Synthesis Lectures on Engineering, Science, and Technology

The focus of this series is general topics, and applications about, and for, engineers and scientists on a wide array of applications, methods and advances. Most titles cover subjects such as professional development, education, and study skills, as well as basic introductory undergraduate material and other topics appropriate for a broader and less technical audience.

Horst R. Beyer

Quantum Spin and Representations of the Poincaré Group, Part I

With a Focus on Physics and Operator Theory

Springer

Horst R. Beyer
Division of Theoretical Astrophysics
University of Tübingen
Tübingen, Germany

ISSN 2690-0300 ISSN 2690-0327 (electronic)
Synthesis Lectures on Engineering, Science, and Technology
ISBN 978-3-031-84139-2 ISBN 978-3-031-84140-8 (eBook)
https://doi.org/10.1007/978-3-031-84140-8

This Springer imprint is published by the registered company Springer Nature Switzerland AG
The registered company address is: Gewerbestrasse 11, 6330 Cham, Switzerland

Acknowledgments

I am indebted to the publisher Springer Nature, in particular, to the Executive Editor Synthesis, Susanne Filler, for the great support. In addition, I am indebted to numerous colleagues, both, from the field of mathematics and physics. The graphics in this text have been created with Wolfram Mathematica® software (www.wolfram.com) and PGF/TikZ software. The text was produced, using the document preparation system LaTeX.

Horst R. Beyer

Introduction

Under changes of the inertial frame of reference, the fields of relativistic quantum field theories must transform under strongly continuous unitary representations of the Poincaré group. The focus of the book, which is divided into two parts, Parts I and II, is the construction of these representations that provide the basis for the formulation of current relativistic quantum field theories of the scalar fields, the Dirac field and the electromagnetic field.

Such construction is tied to the use of the methods of operator theory that also provide the basis for the formulation of quantum mechanics, up to the interpretation of the measurement process. Also because representation spaces of representations of primary interest in quantum theory are infinite dimensional, the use of these methods is essential. Consequently, the text also calculates the generators of relevant strongly continuous one-parameter groups that are associated with the representations and, where appropriate, the corresponding spectrum.

In Part I, we construct a family of strongly continuous unitary representations of the restricted Lorentz group $\mathcal{L}_+^\uparrow$ that are associated with spin 0 and which are the basis for the construction of the representations corresponding to spins $1/2$ and 1 in Part II. Unlike their counterparts for the scalar field, the representations associated with the Dirac field, are only "partially unitary," in the sense that the one-parameter groups that are associated with Lorentz boosts are not unitary. Hence, their generators are not self-adjoint and, as a consequence, cannot function as "observables" in a one-particle theory. Historically, this fact was one of the reasons to abandon the idea of the development of "relativistic quantum mechanics," following the model of quantum mechanics, in favor of the quantization of fields. Still, in the process candidates for relativistic one-particle theories emerged that through the process of "second quantization," appear in the construction of corresponding quantum field theories. The text provides the connections of these one-particle theories with the constructed representations.

The representations associated with the Dirac field are induced by a double cover of $\mathcal{L}_+^\uparrow$, leading to a family of strongly continuous representations of $\mathrm{SL}(2,\mathbb{C})$, the special linear group in 2-dimensions on $\mathbb{C}$, and the concepts of Weyl and Dirac spinor fields.

For the description of non-relativistic spin in quantum mechanics, Part I also gives the construction of a double cover of SO(3), the induced strongly continuous unitary representation of SU(2), calculates the generators and associated spectra corresponding to rotations about the coordinate axes and states the Pauli interaction Hamiltonian.

The book does not assume any knowledge of Lie groups nor differential geometry. The needed algebraic properties of SO(2), SO(3), SU(2), the Lorentz group $\mathscr{L}$, $\mathrm{SL}(2,\mathbb{C})$ and the Poincaré group $\mathcal{P}$ are developed inside the book.

To provide a background on matrix Lie groups, Part I contains a compact introduction to general matrix Lie groups. We choose for this a direct approach that uses the natural embedding of such a group G into the space $\mathrm{M}(n,\mathbb{K})$, of $n \times n$-matrices over $\mathbb{K} \in \{\mathbb{R},\mathbb{C}\}$, where n is some non-zero natural number. Viewed from a differential geometric perspective, this text does not go beyond the definition of tangent spaces to matrix Lie groups, although the approach here allows the definition of differentiable vector fields, in particular, left-invariant vector fields, Lie derivatives, Lie algebras, metrics and related geodesics in matrix Lie groups, essentially leading to a geometric definition of the matrix-valued exponential map. From a perspective of representations of Lie groups in infinite dimensional Banach spaces such geometric considerations are secondary, although occasionally providing interesting insights.[1] In addition to the algebraic requirement that such a representation R is a homomorphism from a matrix Lie group G into the group $\mathrm{GL}(X)$ of bijective bounded linear maps on some Banach space X, R needs to be strongly continuous.[2] Only then R maps one-parameter subgroups and one-parameter semigroups of G to one-parameter subgroups and one-parameter semigroups of $\mathrm{GL}(X)$, respectively, that have a meaningful generator [20, 25, 31, 47, 4]. Since such generators are generically unbounded and commutators of such operators turned out inconclusive, commutators of unbounded operators do not appear in this text. For this reason, Lie algebras play a minor role in this text.

For simplicity, in this text, we identify linear maps from $\mathbb{K}^n$ to $\mathbb{K}^n$ with their matrix representation, with respect to the canonical basis of $\mathbb{K}^n$, where $\mathbb{K} \in \{\mathbb{R},\mathbb{C}\}$ and n is a non-zero natural number. We alert readers from the field of mathematics that, contrary to the standard convention in mathematics, we use in this book the standard convention in

[1] For instance, geometric considerations appear in the text, such as in the motivation of the transformation properties of spinor fields and also in the construction of a family of unitary representations of the restricted Lorentz group.

[2] The proof of strong continuity is merely an application of Lebesgue's dominated convergence theorem.

physics that scalar products are antilinear in the first argument and linear in the second argument. Part I contains 34 solved exercises, to aid comprehension.

Horst R. Beyer

Conventions

For every map f, the symbol Ran$f$ denotes the set consisting of the assumed values. In particular, if f is a linear map between linear spaces, ker$f$ denotes the subspace of the domain of f containing those elements that are mapped to the zero vector. For every non-empty set S, the symbol id_S denotes the identity map on S. We always assume the composition of maps (which includes addition, multiplication etc.) to be maximally defined.

The symbols $\mathbb{N}$, $\mathbb{R}$, $\mathbb{C}$ denote the natural numbers (including zero), all real numbers and all complex numbers, respectively. The symbols $\mathbb{N}^*$, $\mathbb{R}^*$, $\mathbb{C}^*$ denote the corresponding sets without 0. We call $x \in \mathbb{R}$ positive (negative) if $x \geqslant 0$ ($x \leqslant 0$). We call $x \in \mathbb{R}$ strictly positive (strictly negative) if $x > 0$ ($x < 0$). For every $n \in \mathbb{N}^*$, the symbol S^n denotes unit sphere in Euclidean n-space, i.e., $S^n := \{x \in \mathbb{R}^n : |x| = 1\}$.

For $\mathbb{K} \in \{\mathbb{R}, \mathbb{C}\}$, $n, m \in \mathbb{N}^*$, $e_1, \dots, e_n$ denotes the canonical basis of $\mathbb{K}^n$. For every $x \in \mathbb{K}^n$, $|x|$ denotes the canonical norm of x. Further, in connection with matrices, the elements of $\mathbb{K}^n$ are considered as column vectors. Finally, $\text{M}(n \times m, \mathbb{K})$, $(\text{M}(n, \mathbb{K}) := \text{M}(n \times n, \mathbb{K}))$, denotes the vector space of $n \times m$ matrices with entries from $\mathbb{K}$. The $n \times n$ unit matrix is denoted by $E_{n \times n}$ or, if there is no confusion possible, by E. For every $A \in \text{M}(n \times n, \mathbb{K})$, $\det(A)$ denotes its determinant and $\text{Tr}(A)$ its trace. To indicate components of vectors, matrices and coordinates, this text uses only lower indices. In particular, we do not use the Einstein summation convention.

For each $k \in \mathbb{N}$, $n, m \in \mathbb{N}^*$, $\mathbb{K} \in \{\mathbb{R}, \mathbb{C}\}$ and each non-empty open subset Ω of $\mathbb{R}^n$, the symbol $C^k(\Omega, \mathbb{C})$ denotes the linear space of continuous and k-times continuously differentiable complex-valued functions on Ω. Further, $C_0^k(\Omega, \mathbb{C})$ denotes the subspace of $C^k(\Omega, \mathbb{C})$ containing those elements that have a compact support in Ω. If Ω is bounded, $C^k(\overline{\Omega}, \mathbb{C})$ is defined as the subspace of $C^k(\Omega, \mathbb{C})$ consisting of those elements for which there is an extension to an element of $C^k(V, \mathbb{C})$ for some open subset V of $\mathbb{R}^n$ containing Ω. The superscript k is omitted if $k = 0$. For every map $f : U \to \mathbb{K}^m$ which is defined on some subset $U \subset \mathbb{R}^n$ as well as differentiable in $x \in U, f'(x) \in \text{M}(m \times n, \mathbb{K})$ denotes the derivative of f in x defined by

$$f'(x)_{ij} := \frac{\partial f_i}{\partial x_j}(x)$$

for all $i \in \{1, \ldots, m\}$ and $j \in \{1, \ldots, n\}$. In addition, in the case $m = 1$, we define the gradient of f in x by

$$(\nabla f)(x) := \sum_{i=1}^{n} \frac{\partial f}{\partial x_i}(x) e_i .$$

For a differentiable map γ from some non-trivial open interval I of $\mathbb{R}$ into $\mathbb{K}^n$ or $\mathrm{M}(n, \mathbb{K})$, the corresponding derivative is defined component-wise. Further, $BC(\mathbb{R}^n, \mathbb{C})$ denotes the subspace of $C(\mathbb{R}^n, \mathbb{C})$ consisting of those functions which are bounded. $C_\infty(\mathbb{R}^n, \mathbb{C})$ denotes subspace of $C(\mathbb{R}^n, \mathbb{C})$ containing those functions f satisfying

$$\lim_{|x| \to \infty} f(x) = 0.$$

Throughout the text, Lebesgue integration theory is used in the formulation of [54]. Compare also Chap. 1 in [32] and Appendix A in [71]. If not indicated otherwise, the terms *'almost everywhere'* (a.e.), *'measurable'*, *'integrable'*, etc. refer to the Lebesgue measure v^n on $\mathbb{R}^n$, $n \in \mathbb{N}^*$. The appropriate n will be clear from the context. Nevertheless, we often mimic the notation of the Riemann-integral to improve readability. We follow common usage and do not differentiate between a function f which is almost everywhere defined (with respect to the chosen measure) on some set and the associated equivalence class consisting of all functions which are almost everywhere defined on that set and differ from f only on a set of measure zero. In this sense, for $p \geqslant 1$, the symbol $L^p_{\mathbb{C}}(\Omega, \rho)$, where ρ is some a.e. defined and positive real-valued and locally integrable function on Ω, denotes the vector space of all complex-valued measurable functions f which are a.e. defined on Ω and such that $|f|^p$ is integrable with respect to the measure ρv^n. For every such f, we define the L^p-norm $\|f\|_p$ corresponding to f by

$$\|f\|_p := \left(\int_\Omega \rho \, |f|^p \, dv^n \right)^{1/p} .$$

Equipped with $\| \, \|_p$, $L^p_{\mathbb{C}}(\Omega, \rho)$ is a complex Banach space. In addition, we define in the special case $p = 2$ a scalar product $\langle | \rangle_2$ on $L^2_{\mathbb{C}}(\Omega, \rho)$ by

$$\langle f | g \rangle_2 := \int_\Omega \rho f^* g \, dv^n ,$$

for all $f, g \in L^2_{\mathbb{C}}(\Omega, \rho)$, where * denotes complex conjugation on $\mathbb{C}$. As a consequence, $\langle | \rangle_2$ is antilinear in the first argument and linear in its second. This convention is used for sesquilinear forms in general. If ρ is constant of value 1, we omit any reference to ρ in the previous symbols. $L^\infty_{\mathbb{C}}(\Omega)$ denotes the vector space of complex-valued measurable a.e. defined and a.e. bounded functions on Ω. For every $f \in L^\infty_{\mathbb{C}}(\Omega)$ we define $\|f\|_\infty$ as the largest lower bound of $|f|$. Equipped with $\| \|_\infty$, $L^\infty_{\mathbb{C}}(\Omega)$ is a complex Banach space.

Finally, standard results and nomenclature of Operator Theory are used. For this, compare textbooks on Functional Analysis, e.g., [53] Vol. I, [54, 78]. In particular, for every non-trivial normed vector space $(X, \| \|_X)$ and any normed vector space $(Y, \| \|_Y)$ over the same field, we denote by $\mathrm{L}(X, Y)$ the vector space of continuous linear maps from X to Y. Equipped with the operator norm $\| \|_{\mathrm{op},X,Y}$, defined by

$$\|A\|_{\mathrm{op},X,Y} := \sup_{\xi \in X, \|\xi\|_X = 1} \|A\xi\|_Y$$

for all $A \in \mathrm{L}(X, Y)$, $\mathrm{L}(X, Y)$, is a normed vector space which is complete if $(Y, \| \|_Y)$ is complete. Frequently, indices in $\| \|_{\mathrm{op},X,Y}$ are omitted if there is no confusion possible. Finally, for every non-empty subset U of some normed vector space $(X, \| \|_X)$ and any normed vector space $(Y, \| \|_Y)$, the symbol $C(U, Y)$ denotes the vector space of continuous functions from U to Y.

Contents

Background 1

For a reader with basic knowledge in the area of Lie groups and the matrix exponential map, it might suffice to read this chapter quickly to get an overview of the material and proceed to Chap. 2.

The first section provides basic information on general matrix Lie groups. Additional information on particular matrix groups is provided inside subsequent sections. The second section introduces the operator-valued exponential map and a local inverse, the operator-valued logarithm, and provides applications to matrix Lie groups. Again, further information on applications to particular matrix Lie groups is provided inside subsequent sections.

1.1 Matrix Lie Groups

In the following, we are going to give background information on matrix Lie groups. We choose for this a direct approach that uses the natural embedding of such a group G into the space $\mathrm{M}(n, \mathbb{K})$, of $n \times n$-matrices over $\mathbb{K} \in \{\mathbb{R}, \mathbb{C}\}$, where n is some non-zero natural number. The advantage of this direct approach, over an approach via real differentiable manifolds, e.g., see [26, 30, 70], is the simplification that is introduced by its use of a differential structure on G that is induced by its natural embedding into $\mathrm{M}(n, \mathbb{K})$, without losing rigor.[1] In the approach here, the tangent space $T_g G$ at $g \in G$ is defined as the set of all derivatives $c'(0)$ of differentiable paths c in G such that $c(0) = g$, from some open interval I around 0. In this way, $T_g G$ becomes a real subspace of $\mathrm{M}(n, \mathbb{K})$, where we note that $\mathrm{M}(n, \mathbb{C})$ is also a vector space over the real numbers, of dimension $2n^2$. The dimension

[1] The main advantage of an approach via real differentiable manifolds is precisely its independence from any embedding, i.e., its intrinsic nature. On the other hand, normally, this feature becomes important only in questions such as intrinsic vs. extrinsic curvature.

H. R. Beyer, *Quantum Spin and Representations of the Poincaré Group, Part I*,
Synthesis Lectures on Engineering, Science, and Technology,
https://doi.org/10.1007/978-3-031-84140-8_1

of these real vector spaces can be regarded as the equivalent of the dimension of G in an approach via real differentiable manifolds. In the latter approach, the joint dimension of the tangent spaces coincides with the dimension of the manifold, defined earlier in that approach. In the approach here, the dimension of the tangent spaces needs to be determined. Normally, this does not pose problems. At this point the operator-valued exponential map, introduced later, becomes important.

1.1.1 Definition of a Banach Space Structure for $\mathrm{M}(\mathrm{n}, \mathbb{K})$

For the definition of a topology on matrix Lie groups and at the same to connect to the later definition of the operator-valued exponential map, we assume the $\mathbb{K}$-vector space $\mathrm{M}(n, \mathbb{K})$, $n \in \mathbb{N}^*$, $\mathbb{K} \in \{\mathbb{R}, \mathbb{C}\}$, equipped with the operator norm $\| \ \|_{op}$, defined by

$$\|A\|_{\text{op}} := \sup_{x \in \mathbb{K}^n, |x|=1} |A \cdot x| \ ,$$

for every $A \in \mathrm{M}(n, \mathbb{K})$, where $| \ |$ denotes the canonical norm on $\mathbb{K}^n$. We note that $(\mathrm{M}(n, \mathbb{K}), \| \ \|_{op})$ is a n^2-dimensional Banach space over $\mathbb{K}$. Since the operator norm is not always easy to compute, we are also going to use the equivalent maximum norm $\| \ \|_\infty$ on $\mathrm{M}(n, \mathbb{K})$, defined by

$$\|A\|_\infty := \max_{j,k \in \{1,\dots,n\}} |A_{jk}| \ ,$$

for every $A = (A_{jk})_{(j,k) \in \{1,\dots,n\}^2} \in \mathrm{M}(n, \mathbb{K})$. That the norms $\| \ \|_{op}$ and $\| \ \|_\infty$ are equivalent can be seen as follows.[2] For $A = (A_{jk})_{(j,k) \in \{1,\dots,n\}^2} \in \mathrm{M}(n, \mathbb{K})$, we have that

$$\begin{aligned} |A \cdot x| &= |\sum_{k=1}^{n} [\sum_{l=1}^{n} A_{kl} x_l] e_k| \leqslant \sum_{k=1}^{n} \sum_{l=1}^{n} |A_{kl}| \cdot |x_l| \\ &\leqslant \sum_{k=1}^{n} \left[\left(\sum_{l=1}^{n} |A_{kl}|^2 \right)^{1/2} \cdot \left(\sum_{l=1}^{n} |x_l|^2 \right)^{1/2} \right] \leqslant \sum_{k=1}^{n} \left[\left(\sum_{l=1}^{n} \|A\|_\infty^2 \right)^{1/2} \right] \\ &= n^{3/2} \, \|A\|_\infty \ , \end{aligned}$$

for every $x \in \mathbb{K}^n$ satisfying $|x| = 1$, where $e_1, \dots, e_n$ is the canonical basis of $\mathbb{K}^n$. Hence,

$$\|A\|_{\text{op}} \leqslant n^{3/2} \, \|A\|_\infty \ .$$

Further, since

[2] This can also be concluded from a general result, see [39], that all norms defined on a finite dimensional vector space are equivalent.

$$\left(\sum_{k=1}^{2} |A_{kl}|^2\right)^{1/2} = |A \cdot e_l| \leqslant \|A\|_{op} ,$$

we have that

$$\left(\sum_{k=1}^{2} |A_{kl}|^2\right)^{1/2} \leqslant \|A\|_{op} ,$$

for every $l \in \{1, \ldots, n\}$. Hence,

$$\|A\|_\infty^2 \leqslant \sum_{k,l=1}^{2} |A_{kl}|^2 \leqslant n \, \|A\|_{op}^2$$

Hence,

$$\|A\|_{op} \geqslant \frac{1}{n^{1/2}} \|A\|_\infty .$$

and

$$\frac{1}{n^{1/2}} \|A\|_\infty \leqslant \|A\|_{\mathrm{op}} \leqslant n^{3/2} \|A\|_\infty . \tag{1.1}$$

As a consequence, the norms $\| \, \|_{op}$ and $\| \, \|_\infty$ are equivalent and hence induce the same topology on $\mathrm{M}(n, \mathbb{K})$.

In the following, we define the open ball $U_\varepsilon(A)$ of radius $\varepsilon > 0$ around $A \in \mathrm{M}(n, \mathbb{K})$ by

$$U_\varepsilon(A) := \{B \in \mathrm{M}(n, \mathbb{K}) : \|B - A\|_{\mathrm{op}} < \varepsilon\}$$

and the closed ball $B_\varepsilon(A)$ of radius ε around A by

$$B_\varepsilon(A) := \{B \in \| \, \|_{\mathrm{op}} : \|B - A\|_{\mathrm{op}} \leqslant \varepsilon\} .$$

Finally, we define the norm-topology $\mathscr{T}_{\mathrm{M}(n,\mathbb{K})}$ induced by $\| \, \|_{\mathrm{op}}$ on $\mathrm{M}(n, \mathbb{K})$ to consist of arbitrary unions of open balls. Then $(\mathrm{M}(n, \mathbb{K}), \mathscr{T}_{\mathrm{M}(n,\mathbb{K})})$ is a Hausdorff topological space, with a countable basis.

1.1.2 Some Matrix Lie Groups, Important for Applications

In the following, we assume that $n \in \mathbb{N}^*$ and $\mathbb{K} \in \{\mathbb{R}, \mathbb{C}\}$. We define $GL(n, \mathbb{K})$ by

$$GL(n, \mathbb{K}) := \{A \in \mathrm{M}(n, \mathbb{K} : \det(A) \neq 0\} .$$

Since

$$\det(A \cdot B) = \det(A) \cdot \det(B) \neq 0 \ ,$$

for all $A, B \in GL(n, \mathbb{K})$, $GL(n, \mathbb{K})$ is closed under matrix multiplication. Further, we note that the matrix multiplication is associative and, since the $n \times n$ unit matrix E satisfies $\det(E) = 1 \neq 0$, that $E \in GL(n, \mathbb{K})$. Finally, for every $A \in GL(n, \mathbb{K})$, there is $A^{-1} \in \mathrm{M}(n, \mathbb{K})$ such that

$$A \cdot A^{-1} = A^{-1} \cdot A = E \ .$$

The latter implies that $A^{-1} \in GL(n, \mathbb{K})$, since

$$1 = \det(E) = \det(A^{-1} \cdot A) = \det(A^{-1}) \cdot \det(A) \ ,$$

and hence that $\det(A^{-1}) \neq 0$.

> Hence, $GL(n, \mathbb{K})$ equipped with the operation of matrix multiplication is a group, the so called general linear group of degree n over $\mathbb{K}$.

We note that the operation of matrix multiplication is continuous, since if $A, B \in \mathrm{M}(n, \mathbb{K})$ and $A_1, A_2, \ldots$ and $B_1, B_2, \ldots$ are sequences in $\mathrm{M}(n, \mathbb{K})$ that are convergent to A and B, respectively, it follows that

$$\begin{aligned} \|A_\nu \cdot B_\nu - A \cdot B\|_{op} &= \|(A_\nu - A) \cdot B_\nu + A \cdot (B_\nu - B)\|_{op} \\ &= \|(A_\nu - A) \cdot (B_\nu - B) + (A_\nu - A) \cdot B + A \cdot (B_\nu - B)\|_{op} \\ &\leqslant \|A_\nu - A\|_{op} \cdot \|B_\nu - B\|_{op} + \|A_\nu - A\|_{op} \cdot \|B\|_{op} + \|A\|_{op} \cdot \|B_\nu - B\|_{op} \ , \end{aligned}$$

for every $\nu \in \mathbb{N}^*$ and hence that

$$\lim_{\nu \to \infty} \|A_\nu \cdot B_\nu - A \cdot B\|_{op} = 0 \ .$$

Further, also the operation of inversion is continuous. For the proof, let $A \in GL(n, \mathbb{K})$ and $A_1, A_2, \ldots$ be a sequence in $GL(n, \mathbb{K})$ that is convergent to A. Then it follows for $\nu \in \mathbb{N}^*$ that

$$\begin{aligned} &A_\nu^{-1} - A^{-1} = A^{-1}(A - A_\nu)A_\nu^{-1} \\ &= A^{-1}(A - A_\nu)(A_\nu^{-1} - A^{-1}) + A^{-1}(A - A_\nu)A^{-1} \ , \\ &\left[E - A^{-1}(A - A_\nu)\right](A_\nu^{-1} - A^{-1}) = A^{-1}(A - A_\nu)A^{-1} \ . \end{aligned}$$

Since

$$\lim_{\nu \to \infty} \|A_\nu - A\|_{op} = 0 \ ,$$

there is $\nu_0 \in \mathbb{N}^*$ such that

$$\|A^{-1}\|_{op} \cdot \|A_\nu - A\|_{op} \leqslant \frac{1}{2} ,$$

for $\nu \in \mathbb{N}^*$ such that $\nu \geqslant \nu_0$. For such ν, since

$$\|A^{-1}(A - A_\nu)\|_{op} \leqslant \|A^{-1}\|_{op} \cdot \|A_\nu - A\|_{op} ,$$

it follows that

$$\begin{aligned} A_\nu^{-1} - A^{-1} &= \left[E - A^{-1}(A - A_\nu)\right]^{-1} A^{-1}(A - A_\nu)A^{-1} \\ &= \left\{\sum_{k=0}^{\infty} [A^{-1}(A - A_\nu)]^k\right\} A^{-1}(A - A_\nu)A^{-1} \end{aligned}$$

and hence that

$$\begin{aligned} \|A_\nu^{-1} - A^{-1}\|_{op} &\leqslant \left[\sum_{k=0}^{\infty} \|A^{-1}(A - A_\nu)\|_{op}^k\right] \cdot \|A^{-1}(A - A_\nu)A^{-1}\|_{op} \\ &\leqslant 2\,\|A^{-1}\|_{op}^2 \cdot \|A_\nu - A\|_{op} . \end{aligned}$$

As a consequence,

$$\lim_{\nu \to \infty} \|A_\nu^{-1} - A^{-1}\|_{op} = 0 .$$

Hence, as a topological space with a group structure such that the group operations, i.e., multiplication and inversion, are continuous maps, $GL(n, \mathbb{K})$ is a topological group.

Further, since the determinant function, $\det : \mathrm{M}(n, \mathbb{K}) \to \mathbb{K}$ is continuous, $GL(n, \mathbb{K})$ is an open subset of $\mathrm{M}(n, \mathbb{K})$, as the inverse image of the open subset $\mathbb{K}\backslash\{0\}$ of $\mathbb{K}$ under the determinant function. In addition, since $k.E \in GL(n, \mathbb{K})$ and $\|k.E\|_{op} = k$, for every $k \in \mathbb{N}$, $GL(n, \mathbb{K}) \subset \mathrm{M}(n, \mathbb{K})$ is unbounded.

Exercise 1.1 Let X be a non-trivial Banach space over $\mathbb{K} \in \{\mathbb{R}, \mathbb{C}\}$. Show that

$$GL(X) := \{A \in \mathrm{L}(X, X) : A \text{ is bijective}\}$$

is an open subset of $\mathrm{L}(X, X)$, where $\mathrm{L}(X, X)$ is equipped with the operator norm $\|\ \|_{op}$, and that $GL(X)$ equipped with the operations of composition and inversion is a topological group.

In the following, we define a number of important subgroups of $GL(n, \mathbb{K})$ that are assumed equipped with the respective induced topology, i.e., the topology generated by the topology of $GL(n, \mathbb{K})$ through intersection. The group operations of these subgroups are restrictions of the group operations of $GL(n, \mathbb{K})$ and hence continuous with respect to the induced topology. In this way, all these subgroups are topological groups.

We define $GL_+(n, \mathbb{R})$ to consist of all real $n \times n$ matrices M satisfying

$$\det(M) > 0 \ . \tag{1.2}$$

Since

$$\det(M_1 M_2) = \det(M_1)\det(M_2) > 0 \ , \ \det(E) = 1 > 0 \ ,$$
$$\det(M^{-1}) = \frac{1}{\det(M)} > 0 \ ,$$

for all $M, M_1, M_2 \in GL_+(n, \mathbb{R})$, $GL_+(n, \mathbb{R})$ is a subgroup of $GL(n, \mathbb{R})$. Further, since the determinant function, $\det : \mathrm{M}(n, \mathbb{R}) \to \mathbb{R}$ is continuous, $GL_+(n, \mathbb{R})$ is an open subset of $\mathrm{M}(n, \mathbb{R})$, as the inverse image of the open subset $(0, \infty)$ of $\mathbb{R}$ under the determinant function. In addition, since $k.E \in GL_+(n, \mathbb{R})$ and $\|k.E\|_{op} = k$, for every $k \in \mathbb{N}$, $GL_+(n, \mathbb{R}) \subset \mathrm{M}(n, \mathbb{R})$ is unbounded.

We define $\mathrm{O}(n)$ to consist of all real $n \times n$ matrices M satisfying

$$M^* \cdot M = E \ , \tag{1.3}$$

where $\cdot$ denotes matrix multiplication, E the $n \times n$ unit matrix and M^* the transpose of M.

We note that the latter implies that

$$1 = \det(E) = \det(M^* \cdot M) = \det(M^*) \cdot \det(M) = [\det(M)]^2$$

and hence that

$$\det(M) \in \{-1, 1\} \ .$$

Therefore M is invertible and $M^{-1} = M^*$, i.e., M is an orthogonal matrix. Also, according to definition, every orthogonal matrix M is invertible such that $M^{-1} = M^*$ and hence is satisfying (1.3). As a consequence, $\mathrm{O}(n)$ coincides with the set of orthogonal $n \times n$-matrices.

Also, since

$$(M_1 M_2)^* M_1 M_2 = M_2^* M_1^* M_1 M_2 = E \ , \quad E^* E = EE = E \ ,$$
$$\left(M^{-1}\right)^* M^{-1} = \left(M^*\right)^* M^{-1} = M M^{-1} = E \ ,$$

for all $M_1, M_2 \in \mathrm{O}(n)$, $\mathrm{O}(n)$ is a subgroup of the general linear group $GL(n, \mathbb{R})$, the so called orthogonal group of degree n.

We note that from (1.3), it follows that a real $n \times n$-matrix is orthogonal if and only if its column vectors form an orthonormal basis of $\mathbb{R}^n$, equipped with the canonical scalar product.

We note that

$$\begin{aligned}\|M\|_{\mathrm{op}} &:= \sup_{x \in \mathbb{R}^n, |x|=1} |M \cdot x| = \sup_{x \in \mathbb{R}^n, |x|=1} \langle M \cdot x | M \cdot x \rangle^{1/2} \\ &= \sup_{x \in \mathbb{R}^n, |x|=1} \left\langle x | M^* \cdot M \cdot x \right\rangle^{1/2} = 1 \ ,\end{aligned}$$

for every $M \in \mathrm{O}(n)$. Further, if $M_1, M_2, \ldots$ is a sequence in $\mathrm{O}(n)$ that is convergent to $M \in \mathrm{M}(n \times n, \mathbb{R}$, i.e., such that

$$\lim_{\nu \to \infty} \|M_\nu - M\|_{\mathrm{op}} = 0 \ ,$$

it follows for every $\nu \in \mathbb{N}^*$ that

$$\begin{aligned}&\|M_\nu^* M_\nu - M^* M\|_{\mathrm{op}} \\ &= \|(M_\nu - M)^*(M_\nu - M) + M^*(M_\nu - M) + (M_\nu^* - M^*)M\|_{\mathrm{op}} \\ &\leqslant \|(M_\nu - M)^*(M_\nu - M)\|_{\mathrm{op}} + \|M^*(M_\nu - M)\|_{\mathrm{op}} + \|(M_\nu^* - M^*)M\|_{\mathrm{op}} \\ &\leqslant \|M_\nu - M\|_{\mathrm{op}}^2 + 2\,\|M\|_{\mathrm{op}}\,\|M_\nu - M\|_{\mathrm{op}} \ ,\end{aligned}$$

where we used that $\|A^* A\|_{\mathrm{op}} = \|A\|_{\mathrm{op}}^2$, for every $A \in \mathrm{M}(n, \mathbb{R})$, and that the adjoint operation is linear, and hence that $M^* M = E$.

Therefore, $\mathrm{O}(n)$, as a bounded and closed subset of a finite dimensional normed vector space, is compact.

In addition, for every $n \in \mathbb{N}^*$, we define $SO(n)$ to consist of all $M \in \mathrm{O}(n)$ with determinant 1. Since

$$\det(M_1 M_2) = \det(M_1)\det(M_2) = 1 \ , \ \det(E) = 1 \ ,$$
$$\det(M^{-1}) = \det(M^*) = [\det(M)]^* = 1 \ ,$$

for all $M, M_1, M_2 \in SO(n)$, $SO(n)$ is a subgroup of $\mathrm{O}(n)$. Further, since the determinant function $\det : \mathrm{M}(n, \mathbb{R}) \to \mathbb{R}$ is continuous, $SO(n)$ is given by the intersection of $\mathrm{O}(n)$ with the inverse image of the closed subset $\{1\}$ of $\mathbb{R}$ under the determinant. Since the latter is closed,

SO(n), as a bounded and closed subset of a finite dimensional normed vector space, is compact.

Analogously, we define $\mathrm{U}(n)$ to consist of all complex $n \times n$ matrices U satisfying

$$U^* \cdot U = E \ , \tag{1.4}$$

where $\cdot$ denotes matrix multiplication, E the $n \times n$ unit matrix and U^* the adjoint of U.

We note that the latter implies that

$$1 = \det(E) = \det(U^* \cdot U) = \det(U^*) \cdot \det(U) = |\det(U)|^2$$

and hence that

$$\det(U) \in S^1 \ .$$

Therefore U is invertible and $U^{-1} = U^*$, i.e., U is an unitary matrix. Also, according to definition, every unitary matrix U is invertible such that $U^{-1} = U^*$ and hence is satisfying (1.4). As a consequence, $\mathrm{U}(n)$ coincides with the set of unitary $n \times n$-matrices. Also, since

$$(U_1 U_2)^* U_1 U_2 = U_2^* U_1^* U_1 U_2 = E \ , \ E^* E = EE = E \ ,$$
$$\left(U^{-1}\right)^* U^{-1} = \left(U^*\right)^* U^{-1} = UU^{-1} = E \ ,$$

for all $U_1, U_2 \in \mathrm{U}(n)$, $\mathrm{U}(n)$ is a subgroup of the general linear group $GL(n, \mathbb{C})$, the so called unitary group of degree n. In particular, we have that $\mathrm{U}(1)$ is given by S^1, equipped with the operation of complex multiplication.

We note that from (1.4), it follows that a complex $n \times n$-matrix is unitary if and only if its column vectors form an orthonormal basis of $\mathbb{C}^n$, equipped with the canonical scalar product.

Similarly to the case of $\mathrm{O}(n)$, we note that

$$\begin{aligned}\|U\|_{\mathrm{op}} &:= \sup_{x\in\mathbb{C}^n,|x|=1} |U\cdot x| = \sup_{x\in\mathbb{C}^n,|x|=1} \langle U\cdot x|U\cdot x\rangle^{1/2} \\ &= \sup_{x\in\mathbb{C}^n,|x|=1} \left\langle x|U^*\cdot U\cdot x\right\rangle^{1/2} = 1 ,\end{aligned}$$

for every $U \in \mathrm{U}(n)$. Further, if $U_1, U_2, \ldots$ is a sequence in $\mathrm{U}(n)$ that is convergent to $U \in \mathrm{M}(n, \mathbb{C})$, i.e., such that

$$\lim_{\nu\to\infty} \|U_\nu - U\|_{\mathrm{op}} = 0 ,$$

it follows for every $\nu \in \mathbb{N}^*$ that

$$\begin{aligned}&\|U_\nu^* U_\nu - U^* U\|_{\mathrm{op}} \\ &= \|(U_\nu - U)^*(U_\nu - U) + U^*(U_\nu - U) + (U_\nu^* - U^*)U\|_{\mathrm{op}} \\ &\leqslant \|(U_\nu - U)^*(U_\nu - U)\|_{\mathrm{op}} + \|U^*(U_\nu - U)\|_{\mathrm{op}} + \|(U_\nu^* - U^*)U\|_{\mathrm{op}} \\ &\leqslant \|U_\nu - U\|_{\mathrm{op}}^2 + 2\,\|U\|_{\mathrm{op}}\,\|U_\nu - U\|_{\mathrm{op}} ,\end{aligned}$$

where we used that $\|A^*A\|_{\mathrm{op}} = \|A\|_{\mathrm{op}}^2$, for every $A \in \mathrm{M}(n, \mathbb{C})$, and that the adjoint operation is anti-linear, and hence that $U^*U = E$.

Therefore, $\mathrm{U}(n)$, as a bounded and closed subset of a finite dimensional normed vector space, is compact.

In addition, we define $SU(n)$ to consist of all $U \in \mathrm{U}(n)$ with determinant 1. Since

$$\begin{aligned}\det(U_1U_2) &= \det(U_1)\det(U_2) = 1 , \quad \det(E) = 1 , \\ \det(U^{-1}) &= \det(U^*) = [\det(U)]^* = 1 ,\end{aligned}$$

for all $U, U_1, U_2 \in SU(n)$, $SU(n)$ is a subgroup of $\mathrm{U}(n)$. Further, since the determinant function $\det : \mathrm{M}(n, \mathbb{C}) \to \mathbb{C}$ is continuous, $SU(n)$ is given by the intersection of $\mathrm{U}(n)$ with the inverse image of the closed subset $\{1\}$ of $\mathbb{C}$ under the determinant. Since the latter is closed,

> $SU(n)$, as a bounded and closed subset of a finite dimensional normed vector space, is compact.

Further,

> for every $n \in \mathbb{N}^*$, we define the special linear group, $SL(n, \mathbb{K})$, by
>
> $$SL(n, \mathbb{K}) := \{G \in \mathrm{M}(n, \mathbb{K}) : \det(G) = 1\} .$$

Since

$$\det(M_1 M_2) = \det(M_1)\det(M_2) = 1 \ , \ \det(E) = 1 \ ,$$
$$1 = \det(E) = \det(M^{-1} \cdot M) = \det(M^{-1}) \cdot \det(M) = \det(M^{-1}) \ ,$$

for all $M, M_1, M_2 \in SL(n, \mathbb{K})$, $SL(n, \mathbb{K})$ is a subgroup of $GL(n, \mathbb{K})$. Again, since the determinant function $\det : \mathrm{M}(n, \mathbb{C}) \to \mathbb{C}$ is continuous, $SL(n, \mathbb{K})$ is given by the inverse image of the closed subset $\{1\}$ of $\mathbb{K}$ under the determinant. Since the latter is closed, $SL(n, \mathbb{K})$ is closed. On the other hand, for $n \geqslant 2$, $SL(n, \mathbb{K}) \subset \mathrm{M}(n, \mathbb{K})$ is unbounded and hence not compact. For the proof, we note that

$$M_k := \mathrm{diag}(k, 1/k, 0, \ldots, 0) \in SL(n, \mathbb{K}) \ ,$$

for every $k \in \mathbb{N}^*$. Then,

$$|M_k \cdot e_1| = |k\, e_1| = k$$

and hence $\|M_k\|_{op} \geqslant k$.

In the following, we are going to prove that $SO(n)$, $\mathrm{U}(n)$, $SU(n)$ are path connected and that $\mathrm{O}(n)$ has 2 connected components. Later, see Proposition 1.24 and Corollary 1.25, we are going to show that $GL_+(n, \mathbb{R})$, $GL(n, \mathbb{C})$, $SL(2, \mathbb{R})$, $SL(2, \mathbb{C})$ are path connected and that $GL(n, \mathbb{R})$ has 2 connected components.

Lemma 1.1 *Let $n \in \{2, 3, \dots\}$, $e_1, \dots, e_n$ the canonical basis of $\mathbb{R}^n$ and $v \in \mathbb{R}^n$ such that $|v| = 1$. Then there is a continuous path $c : [0, 1] \to SO(n)$ such that $c(0) = E$ and $c(1) \cdot v = e_1$.*

Proof For the proof, we note that if $v = e_1$, then $c := ([0, 1] \to SO(n), \mathbb{R}), t \mapsto E)$ is continuous as well as such that $c(0) = E$ and $c(1) \cdot v = E \cdot e_1 = e_1$. If $v = -e_1$, then we define $f_t \in Hom(\mathbb{R}^n, \mathbb{R}^n)$ as the uniquely determined linear map satisfying

$$f_t(v) := \cos(\pi t)\, v - \sin(\pi t)\, e_2 \ ,$$
$$f_t(e_2) := \sin(\pi t)\, v + \cos(\pi t)\, e_2 \ ,$$
$$f_t e_k := e_k \ ,$$

for all $k \in \{3, \dots, n\}$. In particular, f_t is orthogonal. We note that $f_0 = \mathrm{id}_{\mathbb{R}^n}$ and that $f_1(v) = -v = e_1$. Further, if for every $t \in [0, 1]$, $c(t) \in \mathrm{O}(n)$ is the representation matrix of f_t with respect to the canonical basis of $\mathbb{R}^n$, then $c := ([0, 1] \to \mathrm{O}(n), t \mapsto c(t))$ is continuous as well as such that $c(0) = E$ and $c(1) \cdot v = e_1$. Since, $\det \circ c$ is continuous, $[0, 1]$ is connected and $\det(c(0)) = 1$, we have that $(\det \circ c)([0, 1])$ is connected and hence that $(\det \circ c)([0, 1]) = \{1\}$. As a consequence, $Ran(c) \subset SO(n)$. If $v \in S^n \setminus \{e_1, -e_1\}$, then v and e_1 are linearly independent. Let $v_1, \dots, v_{n-2} \in \mathbb{R}^n$ be an orthonormal basis of the orthogonal complement of $\{v, e_1\}$. We define

$$\hat{e}_1 := \frac{1}{|e_1 - \langle v|e_1\rangle\, v|}\,(e_1 - \langle v|e_1\rangle\, v) = \frac{1}{\sqrt{1 - |\langle v|e_1\rangle|^2}}\,(e_1 - \langle v|e_1\rangle\, v) \ .$$

Then, $v, \hat{e}_1, v_1, \dots, v_{n-2}$ is an orthonormal base of $\mathbb{R}^n$. Further, there is $\tau \in (0, 2\pi]$ such that

$$\cos(\tau)\, v - \sin(\tau)\, \hat{e}_1 = e_1 = \langle v|e_1\rangle\, v + \langle \hat{e}_1|e_1\rangle\, \hat{e}_1 \ .$$

For $t \in [0, 1]$,

$$\cos(\tau t)\, v - \sin(\tau t)\, \hat{e}_1 \ , \ \ \sin(\tau t)\, v + \cos(\tau t)\, \hat{e}_1 \ , \ \ v_1 \ , \ \dots \ , \ v_{n-2}$$

is an orthonormal basis for $\mathbb{R}^n$, and we define $f_t \in Hom(\mathbb{R}^n, \mathbb{R}^n)$ as the uniquely determined linear map satisfying

$$f_t(v) := \cos(\tau t)\, v - \sin(\tau t)\, \hat{e}_1 \ ,$$
$$f_t(\hat{e}_1) := \sin(\tau t)\, v + \cos(\tau t)\, \hat{e}_1 \ ,$$
$$f_t v_k := v_k \ ,$$

for all $k \in \{1, \dots, n-2\}$. In particular, f_t is orthogonal. We note that $f_0 = \mathrm{id}_{\mathbb{R}^n}$ and that $f_1(v) = e_1$. Further, if for every $t \in [0, 1]$, $c(t) \in \mathrm{O}(n)$ is the representation matrix of f_t with respect to the canonical basis of $\mathbb{R}^n$, then $c := ([0, 1] \to \mathrm{O}(n), t \mapsto c(t))$ is continuous as well as such that $c(0) = E$ and $c(1) \cdot v = e_1$. Since, $\det \circ c$ is continuous, $[0, 1]$

is connected and $\det(c(0)) = 1$, we have that $(\det \circ c)([0, 1])$ is connected and hence that $(\det \circ c)([0, 1]) = \{1\}$. As a consequence, $\mathrm{R}an(c) \subset SO(n)$. □

Proposition 1.2 (Path connectedness of $SO(n)$) *Let $n \in \mathbb{N}^*$, then*

(i) $SO(n)$ is path connected.
(ii) We have that

$$\mathrm{O}(n) = SO(n) \mathbin{\dot\cup} \mathrm{O}_-(n) ,$$

where

$$\mathrm{O}_-(n) := \{M \in \mathrm{O}(n) : \det(M) = -1\}$$

is path connected.

Proof 'Part (i)': The proof proceeds by induction on $n \in \mathbb{N}^*$. If $n = 1$, then $SO(n) = SO(1) = \{1\}$. The latter is path connected. In the following, we assume that $SO(n)$ is path connected for $n \in \mathbb{N}^*$. If $R \in SO(n+1)$, then the column vectors $R_{1\cdot}, \dots, R_{(n+1)\cdot}$ form an orthonormal basis of $\mathbb{R}^{n+1}$. From Lemma 1.1, it follows the existence of a continuous path $c : [0, 1] \to SO(n+1)$ such that $c(0) \cdot R_{1\cdot} = e_1$ and $c(1) = E_{(n+1)(n+1)}$. Hence it follows that $c \cdot R : [0, 1] \to SO(n+1)$, defined by $(c \cdot R)(t) := c(t) \cdot R$, for every $t \in [0, 1]$, is continuous such that

$$(c \cdot R)(0) = \begin{pmatrix} 1 & 0 & \cdots & 0 \\ 0 & \tilde{R}_{11} & \cdots & \tilde{R}_{1n} \\ \cdot & \cdot & \cdots & \cdot \\ \cdot & \cdot & \cdots & \cdot \\ \cdot & \cdot & \cdots & \cdot \\ 0 & \tilde{R}_{n1} & \cdots & \tilde{R}_{nn} \end{pmatrix} ,$$

where $\tilde{R} \in SO(n)$, and at the same time such that $(c \cdot R)(1) = R$. Since $SO(n)$ is path connected, there is a continuous path $d : [0, 1] \to SO(n)$ such that $d(0) = E_{n \times n}$ and $d(1) = \tilde{R}$. Hence, $\tilde{c} : [0, 1] \to SO(n+1)$ defined by

$$\tilde{c}(t) := \begin{pmatrix} 1 & 0 & \cdots & 0 \\ 0 & d_{11}(2t) & \cdots & d_{1n}(2t) \\ \cdot & \cdot & \cdots & \cdot \\ \cdot & \cdot & \cdots & \cdot \\ \cdot & \cdot & \cdots & \cdot \\ 0 & d_{n1}(2t) & \cdots & d_{nn}(2t) \end{pmatrix} ,$$

for every $t \in [0, 1/2)$ and $\tilde{c}(t) := (c \cdot R)(2\,(t - (1/2)) \cdot R$, for every $t \in [1/2, 1]$, is continuous and such that $\tilde{c}(0) = E_{(n+1)(n+1)}$ and $\tilde{c}(1) = R$. Hence, it follows that $SO(n+1)$ is path connected.

'Part (ii)': For every $M \in \mathrm{O}(n)$, we have that $\det(M) \in \{-1, 1\}$. Hence

$$O(n) \subset SO(n) \cup O_-(n) \subset O(n) \ .$$

Also, $SO(n)$ and $O_-(n)$ are disjoint. Further, $O_-(n)$ is path connected. For the proof, let $M, M_1, M_2 \in O_-(n)$. Then $M \cdot M_1, M \cdot M_2 \in SO(n)$ and since $SO(n)$ is path connected, there is a continuous path $c : [0, 1] \to SO(n)$ such that $c(0) = M \cdot M_1$ and $c(1) = M \cdot M_2$. Hence, $M^* \cdot c$ is a continuous path in $O_-(n)$ such that $(M^* \cdot c)(0) = M_1$ and $(M^* \cdot c)(1) = M_2$. As a consequence, $O_-(n)$ is path connected. □

Proposition 1.3 (Path connectedness of $U(n)$ and $SU(n)$) *For every $n \in \mathbb{N}^*$, $U(n)$ and $SU(n)$ are both path connected.*

Proof Let $U \in U(n)$. Since U is normal, there is an orthonormal basis $v_1, \dots, v_n \in \mathbb{C}^n$ of eigenvectors of U. In addition, let $e^{i\varphi_1}, \dots, e^{i\varphi_n} \in S^1$, where $\varphi_1, \dots, \varphi_n \in (-\pi, \pi]$, be the corresponding, not necessarily pairwise different, eigenvalues of U, i.e., such that $U \cdot v_j = e^{i\varphi_j}.v_j$, for every $j \in \{1, \dots, n\}$. Hence, if $V \in U(n)$ is the matrix that has the vector v_j as the j-th column vector, $j \in \{1, \dots, n\}$, then

$$U \cdot V = V \cdot \text{diag}(e^{i\varphi_1}, \dots, e^{i\varphi_n}) \ ,$$

where for every sequence $a_1, \dots, a_n \in \mathbb{C}$, $\text{diag}(a_1, \dots, a_n)$ denotes the diagonal matrix $D \in M(n, \mathbb{C})$ satisfying $D_{jj} = a_j$, $j \in \{1, \dots, n\}$. Hence,

$$V^* \cdot U \cdot V = \text{diag}(e^{i\varphi_1}, \dots, e^{i\varphi_n}) \ .$$

Then, $c : [0, 1] \to U(n)$, defined by

$$c(t) := V \cdot \text{diag}(e^{it\varphi_1}, \dots, e^{it\varphi_n}) \cdot V^* \ ,$$

for every $t \in [0, 1]$, is continuous such that $c(0) = E$ and $c(1) = U$. Hence, for $U_1, U_2 \in U(n)$, there is a continuous path $d : [0, 1] \to U(n)$ such that $d(0) = U_1$ and $d(1) = U_2$. Therefore, $U(n)$ is path connected (and hence also connected). We note that if $U \in SU(n)$, then $\varphi_1 + \cdots + \varphi_n = k \cdot 2\pi$, for some $k \in \mathbb{Z}$. Hence, without loss of generality, we can assume that $\varphi_1 + \cdots + \varphi_n = 0$. Then, $Ran(c) \subset SU(n)$. Therefore, also in this case, for $U_1, U_2 \in SU(n)$, there is a continuous path $d : [0, 1] \to U(n)$ such that $d(0) = U_1$ and $d(1) = U_2$. As a consequence, $SU(n)$ is path connected (and hence also connected) (Table 1.1). □

1.1.3 Tangent Spaces of Matrix Lie Groups

In the following, we assume that $n \in \mathbb{N}^*$ and $\mathbb{K} \in \{\mathbb{R}, \mathbb{C}\}$. We are going to use standard definitions of differentiability of Banach space valued maps, see, e.g., [8, 38, 39]. If $c : I \to$

Table 1.1 Topological properties of some matrix Lie groups. See also, Proposition 1.24 and Corollary 1.25

$GL(n, \mathbb{R})$	Open, unbounded	2 connected components
$GL_+(n, \mathbb{R})$	Open, unbounded	Path connected
$GL(n, \mathbb{C})$	Open, unbounded	Path connected
$SL(n, \mathbb{K})$	Closed, unbounded for $n \geqslant 2$	Path connected
$O(n)$	Compact	2 connected components
$SO(n)$	Compact	Path connected
$U(n)$	Compact	Path connected
$SU(n)$	Compact	Path connected

$M(n, \mathbb{K})$, where I is an open interval of $\mathbb{R}$ around 0, then c is differentiable in $t \in I$, with derivative $A = (A_{jk})_{(j,k)\in\{1,\dots,n\}^2} \in M(n, \mathbb{K})$ if

$$\lim_{h\to 0, h\neq 0} \left\| \frac{1}{h} [\, c(t+h) - c(t)\,] - A \right\|_{op} = 0 \,. \tag{1.5}$$

Since the norms $\| \, \|_{\text{op}}$ and $\| \, \|_\infty$ are equivalent, the latter is equivalent to

$$\lim_{h\to 0, h\neq 0} \left\| \frac{1}{h} [\, c(t+h) - c(t)\,] - A \right\|_{\infty} = 0 \,.$$

As a consequence, (1.5) is equivalent to

$$\lim_{h\to 0, h\neq 0} \left| \frac{1}{h} [\, c_{jk}(t+h) - c_{jk}(t)\,] - A_{jk} \right| = 0 \,,$$

for every $(j, k) \in \{1, \dots, n\}^2$, where $c_{jk} := p_{jk} \circ c$ and the projection $p_{jk} : M(n, \mathbb{K}) \to \mathbb{K}$ is defined by $p_{jk}(B) := B_{jk}$, for every $B = (B_{jk})_{(j,k)\in\{1,\dots,n\}^2}$, i.e., (1.5) is equivalent to differentiability in $t \in I$ of every component of c, in the sense of calculus, with derivative A_{jk} for every $(j, k) \in \{1, \dots, n\}^2$.

Assumption 1.4 In the following, let $G \subset M(n, \mathbb{K})$ be such that G equipped with the matrix multiplication forms a group.

Definition 1.5 (*Tangent spaces*) We define for every $g \in G$, the tangent space $T_g G$ of G at g as the set of all tangent vectors $c'(0) \in M(n, \mathbb{K})$ of differentiable paths $c : I \to M(n, \mathbb{K})$ with $\mathrm{R}an(c) \subset G$ and $c(0) = g$, where I is an open interval of $\mathbb{R}$ around 0.

Proposition 1.6 (Elementary properties of tangent spaces) *Let $g \in G$ and e be the unit element of G. Then*

(i) $T_g G = g \cdot T_e G = T_e G \cdot g$;
(ii) $T_g G$ *is a real subspace of* $\mathrm{M}(n, \mathbb{K})$.

Proof 'Part (i)': If $c : I \to \mathrm{M}(n, \mathbb{K})$ is differentiable with $\mathrm{R}an(c) \subset G$ and $c(0) = g$, where I is an open interval of $\mathbb{R}$ around 0, and $h \in G$, then $c_1 := (I \to \mathrm{M}(n, \mathbb{K}), t \mapsto h \cdot c(t))$ and $c_2 := (I \to \mathrm{M}(n, \mathbb{K}), t \mapsto c(t) \cdot h)$ are such that $c_1(0) = h \cdot g$ and $c_2(0) = g \cdot h$, respectively. Further, the range of both paths is contained in G, and both paths are differentiable with derivative $c_1'(0) = h \cdot c'(0)$ and $c_2'(0) = c'(0) \cdot h$, respectively. Hence it follows that

$$h \cdot T_g G \subset T_{hg} G \text{ and } T_g G \cdot h \subset T_{gh} G \ , \tag{1.6}$$

for all $g, h \in G$. From the latter, it follows for $g \in G$ that

$$g \cdot T_e G \subset T_g G \ , \ g^{-1} \cdot T_g G \subset T_e G$$

which implies that

$$g \cdot T_e G \subset T_g G \ , \ T_g G \subset g \cdot T_e G$$

and hence that

$$T_g G = g \cdot T_e G \ .$$

Also, it follows from (1.6) for $g \in G$ that

$$T_e G \cdot g \subset T_g G \ , \ T_g G \cdot g^{-1} \subset T_e G$$

which implies that

$$T_e G \cdot g \subset T_g G \ , \ T_g G \subset T_e G \cdot g$$

and hence that

$$T_g G = T_e G \cdot g \ .$$

'Part (ii)': The constant path $c : \mathbb{R} \to \mathrm{M}(n, \mathbb{K})$, defined by $c(t) := e$, for every $t \in \mathbb{R}$, is differentiable with $\mathrm{R}an(c) \subset G$ and $c(0) = e$. Since $c'(0) = 0$, it follows that $T_e G$ contains the 0-matrix. Further, if $c : I \to \mathrm{M}(n, \mathbb{K})$ is differentiable with $\mathrm{R}an(c) \subset G$ and $c(0) = e$, where I is an open interval of $\mathbb{R}$ around 0, and $\lambda \in \mathbb{R}^*$, then $\tilde{c} : \tilde{I} \to G$, defined by $\tilde{c}(t) := c(\lambda t)$, for every $t \in \lambda^{-1} I$, satisfies $\tilde{c}(0) = e$, the range of $\tilde{c}$ is contained in G and $\tilde{c}$ is differentiable with derivative

$$\tilde{c}'(0) = \lambda \, c'(0) \in T_e G \ .$$

Also, if $d : I \to \mathrm{M}(n, \mathbb{K})$ is differentiable with $\mathrm{R}an(d) \subset G$ and $d(0) = e$, then $c \cdot d : I \to \mathrm{M}(n, \mathbb{K})$, defined by $(c \cdot d)(t) := c(t) \cdot d(t)$, for every $t \in I$, is differentiable with derivative

$$(c \cdot d)'(0) = c'(0) \cdot d(0) + c(0) \cdot d'(0) = c'(0) + d'(0) \in T_e G \ .$$

Table 1.2 Properties of tangent spaces of some matrix Lie groups, where E denotes the $n \times n$-unit matrix

Tangent space	Matrices	Dimension
$T_E GL(n, \mathbb{R}) = T_E GL_+(n, \mathbb{R})$	Real $n \times n$	n^2
$T_E GL(n, \mathbb{C})$	Complex $n \times n$	$2\,n^2$
$T_E SL(n, \mathbb{R})$	Trace-free real $n \times n$	$n^2 - 1$
$T_E SL(n, \mathbb{C})$	Trace-free complex $n \times n$	$2\,(n^2 - 1)$
$T_E \mathrm{O}(n) = T_E SO(n)$	Anti-symmetric real $n \times n$	$\frac{n(n-1)}{2}$
$T_E \mathrm{U}(n)$	Anti-Hermitian complex $n \times n$	n^2
$T_E SU(n)$	Trace-free anti-Hermitian complex $n \times n$	$n^2 - 1$

Hence, $T_e G$ is a real subspace of $\mathrm{M}(n, \mathbb{K})$. Therefore, it follows with the help of (ii) that $T_g G$ is a real subspace of $\mathrm{M}(n, \mathbb{K})$, for every $g \in G$. □

The following gives elementary properties of tangent spaces of some matrix Lie groups. Later, see Propositions 1.16, 1.17, 1.20 and Table 1.2, we are going to give a complete characterization of the tangent spaces of these groups.

Proposition 1.7 (Properties of some tangent spaces) *Let $n \in \mathbb{N}^*$ and E be the $n \times n$ unit matrix.*

(i) If G is a subgroup of $\mathrm{O}(n)$ or $\mathrm{U}(n)$, then the elements of $T_E G$ are anti-symmetric matrices and anti-Hermitian matrices, respectively.
(ii) If G is a subgroup of $\mathrm{GL}(n, \mathbb{K})$, such that $\det(M) = 1$, for every $M \in \mathrm{GL}(n, \mathbb{K})$, then the elements of $T_E G$ are trace-zero matrices.

Proof 'Part (i)': If G is a subgroup of $\mathrm{O}(n)$ or $\mathrm{U}(n)$ and $c : I \to \mathrm{M}(n, \mathbb{K})$ is differentiable with $\mathrm{R}an(c) \subset G$ and $c(0) = E$, where I is an open interval of $\mathbb{R}$ around 0, then

$$\sum_{l=1}^{n} c_{kl}(t)\,(c_{jl}(t))^* = \sum_{l=1}^{n} (c_{jl}(t))^*\,c_{kl}(t) = \delta_{jk}\ ,$$

where

$$\delta_{jk} := \begin{cases} 0 & j \neq k \\ 1 & j = k \end{cases} ,$$

and hence

$$\sum_{l=1}^{n} [(c'_{jl}(t))^*\,c_{kl}(t) + (c_{jl}(t))^*\,c'_{kl}(t)] = 0\ ,$$

for every $t \in I$ and $j, k \in \{1, \ldots, n\}$, where we define $\lambda^* = \lambda$, for every $\lambda \in \mathbb{R}$. Therefore, we have that

$$0 = \sum_{l=1}^{n} [(c'_{jl}(0))^* \, \delta_{kl} + \delta_{jl} \, c'_{kl}(t)] = (c'_{jk}(0))^* + c'_{kj}(0) \ ,$$

and hence that

$$c'_{kj}(0) = -(c'_{jk}(0))^* \ ,$$

for all $j, k \in \{1, \ldots, n\}$. 'Part (ii)': If G is a subgroup of $GL(n, \mathbb{K})$, such that $\det(M) = 1$, for every $M \in GL(n, \mathbb{K})$, and $c : I \to \mathrm{M}(n, \mathbb{K})$ is differentiable with $Ran(c) \subset G$ and $c(0) = E$, where I is an open interval of $\mathbb{R}$ around 0, then

$$\det(c(t)) = \det(c_{1\cdot}(t), \ldots, c_{n\cdot}(t)) = 1 \ ,$$

for all $t \in I$, were $c_{j\cdot}(t)$ denotes the j-th column vector of $c(t)$, $j \in \{1, \ldots, n\}$, and hence

$$\begin{aligned}
0 &= (\det \circ c)'(0) = \det(c'_{1\cdot}(0), \ldots, c_{n\cdot}(0)) + \cdots + \det(c_{1\cdot}(0), \ldots, c'_{n\cdot}(0)) \\
&= \det(c'_{1\cdot}(0), \ldots, e_n) + \cdots + \det(e_1, \ldots, c'_{n\cdot}(0)) \\
&= \det(\langle e_1 | c'_{1\cdot}(0) \rangle \, e_1, \ldots, e_n) + \cdots + \det(e_1, \ldots, \langle e_n | c'_{n\cdot}(0) \rangle \, e_n) \ , \\
&= \sum_{k=1}^{n} \langle e_k | c'_{k\cdot}(0) \rangle = \sum_{k=1}^{n} c'_{kk}(0) = \mathrm{Tr}(c'(0)) \ ,
\end{aligned}$$

where $\langle | \rangle$ denotes the canonical scalar product for $\mathbb{K}^n$. □

1.2 Definition and Properties of the Operator Exponential Map

In the following, we are going to use the exponential map, for bounded linear operators[3] on a non-trivial Banach space $(X, \| \ \|)$ over $\mathbb{K} \in \{\mathbb{R}, \mathbb{C}\}$. For the proof of the following theorem, see Sect. 3.3 of [4]. Here, $\mathrm{L}(X, X)$ denotes the Banach space of bounded[4] linear maps on X, equipped with the operator norm $\| \ \|_{\mathrm{op}}$, defined by

$$\|A\|_{\mathrm{op}} := \sup_{x \in X, \|x\| = 1} \|Ax\| \ ,$$

for every $A \in \mathrm{L}(X, X)$. Here and in the following, wherever possible, we use the standard convention for linear maps that skips the evaluation brackets (), when a linear map is applied to a element from its domain. For instance, for every $x \in \mathbb{K}^n$, we write "Ax" instead

[3] We note that the exponential map can be generalized even further to classes of unbounded linear operators, such as densely-defined, linear and self-adjoint operators in complex Hilbert spaces and generators of strongly continuous semi-groups. For this, we refer to functional analysis texts.

[4] Or equivalently "continuous."

of "$A(x)$." Equipped with the composition of maps as an additional operation, $\mathrm{L}(X, X)$ is an associative Banach algebra with unit element.

In the following, we often consider finite dimensional vector spaces, but these are not always subspaces of $\mathbb{R}^n$ or $\mathbb{C}^n$, for some $n \in \mathbb{N}^*$. In this connection, it needs to be taken into account, e.g., see [39], that all norms defined on a finite dimensional vector space are equivalent, that every finite dimensional normed vector space is complete[5] and that every linear map on a finite dimensional normed vector space is bounded (or equivalently "continuous"). Hence, if X is finite dimensional, there is no difference between the set of all linear maps on X, $Hom(X, X)$, and the set of all bounded (or equivalently "continuous") linear maps, $\mathrm{L}(X, X)$, on X, although strictly speaking, the use of the symbol $\mathrm{L}(X, X)$ implies the choice of a norm for X. For the case $X = \mathbb{K}^n$, where $n \in \mathbb{N}^*$, implicitly, this norm is going to be assumed to be the one that is induced on X by the canonical scalar product, if not specified otherwise.
In connection with matrices, it needs to be taken into account that the map that associates with every element of $\mathrm{L}(\mathbb{K}^n, \mathbb{K}^n)$, where $n \in \mathbb{N}^*$, $\mathbb{K} \in \{\mathbb{R}, \mathbb{C}\}$ and $\mathbb{K}^n$ is assumed equipped with the canonical scalar product, the corresponding representation matrix with respect to the canonical basis of $\mathbb{K}^n$, is an isomorphism between the associative Banach algebras with unit element $(\mathrm{L}(\mathbb{K}^n, \mathbb{K}^n), +, ., \circ, \| \|_{\mathrm{op}})$, here $+$ denotes addition, . denotes scalar multiplication and $\circ$ composition of maps, and $(\mathrm{M}(n, \mathbb{K}), +, ., \cdot, \| \|)$, where "+" denotes component-wise addition, "." denotes component-wise scalar multiplication, "$\cdot$" denotes multiplication of matrices, and $\| \|$ denotes the norm that is induced on $\mathrm{M}(n, \mathbb{K})$ by the operator norm $\| \|_{\mathrm{op}}$ on $\mathrm{L}(\mathbb{K}^n, \mathbb{K}^n)$. In addition, we assume that $\mathrm{M}(n, \mathbb{R})$ is embedded in $\mathrm{M}(n, \mathbb{C})$, i.e., $\mathrm{M}(n, \mathbb{R}) \subset \mathrm{M}(n, \mathbb{C})$.

Theorem 1.8 (Definition and properties of the exponential map) *Let $\mathbb{K} \in \{\mathbb{R}, \mathbb{C}\}$ and $(X, \| \|)$ a non-trivial $\mathbb{K}$-Banach space.*

(i) By

$$\exp(A) := \sum_{k=0}^{\infty} \frac{1}{k!} \,.\, A^k \ ,$$

for every $A \in \mathrm{L}(X, X)$, where $A^0 := id_X$ and $A^{k+1} := A \circ A^k$ for all $k \in \mathbb{N}$, there is defined a map $\exp : \mathrm{L}(X, X) \to \mathrm{L}(X, X)$, *the so called exponential function.*

(ii) For all $A, B \in \mathrm{L}(X, X)$ satisfying $[A, B] = 0$,

$$\exp(A + B) = \exp(A) \circ \exp(B) \ , \tag{1.7}$$

where we define $[C, D] := C \circ D - D \circ C$, for every $C, D \in \mathrm{L}(X, X)$.

(iii) The map $u_A : \mathbb{K} \to \mathrm{L}(X, X)$, defined by

---

[5] As a consequence, every subspace of a finite dimensional normed vector space is a closed subspace.

$$u_A(t) := \exp(t.A) \ ,$$

for every $t \in \mathbb{K}$, is differentiable with derivative

$$u_A'(t) = A \circ u_A(t) \ ,$$

for all $t \in \mathbb{K}$.

Remark 1.9 (*Invertibility of exponentials*) Note that, since

$$1 = \exp(0) = \exp(A + (-A)) = \exp(A) \circ \exp(-A) \ ,$$

$\exp(A)$ is invertible with inverse $\exp(-A)$, for every $A \in \mathrm{L}(X, X)$.

Proposition 1.10 (Continuity of the exponential map) *Let $\mathbb{K} \in \{\mathbb{R}, \mathbb{C}\}$ and $(X, \| \ \|)$ a non-trivial $\mathbb{K}$-Banach space. Then, the exponential function,* $\exp : \mathrm{L}(X, X) \to \mathrm{L}(X, X)$, *is continuous.*

Proof For $A, B \in \mathrm{L}(X, X)$, in a first step, we show by induction that

$$\|A^k - B^k\|_{op} \leqslant (\|A\|_{op} + \|B\|_{op})^{k-1} \cdot \|A - B\|_{op} \ , \tag{1.8}$$

for every $k \in \mathbb{N}^*$. Since,

$$\|A^1 - B^1\|_{op} \leqslant (\|A\|_{op} + \|B\|_{op})^{0} \cdot \|A - B\|_{op} \ ,$$

the statement is true for $k = 1$. For the induction step, we assume that the statement is true for some $k \in \mathbb{N}^*$. Then,

$$\begin{aligned}
&\|A^{k+1} - B^{k+1}\|_{op} = \|A \circ (A^k - B^k) + A \circ B^k - B^{k+1}\|_{op} \\
&= \|A \circ (A^k - B^k) + (A - B) \circ B^k\|_{op} \\
&\leqslant \|A\|_{op} \cdot \|A^k - B^k\|_{op} + \|B\|_{op}^k \cdot \|A - B\|_{op} \\
&\leqslant [\, \|A\|_{op} \cdot (\|A\|_{op} + \|B\|_{op})^{k-1} + \|B\|_{op}^k \,] \cdot \|A - B\|_{op} \\
&\leqslant [\, \|A\|_{op} \cdot (\|A\|_{op} + \|B\|_{op})^{k-1} + \|B\|_{op} \cdot (\|A\|_{op} + \|B\|_{op})^{k-1} \,] \cdot \|A - B\|_{op} \\
&= (\|A\|_{op} + \|B\|_{op})^k \cdot \|A - B\|_{op} \ .
\end{aligned}$$

As a consequence, it follows for $l \in \mathbb{N}^*$ that

$$
\begin{aligned}
\left\| \sum_{k=0}^{l} \frac{1}{k!} \cdot A^k - \sum_{k=0}^{l} \frac{1}{k!} \cdot B^k \right\|_{op} &= \left\| \sum_{k=0}^{l} \frac{1}{k!} \cdot (A^k - B^k) \right\|_{op} \\
&\leqslant \sum_{k=0}^{l} \frac{1}{k!} \cdot \|A^k - B^k\|_{op} = \sum_{k=1}^{l} \frac{1}{k!} \cdot \|A^k - B^k\|_{op} \\
&\leqslant \left[\sum_{k=1}^{l} \frac{1}{k!} \cdot (\|A\|_{op} + \|B\|_{op})^{k-1} \right] \cdot \|A - B\|_{op} \\
&= \left[\sum_{k=0}^{l-1} \frac{1}{(k+1)!} \cdot (\|A\|_{op} + \|B\|_{op})^{k} \right] \cdot \|A - B\|_{op} \\
&\leqslant \exp(\|A\|_{op} + \|B\|_{op}) \cdot \|A - B\|_{op} \ ,
\end{aligned}
$$

and by taking the limit that

$$
\| \exp(A) - \exp(B) \|_{op} \leqslant \exp(\|A\|_{op} + \|B\|_{op}) \cdot \|A - B\|_{op} \ .
$$

From the latter inequality, it follows that $\exp : \mathrm{L}(X, X) \to \mathrm{L}(X, X)$, is continuous. □

Proposition 1.11 (Continuous differentiability of the exponential map) *Let* $\mathbb{K} \in \{\mathbb{R}, \mathbb{C}\}$ *and* $(X, \| \ \|)$ *a non-trivial* $\mathbb{K}$*-Banach space. Then, the exponential function* $\exp : \mathrm{L}(X, X) \to \mathrm{L}(X, X)$ *is continuously differentiable, with derivative*

$$
[\exp'(A)](B) = \sum_{k=0}^{\infty} \frac{1}{(k+1)!} \left(\sum_{l=0}^{k} A^{k-l} \circ B \circ A^l \right) ,
$$

for every $A \in \mathrm{L}(X, X)$ *and* $B \in \mathrm{L}(X, X)$.

Proof For $A, B \in \mathrm{L}(X, X)$, in a first step, we show by induction that

$$
A^k - B^k = \sum_{l=0}^{k-1} A^{k-1-l} \circ (A - B) \circ B^l \ , \tag{1.9}
$$

for every $k \in \mathbb{N}^*$. Since,

$$
\sum_{l=0}^{0} A^{-l} \circ (A - B) \circ B^l = A^0 \circ (A - B) \circ B^0 = A - B \ ,
$$

the statement is true for $k = 1$. For the induction step, we assume that the statement is true for some $k \in \mathbb{N}^*$. Then,

$$
\begin{aligned}
A^{k+1} - B^{k+1} &= A \circ (A^k - B^k) + A \circ B^k - B^{k+1} \\
&= A \circ (A^k - B^k) + (A - B) \circ B^k \\
&= A \circ \left[\sum_{l=0}^{k-1} A^{k-1-l} \circ (A - B) \circ B^l \right] + (A - B) \circ B^k \\
&= \sum_{l=0}^{k-1} A^{k-l} \circ (A - B) \circ B^l + (A - B) \circ B^k \\
&= \sum_{l=0}^{k} A^{k-l} \circ (A - B) \circ B^l \ .
\end{aligned}
$$

Hence, for $A, H \in \mathrm{L}(X, X)$ and $m \in \{2, 3, \dots\}$, we have that

$$
\begin{aligned}
&\sum_{k=0}^{m} \frac{1}{k!} [(A + H)^k - A^k] = \sum_{k=1}^{m} \frac{1}{k!} [(A + H)^k - A^k] \\
&= \sum_{k=1}^{m} \frac{1}{k!} \left[\sum_{l=0}^{k-1} (A + H)^{k-1-l} \circ H \circ A^l \right] = \sum_{k=1}^{m} \frac{1}{k!} \left(\sum_{l=0}^{k-1} A^{k-1-l} \circ H \circ A^l \right) \\
&+ \sum_{k=2}^{m} \frac{1}{k!} \left\{ \sum_{l=0}^{k-2} \left[(A + H)^{k-1-l} - A^{k-1-l} \right] \circ H \circ A^l \right\}
\end{aligned}
$$

and hence that

$$
\begin{aligned}
&\left\| \sum_{k=0}^{m} \frac{1}{k!} [(A + H)^k - A^k] - \sum_{k=1}^{m} \frac{1}{k!} \left(\sum_{l=0}^{k-1} A^{k-1-l} \circ H \circ A^l \right) \right\|_{op} \\
&= \left\| \sum_{k=2}^{m} \frac{1}{k!} \left\{ \sum_{l=0}^{k-2} \left[(A + H)^{k-1-l} - A^{k-1-l} \right] \circ H \circ A^l \right\} \right\|_{op} \\
&\leqslant \|H\|_{op} \sum_{k=2}^{m} \frac{1}{k!} \left[\sum_{l=0}^{k-2} \|(A + H)^{k-1-l} - A^{k-1-l}\|_{op} \cdot \|A\|_{op}^l \right] \\
&\leqslant \|H\|_{op}^2 \sum_{k=2}^{m} \frac{1}{k!} \left[\sum_{l=0}^{k-2} (\|A + H\|_{op} + \|A\|_{op})^{k-1-l} \cdot \|A\|_{op}^l \right] \\
&\leqslant \|H\|_{op}^2 \sum_{k=2}^{m} \frac{1}{k!} \left[\sum_{l=0}^{k-2} (\|A + H\|_{op} + \|A\|_{op})^{k-1} \right] \\
&\leqslant \|H\|_{op}^2 \sum_{k=2}^{m} \frac{1}{(k - 1)!} (\|A + H\|_{op} + \|A\|_{op})^{k-1}
\end{aligned}
$$

$$\begin{aligned}
&\leqslant \|H\|_{op}^2 \sum_{k=1}^{m-1} \frac{1}{k!} (\|A + H\|_{op} + \|A\|_{op})^k \\
&\leqslant \|H\|_{op}^2 \exp(\|A + H\|_{op} + \|A\|_{op}) , \qquad (1.10)
\end{aligned}$$

where we used (1.8). For every $m \in \{2, 3, \ldots\}$, we have that

$$\begin{aligned}
\sum_{k=1}^{m} \frac{1}{k!} \sum_{l=0}^{k-1} \left\| A^{k-1-l} \circ H \circ A^l \right\|_{op} &\leqslant \sum_{k=1}^{m} \frac{1}{k!} \sum_{l=0}^{k-1} \|A\|_{op}^{k-1} \cdot \|H\|_{op} \\
= \|H\|_{op} \sum_{k=1}^{m} \frac{\|A\|_{op}^{k-1}}{(k-1)!} &\leqslant \|H\|_{op} \exp(\|A\|_{op}) .
\end{aligned}$$

Therefore, the family

$$\left(\frac{1}{k!} A^{k-1-l} \circ H \circ A^l \right)_{(k,l) \in I} ,$$

where

$$I := \{(k, l) : k \in \mathbb{N}^* \wedge 0 \leqslant l \leqslant k - 1\} ,$$

is absolutely summable, and

$$\left\| \sum_{(k,l) \in I} \frac{1}{k!} A^{k-1-l} \circ H \circ A^l \right\| \leqslant \|H\|_{op} \exp(\|A\|_{op}) . \qquad (1.11)$$

Hence, it follows from (1.10) that

$$\begin{aligned}
&\lim_{H \to 0, H \neq 0} \frac{\left\| \exp(A + H) - \exp(A) - \sum_{k=1}^{\infty} \frac{1}{k!} \left(\sum_{l=0}^{k-1} A^{k-1-l} \circ H \circ A^l \right) \right\|_{op}}{\|H\|_{op}} \\
&= 0 .
\end{aligned}$$

Further, the map $\lambda : \mathrm{L}(X, X) \to \mathrm{L}(X, X)$, defined by

$$\lambda(H) := \sum_{(k,l) \in I} \frac{1}{k!} A^{k-1-l} \circ H \circ A^l ,$$

for every $H \in \mathrm{L}(X, X)$, is linear, and for every $H \in \mathrm{L}(X, X)$ satisfying $\|H\|_{op} = 1$, it follows from (1.11) that

$$\|\lambda(H)\|_{op} \leqslant \exp(\|A\|_{op}) .$$

Hence, $\lambda \in \mathrm{L}(\mathrm{L}(X, X), \mathrm{L}(X, X))$ with $\|\lambda\|_{op} \leqslant \exp(\|A\|_{op})$. For the proof of the continuity of $\exp'$, for $A_1, A_2, B \in \mathrm{L}(X, X)$, we have that

$$
\begin{aligned}
&[\exp'(A_2)](B) - [\exp'(A_1)](B) \\
&= \sum_{k=2}^{\infty} \frac{1}{k!} \left[\sum_{l=0}^{k-1} (A_2^{k-1-l} \circ B \circ A_2^l - A_1^{k-1-l} \circ B \circ A_1^l) \right] .
\end{aligned}
$$

Further, for $k, l \in \mathbb{N}$ such that $2 \leqslant k, 0 \leqslant l \leqslant k-1$, it follows that

$$
\begin{aligned}
&A_2^{k-1-l} \circ B \circ A_2^l - A_1^{k-1-l} \circ B \circ A_1^l \\
&= (A_2^{k-1-l} - A_1^{k-1-l}) \circ B \circ A_2^l + A_1^{k-1-l} \circ B \circ (A_2^l - A_1^l) \\
&= (A_2^{k-1-l} - A_1^{k-1-l}) \circ B \circ (A_2^l - A_1^l) + (A_2^{k-1-l} - A_1^{k-1-l}) \circ B \circ A_1^l \\
&\quad + A_1^{k-1-l} \circ B \circ (A_2^l - A_1^l)
\end{aligned}
$$

and hence that

$$
\begin{aligned}
&\|A_2^{k-1-l} \circ B \circ A_2^l - A_1^{k-1-l} \circ B \circ A_1^l\|_{op} \\
&\leqslant \Big[\|A_2^{k-1-l} - A_1^{k-1-l}\|_{op} \cdot \|A_2^l - A_1^l\|_{op} \\
&\quad + \|A_2^{k-1-l} - A_1^{k-1-l}\|_{op} \cdot \|A_1\|_{op}^l \\
&\quad + \|A_1\|_{op}^{k-1-l} \cdot \|A_2^l - A_1^l\|_{op} \Big] \|B\|_{op} \\
&\leqslant \|B\|_{op} \cdot \|A_1 - A_2\|_{op} \begin{cases} (\,\|A_1\|_{op} + \|A_2\|_{op})^{k-2} & \text{if } l = 0 \vee l = k-1 \\ 3\,(\,\|A_1\|_{op} + \|A_2\|_{op})^{k-2} & \text{if } 0 < l < k-1 \end{cases} \\
&\leqslant 3\,\|B\|_{op} \cdot (\,\|A_1\|_{op} + \|A_2\|_{op})^{k-2} \cdot \|A_1 - A_2\|_{op} ,
\end{aligned}
$$

where we used (1.8) and that we have that

$$
\begin{aligned}
&\|A_1^k - A_2^k\|_{op} \leqslant (\|A_1\|_{op} + \|A_2\|_{op})^{k-1} \cdot \|A_1 - A_2\|_{op} , \\
&\|A_2^{k-1-l} - A_1^{k-1-l}\|_{op} \cdot \|A_2^l - A_1^l\|_{op} \\
&\leqslant (\|A_1\|_{op} + \|A_2\|_{op})^{k-2-l} \cdot \|A_1 - A_2\|_{op} \cdot (\,\|A_1\|_{op}^l + \|A_2\|_{op}^l) \\
&\leqslant (\|A_1\|_{op} + \|A_2\|_{op})^{k-2} \cdot \|A_1 - A_2\|_{op} , \\
&\|A_2^{k-1-l} - A_1^{k-1-l}\|_{op} \cdot \|A_1\|_{op}^l \\
&\leqslant \|A_1\|_{op}^l \,(\|A_1\|_{op} + \|A_2\|_{op})^{k-2-l} \cdot \|A_1 - A_2\|_{op} \\
&\leqslant (\|A_1\|_{op} + \|A_2\|_{op})^{k-2} \cdot \|A_1 - A_2\|_{op} , \\
&\|A_1\|_{op}^{k-1-l} \cdot \|A_2^l - A_1^l\|_{op} \\
&\leqslant \|A_1\|_{op}^{k-1-l} \cdot (\|A_1\|_{op} + \|A_2\|_{op})^{l-1} \cdot \|A_1 - A_2\|_{op} \\
&\leqslant (\|A_1\|_{op} + \|A_2\|_{op})^{k-2} \cdot \|A_1 - A_2\|_{op} .
\end{aligned}
$$

As a consequence,

$$\begin{aligned}
&\|[\exp'(A_2)](B) - [\exp'(A_1)](B)\|_{op} \\
&\leqslant \sum_{k=2}^{\infty} \frac{1}{k!} \left[\sum_{l=0}^{k-1} 3\,\|B\|_{op} \cdot (\,\|A_1\|_{op} + \|A_2\|_{op})^{k-2} \cdot \|A_1 - A_2\|_{op} \right] \\
&= 3\,\|B\|_{op} \cdot \|A_1 - A_2\|_{op} \cdot \sum_{k=2}^{\infty} \frac{1}{(k-1)!} (\,\|A_1\|_{op} + \|A_2\|_{op})^{k-2} \\
&\leqslant 3\,\|B\|_{op} \cdot \|A_1 - A_2\|_{op} \cdot \sum_{k=2}^{\infty} \frac{1}{(k-2)!} (\,\|A_1\|_{op} + \|A_2\|_{op})^{k-2} \\
&= 3\,\|B\|_{op} \cdot \|A_1 - A_2\|_{op} \exp(\,\|A_1\|_{op} + \|A_2\|_{op}) \ .
\end{aligned}$$

The latter implies that

$$\|\exp'(A_2) - \exp'(A_1)\|_{op} \leqslant 3\,\exp(\,\|A_1\|_{op} + \|A_2\|_{op})\,\|A_1 - A_2\|_{op} \ ,$$

for all $A_1, A_2 \in \mathrm{L}(X, X)$ and hence that $\exp'$ is continuous. □

Corollary 1.12 (Local C^1-invertibility of the exponential map) *Let* $\mathbb{K} \in \{\mathbb{R}, \mathbb{C}\}$ *and* $(X, \|\ \|)$ *a non-trivial* $\mathbb{K}$*-Banach space. Then, the exponential function,* $\exp : \mathrm{L}(X, X) \to \mathrm{L}(X, X)$*, is locally* C^1*-invertible at* 0*, i.e., there is an open set* $U \subset \mathrm{L}(X, X)$ *around* 0 *such that* $\exp(U)$ *is open and such that the restriction of* exp *in domain to* U *and in image to* $\exp(U)$ *is bijective with a* C^1*-inverse.*

Proof We note that

$$[\,\exp'(0)](B) = B \ ,$$

for every $B \in \mathrm{L}(X, X)$, i.e., $\exp'(0)$ is given by the identical map on $\mathrm{L}(\mathrm{L}(X, X), \mathrm{L}(X, X))$ and hence bijective with a bounded linear inverse. Therefore, the statement is a direct consequence of Proposition 1.11 and the inverse mapping theorem, see, e.g., Theorem 1.2 in [38]. □

Proposition 1.13 (Commutators with one-parameter groups of exponentials) *Let* $\mathbb{K} \in \{\mathbb{R}, \mathbb{C}\}$*,* $(X, \|\ \|)$ *a* $\mathbb{K}$*-Banach space,* $A, B \in \mathrm{L}(X, X)$ *and* $\varepsilon > 0$*. Then*

$$[B, \exp(t.A)] = 0 \ ,$$

for every $t \in U_\varepsilon(0)$*, if and only if*

$$[B, A] = 0 \ .$$

Proof For the proof, we define $u_A(t) := \exp(t.A)$, for every $t \in U_\varepsilon(0)$. Then,

$$\left\| \frac{1}{t} [B \circ u_A(t) - B \circ u_A(0)] - B \circ A \right\|_{\text{op}}$$
$$= \left\| B \circ \left\{ \frac{1}{t} [u_A(t) - u_A(0)] - A \circ u_A(0) \right\} \right\|_{\text{op}}$$
$$\leqslant \|B\|_{\text{op}} \cdot \left\| \frac{1}{t} [u_A(t) - u_A(0)] - A \circ u_A(0) \right\|_{\text{op}} ,$$
$$\left\| \frac{1}{t} [u_A(t) \circ B - u_A(0) \circ B] - A \circ B \right\|_{\text{op}}$$
$$= \left\| \left\{ \frac{1}{t} [u_A(t) - u_A(0)] - A \circ u_A(0) \right\} \circ B \right\|_{\text{op}}$$
$$\leqslant \|B\|_{\text{op}} \cdot \left\| \frac{1}{t} [u_A(t) - u_A(0)] - A \circ u_A(0) \right\|_{\text{op}} ,$$

for $t \in U_\varepsilon(0) \setminus \{0\}$. Hence the maps

$$v_1 := (U_\varepsilon(0) \to \mathrm{L}(X, X), t \mapsto B \circ u_A(t)) ,$$
$$v_2 := (U_\varepsilon(0) \to \mathrm{L}(X, X), t \mapsto u_A(t) \circ B)$$

are differentiable in 0, with derivatives

$$v_1'(0) = B \circ A \ , \ \ v_2'(0) = A \circ B \ .$$

If B commutes with every $u_A(t)$, $t \in U_\varepsilon(0)$, then $v_1 = v_2$ and hence

$$[B, A] = 0 \ .$$

If $[B, A] = 0$, we proceed as follows. For $t \in U_\varepsilon(0)$, $\delta > 0$, there is $n_0 \in \mathbb{N}$ such that

$$\left\| \sum_{k=0}^{n_0} \frac{1}{k!} . (t.A)^k - \exp(t.A) \right\|_{\text{op}} \leqslant \delta \ .$$

Hence, it follows that

$$
\begin{aligned}
&\|B \circ \exp(t.A) - \exp(t.A) \circ B\|_{\mathrm{op}} \\
&= \Bigg\| B \circ \exp(t.A) - B \circ \sum_{k=0}^{n_0} \frac{1}{k!} \cdot (t.A)^k \\
&\quad + \left[\sum_{k=0}^{n_0} \frac{1}{k!} \cdot (t.A)^k\right] \circ B - \exp(t.A) \circ B \Bigg\|_{\mathrm{op}} \\
&= \Bigg\| B \circ \left[\exp(t.A) - \sum_{k=0}^{n_0} \frac{1}{k!} \cdot (t.A)^k\right] \\
&\quad + \left\{\left[\sum_{k=0}^{n_0} \frac{1}{k!} \cdot (t.A)^k\right] - \exp(t.A)\right\} \circ B \Bigg\|_{\mathrm{op}} \\
&\leqslant 2\,\|B\|_{\mathrm{op}} \cdot \left\| \sum_{k=0}^{n_0} \frac{1}{k!} \cdot (t.A)^k - \exp(t.A) \right\|_{\mathrm{op}} \leqslant 2\,\|B\|_{\mathrm{op}}\, \delta \ .
\end{aligned}
$$

Since the latter is true for every $\delta > 0$, it follows that

$$[B, \exp(t.A)] = 0 \ ,$$

for every $t \in U_{\varepsilon}(0)$. □

Proposition 1.14 (Invariant subspaces of one-parameter groups of exponentials) *Let* $\mathbb{K} \in \{\mathbb{R}, \mathbb{C}\}$, $(X, \|\ \|)$ *a* $\mathbb{K}$*-Banach space,* $A \in \mathrm{L}(X, X)$*,* Y *a closed subspace of* X *and* $\varepsilon > 0$*. Then* Y *is invariant under* $\exp(t.A)$*, for every* $t \in U_{\varepsilon}(0)$*, if and only if* Y *is invariant under* A*.*

Proof '$\Leftarrow$': If Y is invariant under A, we conclude as follows. For $t \in U_{\varepsilon}(0)$ and $x \in X$, we have that

$$\left\| \sum_{k=0}^{n} \frac{1}{k!} \cdot (t.A)^k x - \exp(t.A)x \right\| \leqslant \left\| \sum_{k=0}^{n} \frac{1}{k!} \cdot (t.A)^k - \exp(t.A) \right\|_{\mathrm{op}} \cdot \|x\| \ ,$$

for every $n \in \mathbb{N}^*$, and hence that

$$\lim_{n \to \infty} \sum_{k=0}^{n} \frac{1}{k!} \cdot (t.A)^k x = \exp(t.A)x \ .$$

In particular, if $x \in Y$, then

$$\sum_{k=0}^{n} \frac{1}{k!} \cdot (t.A)^k x \in Y \ ,$$

for every $n \in \mathbb{N}^*$, and

$$\exp(t.A)x = \lim_{n\to\infty} \sum_{k=0}^{n} \frac{1}{k!} . (t.A)^k x \in Y .$$

Hence, Y is invariant under $\exp(t.A)$, for every $t \in U_\varepsilon(0)$.
'$\Rightarrow$': If Y is invariant under $\exp(t.A)$, for every $t \in U_\varepsilon(0)$, we define $u_A(t) := \exp(t.A)$, for every $t \in U_\varepsilon(0)$. For $x \in X$ and $t \in U_\varepsilon(0)\backslash\{0\}$, we have that

$$\begin{aligned} \left\| \frac{1}{t} [u_A(t)x - x] - Ax \right\| &= \left\| \left[\frac{1}{t} [u_A(t) - u_A(0)] - A \right] x \right\| \\ &\leqslant \left\| \left[\frac{1}{t} [u_A(t) - u_A(0)] - A \circ u_A(0) \right] \right\|_{\text{op}} \cdot \|x\| \end{aligned}$$

and hence that $(U_\varepsilon(0) \to X, t \mapsto u_A(t)x)$ is differentiable in 0, with derivative Ax, i.e., that

$$\lim_{t\to 0, t\neq 0} \frac{1}{t} [u_A(t)\, x - x] = Ax .$$

In particular, if $x \in Y$, then

$$\frac{1}{t} [u_A(t)\, x - x] \in Y ,$$

for every $t \in U_\varepsilon(0)\backslash\{0\}$, and

$$Ax = \lim_{t\to 0, t\neq 0} \frac{1}{t} [u_A(t)\, x - x] \in Y .$$

Since this is true for every $x \in Y$, A leaves Y invariant. □

Proposition 1.15 (Determinants of matrix exponentials) *Let* $\mathbb{K} \in \{\mathbb{R}, \mathbb{C}\}$, $n \in \mathbb{N}^*$ *and* $A \in \mathrm{M}(n, \mathbb{K})$. *Then*

$$\det(\exp(A)) = \exp(Tr(A)) .$$

Proof If $\mathbb{K} = \mathbb{C}$ and $A \in \mathrm{M}(n, \mathbb{C})$ has n pairwise different eigenvalues $\lambda_1, \dots, \lambda_n \in \mathbb{C}$, with corresponding eigenvectors $v_1, \dots, v_n \in \mathbb{C}^n$, then

$$\left(\sum_{k=0}^{n} \frac{1}{k!} . A^k \right) \cdot v_j = \sum_{k=0}^{n} \frac{1}{k!} . (A^k \cdot v_j) = \left(\sum_{k=0}^{n} \frac{\lambda_j^k}{k!} \right) v_j ,$$

and hence, using that for every $B \in \mathrm{M}(n, \mathbb{C})$, we have that $|B \cdot v| \leqslant \|B\|_{op} \cdot |v|$, for every $\mathbb{C}^n$, where $|\ |$ is the canonical norm on $\mathbb{C}^n$, that

$$\exp(A) \cdot v_j = \exp(\lambda_j).v_j ,$$

for every $j \in \{1, \dots, n\}$. Hence, if $S \in \mathrm{M}(n, \mathbb{C})$ is the matrix that has the vector v_j as the j-th column vector, $j \in \{1, \dots, n\}$, then

$$A \cdot S = S \cdot \text{diag}(\lambda_1, \ldots, \lambda_n) \ , \ \ \exp(A) \cdot S = S \cdot \text{diag}(\exp(\lambda_1), \ldots, \exp(\lambda_n)) \ ,$$

where for every sequence $a_1, \ldots, a_n \in \mathbb{C}$, $\text{diag}(a_1, \ldots, a_n)$ denotes the diagonal matrix $D \in \text{M}(n, \mathbb{C})$ satisfying $D_{jj} = a_j$, $j \in \{1, \ldots, n\}$. Hence,

$$S^{-1} \cdot A \cdot S = \text{diag}(\lambda_1, \ldots, \lambda_n) \ , \ \ S^{-1} \cdot \exp(A) \cdot S = \text{diag}(\exp(\lambda_1), \ldots, \exp(\lambda_n)) \ ,$$

where we note that $\det(S) \neq 0$, since $v_1, \ldots, v_n$ are linearly independent, and

$$\begin{aligned} \det(\exp(A)) &= \det(S^{-1} \cdot \exp(A) \cdot S) = \prod_{j=1}^{n} \exp(\lambda_j) = \exp\left(\sum_{j=1}^{n} \lambda_j \right) \\ &= \exp\big(\text{Tr}(S^{-1} \cdot A \cdot S)\big) = \exp(\text{Tr}(A)) \ . \end{aligned}$$

If $A \in \text{M}(n, \mathbb{C})$, according to linear algebra, there is $S \in \text{M}(n, \mathbb{C})$ such that $S^{-1} \cdot A \cdot S$ is an upper triangular matrix. For every $k \in \{1, \ldots, n\}$, we define $a_k := (S^{-1} \cdot A \cdot S)_{kk}$. Since $S^{-1} \cdot A \cdot S$ is upper triangular, $a_1, \ldots, a_n$ are eigenvalues of $S^{-1} \cdot A \cdot S$ (and hence also of A). Then, there is a sequence $\varepsilon_1, \varepsilon_2, \ldots$ in $(0, \infty)$ that is converging to 0 and such that $a_1 - \varepsilon_\nu, \ldots, a_n - \nu\varepsilon_\nu$ are pairwise different for every $\nu \in \mathbb{N}^*$. Then, $A_1, A_2, \ldots$, defined by

$$A_\nu := A - \varepsilon_\nu \, S \, \text{diag}(1, \ldots, n) \, S^{-1} \ ,$$

for every $\nu \in \mathbb{N}^*$, is a sequence in $\text{M}(n, \mathbb{C})$ that is convergent to A and whose members have pairwise different eigenvalues. Hence

$$\det(\exp(A_\nu)) = \exp(\text{Tr}(A_\nu)) \ ,$$

for every $\nu \in \mathbb{N}^*$, and, since the involved maps are continuous,

$$\det(\exp(A)) = \lim_{\nu \to \infty} \det(\exp(A_\nu)) = \lim_{\nu \to \infty} \exp(\text{Tr}(A_\nu)) = \exp(\text{Tr}(A)) \ .$$

Therefore, it follows that $\det(\exp(A)) = \exp(\text{Tr}(A))$, for every $A \in \text{M}(n, \mathbb{C})$ and hence also that $\det(\exp(A)) = \exp(\text{Tr}(A))$, for every $A \in \text{M}(n, \mathbb{R})$. □

Proposition 1.16 (Tangent spaces of $GL(n, \mathbb{K})$, $GL_+(n, \mathbb{R})$) *Let $\mathbb{K} \in \{\mathbb{R}, \mathbb{C}\}$, $n \in \mathbb{N}^*$ and E be the $n \times n$ unit matrix. Then,*

$$T_E\text{GL}(n, \mathbb{K}) = \text{M}(n, \mathbb{K}) \ , \ \ T_E\text{GL}_+(n, \mathbb{R}) = \text{M}(n, \mathbb{R})$$

In particular, the real vector spaces $T_E\text{GL}(n, \mathbb{R}) = T_E\text{GL}_+(n, \mathbb{R})$ and $T_E\text{GL}(n, \mathbb{C})$ are of dimension n^2 and $2\,n^2$, respectively.

Proof According to Definition 1.5, we have that

$$T_E GL(n, \mathbb{K}) \subset \mathrm{M}(n, \mathbb{K}) \ .$$

Further, if $A \in \mathrm{M}(n \times n, \mathbb{K})$, then $u_A : \mathbb{R} \to \mathrm{M}(n, \mathbb{C})$, defined by $u_A(t) := \exp(tA)$, for every $t \in \mathbb{R}$, is a differentiable one-parameter subgroup of $GL(n, \mathbb{K})$ such that $u'_A(0) = A$. Further, according to Proposition 1.15, we have that

$$\det(u_A(t)) = \det(\exp(tA)) = \exp(\mathrm{T}r(tA)) > 0 \ ,$$

for every $t \in \mathbb{R}$. Hence,

$$T_E GL(n, \mathbb{K}) = \mathrm{M}(n, \mathbb{K}) \ , \ \ T_E GL_+(n, \mathbb{R}) = \mathrm{M}(n, \mathbb{R}) \ .$$

□

Proposition 1.17 (Tangent spaces of $SL(n, \mathbb{K})$) *Let $\mathbb{K} \in \{\mathbb{R}, \mathbb{C}\}$, $n \in \mathbb{N}^*$ and E be the $n \times n$ unit matrix. Then,*

$$T_E SL(n, \mathbb{K}) = \{A \in \mathrm{M}(n, \mathbb{K}) : \mathrm{T}r(A) = 0\} \ .$$

In particular, the real vector spaces $T_E SL(n, \mathbb{R})$ and $T_E SL(n, \mathbb{C})$ are of dimension $n^2 - 1$ and $2\,(n^2 - 1)$, respectively.

Proof According to Lemma 1.7, we have that

$$T_E SL(n, \mathbb{K}) \subset \{A \in \mathrm{M}(n, \mathbb{K}) : \mathrm{T}r(A) = 0\} \ .$$

Further, if $A \in \mathrm{M}(n \times n, \mathbb{K})$ is a trace zero matrix, then $u_A : \mathbb{R} \to \mathrm{M}(n, \mathbb{C})$, defined by $u_A(t) := \exp(tA)$, for every $t \in \mathbb{R}$, is a differentiable one-parameter subgroup of $GL(n, \mathbb{K})$ such that $u'_A(0) = A$. Further, according to Proposition 1.15, we have that

$$\det(u_A(t)) = \det(\exp(tA)) = \exp(\mathrm{T}r(tA)) = \exp(t\mathrm{T}r(A)) = \exp(0) = 1 \ ,$$

for every $t \in \mathbb{R}$. Hence, $\mathrm{R}an(u_A) \subset SL(n, \mathbb{K})$ and, according to Definition 1.5, $A \in T_E SL(n, \mathbb{K})$. □

Proposition 1.18 (Calculation of matrix exponentials I) *Let $n \in \mathbb{N}^*$, $H \in \mathrm{M}(n, \mathbb{C})$ be Hermitian and $v_1, \ldots, v_n$ a corresponding basis of eigenvectors. In addition, let $\lambda_1, \ldots, \lambda_n \in \mathbb{R}$ be the corresponding, not necessarily pairwise different, eigenvalues of H, i.e., such that $H \cdot v_j = \lambda_j . v_j$, for every $j \in \{1, \ldots, n\}$. Then,*

$$\exp(zH) = U \cdot \mathrm{diag}(\exp(z\lambda_1), \ldots, \exp(z\lambda_n)) \cdot U^* \ ,$$

for every $z \in \mathbb{C}$*, where* $U \in U(n)$ *is the matrix that has the vector* v_j *as the* j*-th column vector and* $\operatorname{diag}(\exp(z\lambda_1), \ldots, \exp(z\lambda_n))$ *denotes the diagonal matrix* $D \in \mathrm{M}(n, \mathbb{C})$ *satisfying* $D_{jj} = \exp(z\lambda_j)$*, for* $j \in \{1, \ldots, n\}$.

Proof In the following, $\langle | \rangle$ denotes the canonical scalar product for $\mathbb{C}^n$. Since H is Hermitian, there is an orthonormal basis $v_1, \ldots, v_n$ of eigenvectors of H. Let $\lambda_1, \ldots, \lambda_n \in \mathbb{R}$ be the corresponding, not necessarily pairwise different, eigenvalues of H, i.e., such that $H \cdot v_j = \lambda_j . v_j$, for every $j \in \{1, \ldots, n\}$. Then,

$$\sum_{k=0}^{m} \frac{1}{k!} \cdot (zH)^k \cdot v_j = \left[\sum_{k=0}^{m} \frac{(z\lambda_j)^k}{k!} \right] . v_j \ ,$$

for every $m \in \mathbb{N}$ and hence

$$\exp(zH) \cdot v_j = \exp(z\lambda_j) . v_j \ ,$$

for $z \in \mathbb{C}$. Hence, if $U \in U(n)$ is the matrix that has the vector v_j as the j-th column vector, $j \in \{1, \ldots, n\}$, then

$$\exp(zH) \cdot U = U \cdot \operatorname{diag}(\exp(z\lambda_1), \ldots, \exp(z\lambda_n)) \ ,$$

where and $\operatorname{diag}(\exp(z\lambda_1), \ldots, \exp(z\lambda_n))$ denotes the diagonal matrix $D \in \mathrm{M}(n, \mathbb{C})$ satisfying $D_{jj} = \exp(z\lambda_j)$, for $j \in \{1, \ldots, n\}$, and therefore

$$\exp(zH) = U \cdot \operatorname{diag}(\exp(z\lambda_1), \ldots, \exp(z\lambda_n)) \cdot U^* \ .$$

□

Proposition 1.19 (Differentiable one-parameter subgroups of $\mathrm{U}(n)$ and $SO(n)$) *Let* $n \in \mathbb{N}^*$*. Then, the following is true.*

(i) If $A \in \mathrm{M}(n, \mathbb{C})$ *is anti-Hermitian, then* $u_A : \mathbb{R} \to \mathrm{M}(n, \mathbb{C})$*, defined by* $u_A(t) := \exp(tA)$*, for every* $t \in \mathbb{R}$*, is a differentiable one-parameter subgroup of* $U(n)$ *with derivative* $u_A'(t) = A \cdot u_A(t)$*, for every* $t \in \mathbb{R}$*. If, in addition,* $\operatorname{Tr}(A) = 0$*, then* $\operatorname{Ran}(u_A) \subset SU(n)$*.*
(ii) If $A \in \mathrm{M}(n, \mathbb{R})$ *is anti-symmetric, then* $u_A : \mathbb{R} \to \mathrm{M}(n, \mathbb{R})$*, defined by* $u_A(t) := \exp(tA)$*, for every* $t \in \mathbb{R}$*, is a differentiable one-parameter subgroup of* $SO(n)$ *with derivative* $u_A'(t) = A \cdot u_A(t)$*, for every* $t \in \mathbb{R}$*.*

Proof If $t \in \mathbb{R}$ and $A \in \mathrm{M}(n, \mathbb{C})$ is anti-Hermitian, then iA is Hermitian, since $(iA)^* = -i\,(-A) = iA$. If $v_1, \ldots, v_n$ is a basis of eigenvectors of iA and $\lambda_1, \ldots, \lambda_n \in \mathbb{R}$ are the corresponding, not necessarily pairwise different, eigenvalues, i.e., such that $iA \cdot v_j = \lambda_j . v_j$, for every $j \in \{1, \ldots, n\}$. Then,

$$\exp(tA) = U \cdot \text{diag}(\exp(-it\lambda_1), \ldots, \exp(-it\lambda_n)) \cdot U^* ,$$

where $U \in U(n)$ is the matrix that has the vector v_j as the j-th column vector and $\text{diag}(\exp(-it\lambda_1), \ldots, \exp(-it\lambda_n))$ denotes the diagonal matrix $D \in \text{M}(n, \mathbb{C})$ satisfying $D_{jj} = \exp(-it\lambda_j)$, for $j \in \{1, \ldots, n\}$. Since the column vectors of $\text{diag}(\exp(-it\lambda_1), \ldots,$ $\exp(-it\lambda_n))$ form an orthonormal basis of $\mathbb{C}^n$, equipped with the canonical scalar product, $\text{diag}(\exp(-it\lambda_1), \ldots, \exp(-it\lambda_n)) \in U(n)$ and hence $\exp(tA) \in U(n)$. As a consequence, $u_A : \mathbb{R} \to \text{M}(n, \mathbb{C})$, defined by $u_A(t) := \exp(tA)$, for every $t \in \mathbb{R}$, is a differentiable one-parameter subgroup of $U(n)$ with derivative $u_A'(t) = A \circ u_A(t)$, for every $t \in \mathbb{R}$. If, in addition, $\text{T}r(A) = 0$, it follows from Proposition 1.15 that

$$\det(\exp(tA)) = \exp(\text{T}r(tA)) = \exp(t\text{T}r(A)) = \exp(0) = 1$$

and hence that $\text{R}an(u_A) \subset SU(n)$. In the special case that $A \in \text{M}(n, \mathbb{R})$ is anti-symmetric, $t \in \mathbb{R}$, we have that $\exp(tA) \in \text{M}(n, \mathbb{R})$ and that the column vectors of $\exp(tA)$ form a orthonormal basis of $\mathbb{C}^n$, equipped with the canonical scalar product. Since these column vectors are elements of $\mathbb{R}^n$, the column vectors of $\exp(tA)$ also form a orthonormal basis of $\mathbb{R}^n$, equipped with the canonical scalar product. Hence, $\exp(tA) \in \text{O}(n)$. Further, it follows from Proposition 1.15 that

$$\det(\exp(tA)) = \exp(\text{T}r(tA)) = \exp(t\text{T}r(A)) = \exp(0) = 1$$

and hence $\exp(tA) \in SO(n)$. As a consequence, $u_A : \mathbb{R} \to \text{M}(n, \mathbb{R})$, defined by $u_A(t) :=$ $\exp(tA)$, for every $t \in \mathbb{R}$, is a differentiable one-parameter subgroup of $SO(n)$ with derivative $u_A'(t) = A \circ u_A(t)$, for every $t \in \mathbb{R}$. □

Proposition 1.20 (Tangent spaces of $SO(n)$, $\text{O}(n)$, $SU(n)$ and $\text{U}(n)$) *Let $n \in \mathbb{N}^*$ and E be the $n \times n$ unit matrix. Then,*

$$T_E SO(n) = T_E\text{O}(n) = \{A \in \text{M}(n, \mathbb{K}) : A \textit{ is anti-symmetric}\} ,$$

$$T_E SU(n) = \{A \in \text{M}(n, \mathbb{K}) : A \textit{ is anti-Hermitian and} \, \text{Tr}(A) = 0\} ,$$

$$T_E\text{U}(n) = \{A \in \text{M}(n, \mathbb{K}) : A \textit{ is anti-Hermitian}\} .$$

In particular, the real vector spaces $T_E SO(n)$, $T_E\text{O}(n)$ are of dimension $\frac{n(n-1)}{2}$; the real vector spaces $T_E SU(n)$ and $T_E\text{U}(n)$ are of dimension $n^2 - 1$ and n^2, respectively.

Proof The statement is a direct consequence of Lemma 1.7 and Proposition 1.19. □

Theorem 1.21 (Spectral decomposition of normal real linear maps) *Let $M \in \text{M}(n, \mathbb{R})$ be normal. Then, there is an orthonormal basis $x_1, \ldots, x_m, y_{11}, y_{12}, \ldots, y_{m'1}, y_{m'2}$ of $\mathbb{R}^n$, where m and m' are natural numbers such that $m + 2m' = n$, with the following properties.*

(i) *The (possibly empty) sequence* $x_1, \ldots, x_m$, *with corresponding sequence of eigenvalues* $\Lambda_1, \ldots, \Lambda_m$, *is a basis of the span of the eigenvectors of* M.

(ii) *There is a (possibly empty) sequence* $(\lambda_1, \mu_1), \ldots, (\lambda_{m'}, \mu_{m'})$ *in* $\mathbb{R} \times (0, \infty)$ *such that for every* $k \in \{1, \ldots, m'\}$

$$M \cdot y_{k1} = \lambda_k y_{k1} - \mu_k y_{k2} \ , \ \ M \cdot y_{k2} = \mu_k y_{k1} + \lambda_k y_{k2} \ .$$

(iii)

$$\det(M) = \left[\prod_{k=1}^{m} \Lambda_k \right] \cdot \left[\prod_{k=1}^{m'} (\lambda_k^2 + \mu_k^2) \right] \ ,$$

where we use the convention that empty products are equal to 1.

Proof We consider $\mathrm{M}(n, \mathbb{R})$ as a subset of $\mathrm{M}(n, \mathbb{C})$ and $\mathbb{C}^n$ as equipped with the canonical scalar product $\langle | \rangle$. Since M is a normal real matrix, and taking into account that the coefficients of its characteristic polynomial are real, it follows the existence of an orthonormal basis $x_1, \ldots, x_m, z_1, z_1^*, \ldots, z_{m'}, z_{m'}^*$ of eigenvectors of M, where $x_1, \ldots, x_m \in \mathbb{R}^n$ is a basis of the span of the eigenvectors of A corresponding to real eigenvalues $\Lambda_1, \ldots, \Lambda_m$ of M and $z_1, z_1^*, \ldots, z_{m'}, z_{m'}^*$ is a basis of the span of eigenvectors of M corresponding to non-real eigenvalues $\kappa_1, \kappa_1^*, \ldots, \kappa_{m'}, \kappa_{m'}^*$ of M, where m and m' are natural numbers such that $m + 2m' = n$. If $m = 0$, then the sequence $x_1, \ldots, x_m$ is empty. Also, if $m' = 0$, then the sequences $z_1, z_1^*, \ldots, z_{m'}, z_{m'}^*$ and $\kappa_1, \kappa_1^*, \ldots, \kappa_{m'}, \kappa_{m'}^*$ are empty. For $k \in \{1, \ldots, m'\}$, $z_k = y_{k1} + i y_{k2}$, where $y_{k1}, y_{k2} \in \mathbb{R}^n$, and $\kappa_k = \lambda_k + i\mu_k$, where $\lambda_k \in \mathbb{R}$, $\mu_k \in \mathbb{R}^*$, and we can assume without loss of generality that $\mu_k > 0$. Then,

$$\begin{aligned} 1 &= \langle y_{k1} + i y_{k2} | y_{k1} + i y_{k2} \rangle = \langle y_{k1} | y_{k1} \rangle - i \langle y_{k2} | y_{k1} \rangle + i \langle y_{k1} | y_{k2} \rangle + \langle y_{k2} | y_{k2} \rangle \\ &= \langle y_{k1} | y_{k1} \rangle + \langle y_{k2} | y_{k2} \rangle \ , \\ 0 &= \langle y_{k1} + i y_{k2} | y_{k1} - i y_{k2} \rangle = \langle y_{k1} | y_{k1} \rangle - i \langle y_{k2} | y_{k1} \rangle - i \langle y_{k1} | y_{k2} \rangle - \langle y_{k2} | y_{k2} \rangle \\ &= \langle y_{k1} | y_{k1} \rangle - 2i \langle y_{k1} | y_{k2} \rangle - \langle y_{k2} | y_{k2} \rangle \ , \end{aligned}$$

it follows that

$$\langle y_{k1} | y_{k1} \rangle = \langle y_{k2} | y_{k2} \rangle = 1 \ , \ \ \langle y_{k1} | y_{k2} \rangle = 0 \ ,$$

i.e., that $y_{k1}, y_{k2} \in \mathbb{R}^n$ are orthonormal, where $\mathbb{R}^n$ is equipped with the canonical scalar product $\langle | \rangle$. Further, we have that

$$\begin{aligned} M \cdot y_{k1} + i M \cdot y_{k2} &= M \cdot (y_{k1} + i y_{k2}) \\ &= (\lambda_k + i\mu_k)(y_{k1} + i y_{k2}) = \lambda_k y_{k1} - \mu_k y_{k2} + i(\mu_k y_{k1} + \lambda_k y_{k2}) \ , \end{aligned}$$

and hence that

Table 1.3 Ranges of the matrix-valued exponential map. The proof of the 3 statements at the bottom of the table is part of Exercises 1.2 and 1.3

$\exp(T_E GL(n, \mathbb{C})) = GL(n, \mathbb{C})$
$\exp(T_E \mathrm{O}(n)) = \exp(T_E SO(n)) = SO(n)$
$\exp(T_E \mathrm{U}(n)) = \mathrm{U}(n)$
$\exp(T_E SU(n)) = SU(n)$
$\exp(T_E SL(2, \mathbb{R})) \subsetneqq SL(2, \mathbb{R})$
$\exp(T_E SL(2, \mathbb{C})) \subsetneqq SL(2, \mathbb{C})$
$\exp(T_E GL_+(2, \mathbb{R})) \subsetneqq GL_+(2, \mathbb{R})$

$$M \cdot y_{k1} = \lambda_k y_{k1} - \mu_k y_{k2} \ , \quad M \cdot y_{k2} = \mu_k y_{k1} + \lambda_k y_{k2} \ .$$

For the proof of (iii), let $U \in \mathrm{M}(n, \mathbb{C})$ be the matrix that has $x_1, \ldots, x_m, z_1, z_1^*, \ldots, z_{m'}, z_{m'}^*$, in this order, as column vectors. Since, $x_1, \ldots, x_m, z_1, z_1^*, \ldots, z_{m'}, z_{m'}^*$ is an orthogonal basis of $\mathbb{C}^n$, U is unitary. In addition, we have that

$$(M \cdot U)^t = \mathrm{diag}(\lambda_1, \ldots, \lambda_m, \kappa_1, \ldots, \kappa_{m'}) \cdot U^t$$

and hence that

$$\begin{aligned} U^{-1} \cdot M \cdot U &= U^{-1} \cdot U \cdot \mathrm{diag}(\lambda_1, \ldots, \lambda_m, \kappa_1, \ldots, \kappa_{m'}) \\ &= \mathrm{diag}(\lambda_1, \ldots, \lambda_m, \kappa_1, \ldots, \kappa_{m'}) \ . \end{aligned}$$

As a consequence,

$$\begin{aligned} \det(M) &= \det(U^{-1} \cdot M \cdot U) = \det(\mathrm{diag}(\lambda_1, \ldots, \lambda_m, \kappa_1, \ldots, \kappa_{m'})) \\ &= \left[\prod_{k=1}^{m} \Lambda_k \right] \cdot \left[\prod_{k=1}^{m'} (\lambda_k^2 + \mu_k^2) \right] \ . \end{aligned}$$

□

In the following, we are going to investigate the surjectivity of the matrix-valued exponential map. For the results, compare Table 1.3.

Proposition 1.22 (Surjectivity of the matrix exponential map I) *Let $n \in \mathbb{N}^*$. For every $R \in SO(n)$, there is an anti-symmetric $A \in \mathrm{M}(n, \mathbb{R})$ such that $R = \exp(A)$.*

Proof For this purpose, let $\mathbb{R}^n$ be equipped with the canonical scalar product $\langle | \rangle$, $R \in SO(n)$ and $\rho : \mathbb{R}^n \to \mathbb{R}^n$, defined by $\rho(x) := R \cdot x$, for every $x \in \mathbb{R}^n$. Then, according to Proposition 1.21, there is an orthonormal basis $x_1, \ldots, x_m, y_{11}, y_{12}, \ldots, y_{m'1}, y_{m'2}$ of $\mathbb{R}^n$, where m and m' are natural numbers such that $m + 2m' = n$; for every $k \in \{1, \ldots, m\}$, x_k is an eigenvector of R corresponding to a real eigenvalue Λ_k; and for every $k \in \{1, \ldots, m'\}$, there are $\lambda_k \in \mathbb{R}$ and $\mu_k > 0$ such that

$$R \cdot y_{k1} = \lambda_k y_{k1} - \mu_k y_{k2} \ , \quad R \cdot y_{k2} = \mu_k y_{k1} + \lambda_k y_{k2} \ .$$

In the latter case, since

$$\begin{aligned} 1 &= \langle y_{k1} | y_{k1} \rangle = \langle R \cdot y_{k1} | R \cdot y_{k1} \rangle \\ &= \langle \lambda_k y_{k1} - \mu_k y_{k2} | \lambda_k y_{k1} - \mu_k y_{k2} \rangle = \lambda_k^2 + \mu_k^2 \ , \end{aligned}$$

we have that

$$\lambda_k^2 + \mu_k^2 = 1 \ .$$

Then, we define $V_k := Span\{y_{k1}, y_{k2}\}$ and $B_k : V_k \to V_k$ as the unique linear map satisfying

$$B_k y_{k1} = -y_{k2} \ , \quad B_k y_{k2} = y_{k1} \ .$$

We note that B_k is anti-symmetric, since the representation matrix of B_k with respect to the basis $\{y_{k1}, y_{k2}\}$ of V_k is given by the anti-symmetric 2×2-matrix

$$\begin{pmatrix} 0 & -1 \\ 1 & 0 \end{pmatrix} .$$

Further, we have that

$$\begin{aligned} B_k^{2l} y_{k1} &= (-1)^l y_{k1} \ , \quad B_k^{2l} y_{k2} = (-1)^l y_{k2} \ , \\ B_k^{2l+1} y_{k1} &= -(-1)^l y_{k2} \ , \quad B_k^{2l+1} y_{k2} = (-1)^l y_{k1} \ , \end{aligned}$$

for every $l \in \mathbb{N}$. The proof proceeds by induction on $l \in \mathbb{N}$. For $l = 0$, the statement is obviously satisfied. If the statement is true for $l \in \mathbb{N}$, then

$$\begin{aligned} B_k^{2(l+1)} y_{k1} &= -(-1)^l B_k y_{k2} = (-1)^{l+1} y_{k1} \ , \\ B_k^{2(l+1)} y_{k2} &= (-1)^l B_k y_{k1} = (-1)^{l+1} y_{k2} \ , \\ B_k^{2(l+1)+1} y_{k1} &= (-1)^{l+1} B_k y_{k1} = -(-1)^{l+1} y_{k2} \ , \\ B_k^{2(l+1)+1} y_{k2} &= (-1)^{l+1} B_k y_{k2} = (-1)^{l+1} y_{k1} \ . \end{aligned}$$

Hence, the statement is true for $l + 1$. In the next step, we conclude that

$$\begin{aligned}
\exp(\varphi B_k)\, y_{k1} &= \sum_{l=0}^{\infty} \frac{\varphi^l}{l!} B_k^l\, y_{k1} = \sum_{l=0}^{\infty} \frac{\varphi^{2l}}{(2l)!} B_k^{2l}\, y_{k1} + \sum_{l=0}^{\infty} \frac{\varphi^{2l+1}}{(2l+1)!} B_k^{2l+1}\, y_{k1} \\
&= \left[\sum_{l=0}^{\infty} (-1)^l \frac{\varphi^{2l}}{(2l)!}\right] y_{k1} - \left[\sum_{l=0}^{\infty} (-1)^l \frac{\varphi^{2l+1}}{(2l+1)!}\right] y_{k2} \\
&= \cos(\varphi)\, y_{k1} - \sin(\varphi)\, y_{k2} \ , \\
\exp(\varphi B_k)\, y_{k2} &= \sum_{l=0}^{\infty} \frac{\varphi^l}{l!} B_k^l\, y_{k2} = \sum_{l=0}^{\infty} \frac{\varphi^{2l}}{(2l)!} B_k^{2l}\, y_{k2} + \sum_{l=0}^{\infty} \frac{\varphi^{2l+1}}{(2l+1)!} B_k^{2l+1}\, y_{k2} \\
&= \left[\sum_{l=0}^{\infty} (-1)^l \frac{\varphi^{2l}}{(2l)!}\right] y_{k2} + \left[\sum_{l=0}^{\infty} (-1)^l \frac{\varphi^{2l+1}}{(2l+1)!}\right] y_{k1} \\
&= \sin(\varphi)\, y_{k1} + \cos(\varphi)\, y_{k2} \ ,
\end{aligned}$$

for every $\varphi \in \mathbb{R}$. Then $\alpha_k := \arccos(\lambda_k) \in [0, \pi]$ satisfies $\cos(\alpha_k) = \lambda_k$. Further, since

$$\mu_k^2 = 1 - \lambda_k^2 = 1 - \cos^2(\alpha_k) = \sin^2(\alpha_k) \ ,$$

and $\mu_k > 0$, $\sin(\alpha_k) \geqslant 0$, we have that $\sin(\alpha_k) = \mu_k$. As a consequence,

$$\begin{aligned}
R \cdot y_{k1} &= \lambda_k y_{k1} - \mu_k y_{k2} = \cos(\alpha_k)\, y_{k1} - \sin(\alpha_k)\, y_{k2} = \exp(\alpha_k B_k)\, y_{k1} \ , \\
R \cdot y_{k2} &= \mu_k y_{k1} + \lambda_k y_{k2} = \sin(\alpha_k)\, y_{k1} + \cos(\alpha_k)\, y_{k2} = \exp(\alpha_k B_k)\, y_{k2} \ .
\end{aligned}$$

Further, if $k \in \{1, \ldots, m\}$, we define $V_k := \mathbb{R}.x_k$. Since

$$R^* R = E \ ,$$

where E denotes the $n \times n$ unit matrix,

$$1 = \langle x_k | x_k \rangle = \langle R x_k | R x_k \rangle = \langle \Lambda_k x_k | \Lambda_k x_k \rangle = \Lambda_k^2 \langle x_k | x_k \rangle = \Lambda_k^2 \ ,$$

where $\Lambda_k \in \mathbb{R}$ is the eigenvalue corresponding to x_k. Hence, we have that $\Lambda_k \in \{-1, 1\}$. If $\Lambda_k = 1$, we define $B_k : V_k \to V_k$ by $B_k x := 0$, for every $x \in V_k$. We note that B_k is anti-symmetric, since the representation matrix of B_k with respect to the basis $\{x_k\}$ of V_k is given by the anti-symmetric 1×1-matrix $\big(0\big)$. Further,

$$R \cdot x = \Lambda_k\, x = x = \exp(\alpha_k B_k)\, x \ ,$$

for every $x \in V_k$, where $\alpha_k := 0$. On the other hand, if $\Lambda_k = -1$, we note that it follows from Proposition 1.21 (iii) that

$$1 = \det(R) = \prod_{k=1}^{m} \Lambda_k \ .$$

and hence that there is another $\tilde{k} \in \{1, \ldots, m\}$ such that $\Lambda_{\tilde{k}} = -1$. Then $V_k \otimes V_{\tilde{k}}$ is a 2-dimensional invariant subspace of ρ and applying the analogous reasoning as above, we arrive at

$$R \cdot x_k = -x_k = \cos(\alpha_k)\, x_k = \exp(\alpha_k B_k)\, x_k \ ,$$
$$R \cdot x_{\tilde{k}} = -x_k = \cos(\alpha_k)\, x_{\tilde{k}} = \exp(\alpha_k B_k)\, x_{\tilde{k}} \ ,$$

where $B_k : V_k \otimes V_{\tilde{k}} \to V_k \otimes V_{\tilde{k}}$ as the unique linear map satisfying

$$B_k x_k = -x_{\tilde{k}} \ , \ \ B_k x_{\tilde{k}} = x_k \ .$$

and $\alpha_k := \pi$. We note that B_k is anti-symmetric, since the representation matrix of B_k with respect to the basis $\{x_k, x_{\tilde{k}}\}$ of $V_k \otimes V_{\tilde{k}}$ is given by the anti-symmetric 2×2-matrix

$$\begin{pmatrix} 0 & -1 \\ 1 & 0 \end{pmatrix} .$$

Finally, in this way, we arrive at a representation $\rho = \exp(A)$, where $A \in L(\mathbb{R}^n, \mathbb{R}^n)$ is the direct sum of the operators $\alpha_k B_k$. By construction, A is an anti-symmetric operator, i.e., $A^* = -A$. □

Proposition 1.23 (Surjectivity of the matrix exponential map II) *Let $n \in \mathbb{N}^*$.*

(i) For every $U \in \mathrm{U}(n)$, there is an anti-Hermitian $A \in \mathrm{M}(n, \mathbb{C})$ such that $U = \exp(A)$.
(ii) For every $U \in SU(n)$, there is an anti-Hermitian traceless $A \in \mathrm{M}(n, \mathbb{C})$ such that $U = \exp(A)$.

Proof For this purpose, let $\mathbb{C}^n$ be equipped with the canonical scalar product $\langle | \rangle$. Let $U \in \mathrm{U}(n)$. Since U is normal, there is an orthonormal basis $v_1, \ldots, v_n \in \mathbb{C}^n$ of eigenvectors of U. In addition, let $e^{i\varphi_1}, \ldots, e^{i\varphi_n} \in S^1$, where $\varphi_1, \ldots, \varphi_n \in (-\pi, \pi]$, be the corresponding, not necessarily pairwise different, eigenvalues of U, i.e., such that $U \cdot v_j = e^{i\varphi_j}.v_j$, for every $j \in \{1, \ldots, n\}$. Hence, if $V \in \mathrm{U}(n)$ is the matrix that has the vector v_j as the j-th column vector, $j \in \{1, \ldots, n\}$, then

$$U \cdot V = V \cdot \mathrm{diag}(e^{i\varphi_1}, \ldots, e^{i\varphi_n}) \ ,$$

where for every sequence $a_1, \ldots, a_n \in \mathbb{C}$, $\mathrm{diag}(a_1, \ldots, a_n)$ denotes the diagonal matrix $D \in \mathrm{M}(n, \mathbb{C})$ satisfying $D_{jj} = a_j$, $j \in \{1, \ldots, n\}$. Hence,

$$\begin{aligned} U &= V \cdot \mathrm{diag}(e^{i\varphi_1}, \ldots, e^{i\varphi_n}) \cdot V^* = V \cdot \exp(i\, \mathrm{diag}(\varphi_1, \ldots, \varphi_n)) \cdot V^* \\ &= \exp(i\, V \cdot \mathrm{diag}(\varphi_1, \ldots, \varphi_n) \cdot V^*) \ . \end{aligned}$$

We note that

$$[i\,V \cdot \mathrm{diag}(\varphi_1, \ldots, \varphi_n) \cdot V^*]^* = -i\,V \cdot \mathrm{diag}(\varphi_1, \ldots, \varphi_n) \cdot V^* \ ,$$

i.e., that $i\,V \cdot \mathrm{diag}(\varphi_1, \ldots, \varphi_n) \cdot V^*$ is anti-Hermitian. Further, in the special case that $U \in SU(n)$, then $\varphi_1 + \cdots + \varphi_n = k \cdot 2\pi$, for some $k \in \mathbb{Z}$. Hence, without loss of generality, we can assume that $\varphi_1 + \cdots + \varphi_n = 0$. Then,

$$\mathrm{Tr}\big(i\,V \cdot \mathrm{diag}(\varphi_1, \ldots, \varphi_n) \cdot V^*\big) = i\mathrm{Tr}(\mathrm{diag}(\varphi_1, \ldots, \varphi_n)) = 0 \ .$$

□

Combining the results of Exercises 1.2 and 1.3, Proposition 1.19, of the proof of Proposition 1.17 as well as of Propositions 1.22, 1.23 and 1.30, we arrive at the following table.

Exercise 1.2 (a) Using the Cayley-Hamilton theorem, explicitly calculate $\exp(A)$, for every trace-free $A \in \mathrm{M}(2, \mathbb{C})$.

(b) Show that there is no trace-free $A \in \mathrm{M}(2, \mathbb{C})$ such that

$$\exp(A) = \begin{pmatrix} -1 & 1 \\ 0 & -1 \end{pmatrix} .$$

(c) Conclude that

$$\exp(T_E SL(2, \mathbb{R})) \subsetneq SL(2, \mathbb{R}) \ , \ \exp(T_E SL(2, \mathbb{C})) \subsetneq SL(2, \mathbb{C}) \ .$$

Exercise 1.3 (a) Using the Cayley-Hamilton theorem, show that there is no $B \in \mathrm{M}(2, \mathbb{R})$ such that

$$B^2 = \begin{pmatrix} -1 & 1 \\ 0 & -1 \end{pmatrix} .$$

(b) Show that there is no $B \in \mathrm{M}(2, \mathbb{R})$ such that

$$\exp(B) = \begin{pmatrix} -1 & 1 \\ 0 & -1 \end{pmatrix} .$$

(c) Conclude that

$$\exp(T_E \mathrm{GL}_+(2, \mathbb{R})) \subsetneq \mathrm{GL}_+(2, \mathbb{R}) \ .$$

The proof that $SL(n, \mathbb{K})$, where $n \in \mathbb{N}^*$ and $\mathbb{K} \in \{\mathbb{R}, \mathbb{C}\}$, is path-connected uses the polar decomposition of elements of $\mathrm{M}(n, \mathbb{K})$. The proof of the existence and uniqueness of this decomposition is given in the Appendix, in a more general setting, see Theorem A.4.

Proposition 1.24 (Path-connectedness of $SL(n, \mathbb{K})$) *Let $\mathbb{K} \in \{\mathbb{R}, \mathbb{C}\}$ and $n \in \mathbb{N}^*$. Then, $SL(n, \mathbb{K})$ is path connected.*

Proof For this purpose, let $\mathbb{K}^n$ be equipped with the canonical scalar product $\langle | \rangle$. According to Theorem A.4, for $M \in SL(n, \mathbb{K})$, there is $U \in \mathrm{O}(n)/\mathrm{U}(n)$ such that

$$M = U \cdot (M^*M)^{1/2} \ .$$

We note that

$$\begin{aligned}1 &= [\det(M)]^* \cdot \det(M) = \det(M^*) \cdot \det(M) = \det(M^*M) \\ &= \det((M^*M)^{1/2}\,(M^*M)^{1/2}) = [\det((M^*M)^{1/2})]^2\end{aligned}$$

and hence, since $(M^*M)^{1/2}$ is positive self-adjoint, that

$$\det((M^*M)^{1/2}) = 1 \ .$$

As a consequence, $U \in SO(n)/SU(n)$, and, according to Propositions 1.22 and 1.23, there is an anti-symmetric traceless/anti-Hermitian traceless $A \in \mathrm{M}(n, \mathbb{K})$ such that $U = \exp(A)$. Further, since $(M^*M)^{1/2}$ is positive self-adjoint with non-vanishing determinant, there is a positive self-adjoint $B \in \mathrm{M}(n, \mathbb{K})$ such that

$$(M^*M)^{1/2} = \exp(B) \ .$$

Hence, $d : [0, 1] \to \mathrm{M}(n, \mathbb{K})$, defined by

$$d(t) := \exp(tA)\exp(tB) \ ,$$

for every $t \in [0, 1]$, is continuous such that

$$\begin{aligned}&d(0) = E \ , \ d(1) = \exp(A)\exp(B) = U \cdot (M^*M)^{1/2} = M \ , \\ &\det(d(0)) = \det(E) = 1 \ , \ \det(d(1)) = \det(M) = 1 \ , \\ &\det(d(t)) = \exp(\mathrm{Tr}(tA))\exp(\mathrm{Tr}(tB)) = \exp(t\mathrm{Tr}(A))\exp(t\mathrm{Tr}(B)) \\ &= \exp(t\mathrm{Tr}(B)) > 0 \ .\end{aligned}$$

Therefore, $c : [0, 1] \to SL(n, \mathbb{K})$, defined by $c(t) := [\det(d(t))]^{-1/n} d(t)$, for every $t \in [0, 1]$, is continuous such that $c(0) = E$ and $c(1) = M$. □

Corollary 1.25 (Path-connectedness of $GL(n, \mathbb{C})$ and $GL_+(n, \mathbb{R})$) *Let and $n \in \mathbb{N}^*$.*

(i) Then, $GL(n, \mathbb{C})$ and $GL_+(n, \mathbb{R})$ are path connected.
(ii) We have that

$$GL(n, \mathbb{R}) = GL_+(n, \mathbb{R}) \,\dot\cup\, GL_-(n, \mathbb{R})(n) \ ,$$

where

$$GL_-(n,\mathbb{R})(n) := \{M \in \mathrm{M}(n,\mathbb{R}) : \det(M) < 0\}$$

is path connected.

Proof Part (i): If $\mathbb{K} \in \{\mathbb{R}, \mathbb{C}\}$ and $M \in GL(n,\mathbb{K})$, it follows from Proposition 1.24 the existence of a continuous path $c_2 : [0,1] \to GL(n,\mathbb{K})$, such that $\det(c(t)) = \det(M)$, for every $t \in [0,1]$, $c_2(0) = [\det(M)].E$ and $c_2(1) = M$. We consider cases. If in particular, $M \in GL_+(n,\mathbb{R})$, then the range of c_2 is contained in $GL_+(n,\mathbb{R})$ and $c_1 : [0,1] \to GL_+(n,\mathbb{R})$, defined by $c_1(t) = [1 - t + t\ \det(M)].E$, for every $t \in [0,1]$, is continuous and such that $c_1(0) = E$ and $c_1(1) = [\det(M)].E$. If in particular, $M \in GL(n,\mathbb{C})$, then $c_1 : [0,1] \to GL(n,\mathbb{C})$, defined by $c_1(t) = \exp(\ln(\det(M))\,t).E$, for every $t \in [0,1]$, is continuous and such that $c_1(0) = E$ and $c_1(1) = [\det(M)].E$. Here $\ln(\det(M)) \in \mathbb{C}$ is such that $\exp(\ln(\det(M))) = \det(M)$. Hence in both cases, $c : [0,1] \to GL(n,\mathbb{K})$, defined by

$$c(t) := \begin{cases} c_1(2t) & \text{if } t \in [0,1/2) \\ c_2(2(t-\frac{1}{2})) & \text{if } t \in [1/2,1) \end{cases},$$

for every $t \in [0,1]$, is a continuous path such that $c(0) = E$ and $c(1) = M$.
Part (ii): For every $M \in GL(n,\mathbb{R})$, we have either $\det(M) > 0$ or $\det(M) < 0$. Hence

$$GL(n,\mathbb{R}) \subset GL_+(n,\mathbb{R}) \,\dot\cup\, GL_-(n,\mathbb{R}) \subset GL(n,\mathbb{R}) \ .$$

Further, $GL_-(n,\mathbb{R})$ is path connected. For the proof, let $M, M_1, M_2 \in GL_-(n,\mathbb{R})$. Then $M \cdot M_1, M \cdot M_2 \in GL_+(n,\mathbb{R})$ and since $GL_+(n,\mathbb{R})$ is path connected, there is a continuous path $c : [0,1] \to GL_+(n,\mathbb{R})$ such that $c(0) = M \cdot M_1$ and $c(1) = M \cdot M_2$. Hence, $M^{-1} \cdot c$ is a continuous path in $GL_-(n)$ such that $(M^{-1} \cdot c)(0) = M_1$ and $(M^{-1} \cdot c)(1) = M_2$. As a consequence, $GL_-(n)$ is path connected. □

1.2.1 Definition and Properties of the Operator Logarithm

Proposition 1.26 (Definition and continuity of the logarithm) *Let* $\mathbb{K} \in \{\mathbb{R}, \mathbb{C}\}$ *and* $(X, \|\ \|)$ *a non-trivial* $\mathbb{K}$*-Banach space. By*

$$\ln(A) := \sum_{k=1}^{\infty} \frac{(-1)^{k+1}}{k}\,(A - 1)^k \ ,$$

for every $A \in U_1(1)$*, there is defined a continuous map* $\ln : U_1(1) \to \mathrm{L}(X,X)$.

Proof In the following, we use (1.9) from the proof of Proposition 1.11 that

$$A^k - B^k = \sum_{l=0}^{k-1} A^{k-1-l} \circ (A - B) \circ B^l \ ,$$

for every $A, B \in \mathrm{L}(X, X)$ and $k \in \mathbb{N}^*$. Hence, it follows that

$$\begin{aligned} \|A^k - B^k\|_{op} &\leqslant \left[\sum_{l=0}^{k-1} \|A\|_{op}^{k-1-l} \|B\|_{op}^{l}\right] \cdot \|A - B\|_{op} \\ &\leqslant k\, M_{A,B}^{k-1} \|A - B\|_{op} \,, \end{aligned} \tag{1.12}$$

where

$$M_{A,B} := \max\{\|A\|_{op}, \|B\|_{op}\} \,,$$

and $k \in \mathbb{N}^*$. If $A, B \in U_1(0)$, it follows further that

$$\| \ln(1 + A) - \ln(1 + B)\|_{op} = \| \sum_{k=1}^{\infty} \frac{(-1)^{k+1}}{k} (A^k - B^k)\|_{op}$$

$$\leqslant \sum_{k=1}^{\infty} \frac{1}{k} \|A^k - B^k\|_{op} \leqslant \left(\sum_{k=1}^{\infty} \frac{1}{k}\, k\, M_{A,B}^{k-1}\right) \cdot \|A - B\|_{op}$$

$$= \left(\sum_{k=1}^{\infty} M_{A,B}^{k-1}\right) \cdot \|A - B\|_{op} = \frac{1}{1 - M_{A,B}} \|A - B\|_{op}$$

and hence that ln is continuous. □

Proposition 1.27 (Continuous differentiability of the logarithm) *Let* $\mathbb{K} \in \{\mathbb{R}, \mathbb{C}\}$ *and* $(X, \| \, \|)$ *a non-trivial* $\mathbb{K}$*-Banach space. Then, the natural logarithm* $\ln : U_1(1) \to \mathrm{L}(X, X)$ *is continuously differentiable, with derivative*

$$[\ln'(A)](B) = \sum_{k=1}^{\infty} \frac{(-1)^{k+1}}{k} \left[\sum_{l=0}^{k-1} (A - 1)^{k-1-l} \circ B \circ (A - 1)^{l}\right] ,$$

for every $A \in U_1(1)$ *and* $B \in \mathrm{L}(X, X)$.

Proof If $A \in U_1(0)$ and $H \in \mathrm{L}(X, X)$ are such that $A + H \in U_1(0)$, then

$$\ln(1 + A + H) - \ln(1 + A) = \sum_{k=1}^{\infty} \frac{(-1)^{k+1}}{k} [(A + H)^k - A^k]$$

$$= \sum_{k=1}^{\infty} \frac{(-1)^{k+1}}{k} \sum_{l=0}^{k-1} (A + H)^{k-1-l} \circ H \circ A^l \,,$$

where we used (1.9) from the proof of Proposition 1.11. Hence,

$$\begin{aligned}
&\| \ln(1+A+H) - \ln(1+A) - \sum_{k=1}^{\infty} \frac{(-1)^{k+1}}{k} \sum_{l=0}^{k-1} A^{k-1-l} \circ H \circ A^l \|_{op} \\
&= \| \sum_{k=2}^{\infty} \frac{(-1)^{k+1}}{k} \sum_{l=0}^{k-1} [(A+H)^{k-1-l} - A^{k-1-l}] \circ H \circ A^l \|_{op} \\
&\leqslant \|H\|_{op} \sum_{k=2}^{\infty} \frac{1}{k} \sum_{l=0}^{k-1} \|(A+H)^{k-1-l} - A^{k-1-l}\|_{op} \cdot \|A\|_{op}^l \\
&\leqslant \|H\|_{op}^2 \sum_{k=2}^{\infty} \frac{1}{k} \sum_{l=0}^{k-1} (k-1-l)\,[\max\{\|A+H\|_{op}, \|A\|_{op}\}]^{k-2-l} \cdot \|A\|_{op}^l \\
&\leqslant \|H\|_{op}^2 \sum_{k=2}^{\infty} \frac{k-1}{k} \sum_{l=0}^{k-1} [\max\{\|A+H\|_{op}, \|A\|_{op}\}]^{k-2-l} \cdot \|A\|_{op}^l \\
&\leqslant \|H\|_{op}^2 \sum_{k=2}^{\infty} (k-1)\,[\max\{\|A+H\|_{op}, \|A\|_{op}\}]^{k-2} \\
&= \frac{\|H\|_{op}^2}{(1-\max\{\|A+H\|_{op}, \|A\|_{op}\})^2}\,,
\end{aligned} \tag{1.13}$$

where we used (1.12) from the proof of Proposition 1.26. Further, for every $m \in \mathbb{N}^*$, we have that

$$\begin{aligned}
&\sum_{k=1}^{m} \frac{(-1)^{k+1}}{k} \sum_{l=0}^{k-1} \left\| A^{k-1-l} \circ H \circ A^l \right\|_{op} \leqslant \sum_{k=1}^{m} \frac{1}{k} \sum_{l=0}^{k-1} \|A\|_{op}^{k-1} \cdot \|H\|_{op} \\
&= \|H\|_{op} \sum_{k=1}^{m} \|A\|_{op}^{k-1} \leqslant \frac{\|H\|_{op}}{1-\|A\|_{op}}\,.
\end{aligned}$$

Therefore, the family

$$\left(\frac{(-1)^{k+1}}{k} A^{k-1-l} \circ H \circ A^l \right)_{(k,l)\in I}\,,$$

where

$$I := \{(k,l) : k \in \mathbb{N}^* \wedge 0 \leqslant l \leqslant k-1\}\,,$$

is absolutely summable, and

$$\left\| \sum_{(k,l)\in I} \frac{(-1)^{k+1}}{k} A^{k-1-l} \circ H \circ A^l \right\| \leqslant \frac{\|H\|_{op}}{1-\|A\|_{op}}\,. \tag{1.14}$$

Hence, it follows from (1.13) that

$$\lim_{H\to 0, H\neq 0} \frac{\left\| \ln(1+A+H) - \ln(1+A) - \sum_{k=1}^{\infty} \frac{(-1)^{k+1}}{k} \left(\sum_{l=0}^{k-1} A^{k-1-l} \circ H \circ A^l \right) \right\|_{op}}{\|H\|_{op}}$$
$$= 0 \; .$$

Further, the map $\lambda : \mathrm{L}(X, X) \to \mathrm{L}(X, X)$, defined by

$$\lambda(H) := \sum_{(k,l)\in I} \frac{(-1)^{k+1}}{k} A^{k-1-l} \circ H \circ A^l \; ,$$

for every $H \in \mathrm{L}(X, X)$, is linear, and for every $H \in \mathrm{L}(X, X)$ satisfying $\|H\|_{op} = 1$, it follows from (1.14) that

$$\|\lambda(H)\|_{op} \leqslant \frac{1}{1 - \|A\|_{op}} \; .$$

Hence, $\lambda \in \mathrm{L}(\mathrm{L}(X, X), \mathrm{L}(X, X))$ with $\|\lambda\|_{op} \leqslant (1 - \|A\|_{op})^{-1}$. For the proof of the continuity of $\ln'$, for $A_1, A_2 \in U_1(0)$ and $B \in \mathrm{L}(X, X)$, we have that

$$[\ln'(1 + A_2)](B) - [\ln'(1 + A_1)](B)$$
$$= \sum_{k=2}^{\infty} \frac{(-1)^{k+1}}{k} \left[\sum_{l=0}^{k-1} (A_2^{k-1-l} \circ B \circ A_2^l - A_1^{k-1-l} \circ B \circ A_1^l) \right] .$$

Further, for $k, l \in \mathbb{N}$ such that $2 \leqslant k, 0 \leqslant l \leqslant k - 1$, it follows that

$$A_2^{k-1-l} \circ B \circ A_2^l - A_1^{k-1-l} \circ B \circ A_1^l$$
$$= (A_2^{k-1-l} - A_1^{k-1-l}) \circ B \circ A_2^l + A_1^{k-1-l} \circ B \circ (A_2^l - A_1^l)$$
$$= (A_2^{k-1-l} - A_1^{k-1-l}) \circ B \circ (A_2^l - A_1^l) + (A_2^{k-1-l} - A_1^{k-1-l}) \circ B \circ A_1^l$$
$$+ A_1^{k-1-l} \circ B \circ (A_2^l - A_1^l)$$

and hence that

$$\|A_2^{k-1-l} \circ B \circ A_2^l - A_1^{k-1-l} \circ B \circ A_1^l\|_{op}$$
$$\leqslant \Big[\|A_2^{k-1-l} - A_1^{k-1-l}\|_{op} \cdot \|A_2^l - A_1^l\|_{op}$$
$$+ \|A_2^{k-1-l} - A_1^{k-1-l}\|_{op} \cdot \|A_1\|_{op}^l$$
$$+ \|A_1\|_{op}^{k-1-l} \cdot \|A_2^l - A_1^l\|_{op} \Big] \|B\|_{op}$$
$$\leqslant [2l\,(k - l - 1) + k - l - 1 + l]\, M_{A_1,A_2}^{k-2}\, \|A_1 - A_2\|_{op}\, \|B\|_{op}$$
$$\leqslant 2k(k - 1)\, M_{A_1,A_2}^{k-2}\, \|A_1 - A_2\|_{op}\, \|B\|_{op} \; ,$$

where we used (1.12) from the proof of Proposition 1.26, and that we have that

$$
\begin{aligned}
&\|A_2^{k-1-l} - A_1^{k-1-l}\|_{op} \cdot \|A_2^l - A_1^l\|_{op} \\
&\leqslant (k-l-1)\, M_{A_1,A_2}^{k-l-2}\, \|A_1 - A_2\|_{op} \cdot l\, M_{A_1,A_2}^{l-1}\, \|A_1 - A_2\|_{op} \\
&= l\,(k-l-1) M_{A_1,A_2}^{k-3}\, \|A_1 - A_2\|_{op}^2 \leqslant 2l\,(k-l-1)\, M_{A_1,A_2}^{k-2}\, \|A_1 - A_2\|_{op} \,, \\
&\|A_2^{k-1-l} - A_1^{k-1-l}\|_{op} \cdot \|A_1\|_{op}^l \\
&\leqslant \|A_1\|_{op}^l\, (k-l-1)\, M_{A_1,A_2}^{k-l-2}\, \|A_1 - A_2\|_{op} \\
&\leqslant (k-l-1)\, M_{A_1,A_2}^{k-2}\, \|A_1 - A_2\|_{op} \,, \\
&\|A_1\|_{op}^{k-1-l} \cdot \|A_2^l - A_1^l\|_{op} \leqslant M_{A_1,A_2}^{k-l-1} \cdot l\, M_{A_1,A_2}^{l-1}\, \|A_1 - A_2\|_{op} \\
&= l\, M_{A_1,A_2}^{k-2}\, \|A_1 - A_2\|_{op} \,.
\end{aligned}
$$

As a consequence,

$$
\begin{aligned}
&\|[\ln'(1 + A_2)](B) - [\ln'(1 + A_1)](B)\|_{op} \\
&\leqslant 2\|A_1 - A_2\|_{op}\, \|B\|_{op} \sum_{k=2}^{\infty} k(k-1) M_{A_1,A_2}^{k-2} \\
&\leqslant \frac{4}{(1 - M_{A_1,A_2})^3}\, \|A_1 - A_2\|_{op}\, \|B\|_{op} \,.
\end{aligned}
$$

The latter implies that

$$
\|\ln'(1 + A_2) - \ln'(1 + A_1)\|_{op} \leqslant \frac{4}{(1 - M_{A_1,A_2})^3}\, \|A_1 - A_2\|_{op} \,,
$$

for all $A_1, A_2 \in U_1(0)$ and hence that $\ln'$ is continuous. □

Corollary 1.28 (Local C^1-invertibility of the logarithm) *Let $\mathbb{K} \in \{\mathbb{R}, \mathbb{C}\}$ and $(X, \|\ \|)$ a non-trivial $\mathbb{K}$-Banach space. Then, the logarithm function, $\ln : U_1(1) \to \mathrm{L}(X, X)$, is locally C^1-invertible at 1, i.e., there is an open set $U \subset \mathrm{L}(X, X)$ around 1 such that $\ln(U)$ is open and such that the restriction of $\ln$ in domain to U and in image to $\ln(U)$ is bijective with a C^1-inverse.*

Proof We note that

$$
[\ln'(1)](B) = B \,,
$$

for every $B \in \mathrm{L}(X, X)$, i.e., $\ln'(1)$ is given by the identical map on $\mathrm{L}(\mathrm{L}(X, X), \mathrm{L}(X, X))$ and hence bijective with a bounded linear inverse. Therefore, the statement is a direct consequence of Proposition 1.11 and the inverse mapping theorem, see, e.g., Theorem 1.2 in [38]. □

Proposition 1.29 (The logarithm as inverse of the exponential map) *Let* $\mathbb{K} \in \{\mathbb{R}, \mathbb{C}\}$ *and* $(X, \| \, \|)$ *a non-trivial* $\mathbb{K}$*-Banach space. Then,*

(i) $\ln(\exp(A)) = A$, *for every* $A \in U_{\ln(2)}(0)$,
(ii) $\exp(\ln(A)) = A$, *for every* $A \in U_1(1)$.

Proof "Part (i)": We note that if $A \in U_{\ln(2)}(0)$, it follows that

$$\| \exp(A) - 1\|_{op} \leqslant \sum_{n=1}^{\infty} \frac{1}{n!} \|A\|_{op}^n = e^{\|A\|_{op}} - 1 < e^{\ln(2)} - 1 = 2 - 1 = 1 ,$$

and hence that $\| \exp(A) - 1\|_{op} < 1$. We define h by

$$h(z) := z\, A ,$$

for every $z \in U_{\ln(2)\, \|A\|_{op}^{-1}}(0)$. We note that, since

$$\|z\, A\|_{op} = |z| \cdot \|A\|_{op} < \ln(2) ,$$

for every $z \in U_{\ln(2)\, \|A\|_{op}^{-1}}(0)$, we have that $Ran(h) \subset U_{\ln(2)}(0)$. In addition, since $\ln(2)\, \|A\|_{op}^{-1} > 1$, we have that $1 \in U_{\ln(2)\, \|A\|_{op}^{-1}}(0)$. Further, from Theorem 1.11, Proposition 1.27 and the chain rule, along with some calculation, it follows that

$$\ln \circ \exp \circ h : U_{\ln(2)\, \|A\|_{op}^{-1}}(0) \to \mathrm{L}(X, X)$$

is differentiable, with derivative

$$(\ln \circ \exp \circ h)'(z) = A ,$$

for every $z \in U_{\ln(2)\, \|A\|_{op}^{-1}}(0)$. Hence, if $\omega \in \mathrm{L}(\mathrm{L}(X, X), \mathbb{K})$, then $\omega \circ (\ln \circ \exp \circ h - h)$ is differentiable and holomorphic on $U_{\ln(2)\, \|A\|_{op}^{-1}}(0)$, respectively, with a derivative that is constant of value 0. Hence, since $[\, \omega \circ (\ln \circ \exp \circ h - h)](0) = 0$, it follows that

$$\omega \circ \ln \circ \exp \circ h = \omega \circ h .$$

Since $\mathrm{L}(X, \mathbb{K})$ separates the points of $\mathrm{L}(X, X)$, the latter implies that

$$\ln \circ \exp \circ h = h$$

and in particular that

$$\ln(\exp(A)) = (\ln \circ \exp \circ h)(1) = h(1) = A .$$

"Part (ii)": If $A \in U_1(1)$, we define h by

$$h(z) := 1 + z\,(A - 1)\ ,$$

for every $z \in U_{\|A-1\|_{op}^{-1}}(0)$. We note that, since

$$\|z\,(A - 1)\|_{op} = |z| \cdot \|A - 1\|_{op} < 1\ ,$$

for every $z \in U_{\|A-1\|_{op}^{-1}}(0)$, we have that $\mathrm{R}an(h) \subset U_1(1)$. In addition, since $\|A - 1\|_{op}^{-1} > 1$, we have that $1 \in U_{\|A-1\|_{op}^{-1}}(0)$. Further, from Theorem 1.11, Proposition 1.27 and the chain rule, along with some calculation, it follows that

$$\exp \circ \ln \circ h : U_{\|A-1\|_{op}^{-1}}(0) \to \mathrm{L}(X, X)$$

is differentiable, with derivative

$$(\exp \circ \ln \circ h)'(z) = (A - 1) \circ [h(z)]^{-1} \circ (\exp \circ \ln \circ h)(z)\ ,$$

for every $z \in U_{\|A-1\|_{op}^{-1}}(0)$. Hence, if $\omega \in \mathrm{L}(\mathrm{L}(X, X), \mathbb{K})$, then $\omega \circ (\exp \circ \ln \circ h - h)$ is differentiable and holomorphic on $U_{\|A-1\|_{op}^{-1}}(0)$, respectively, with a derivative that is constant of value 0. Hence, since $[\omega \circ (\exp \circ \ln \circ h - h)](0) = 0$, it follows that

$$\omega \circ \exp \circ \ln \circ h = \omega \circ h\ .$$

Since $\mathrm{L}(X, \mathbb{K})$ separates the points of $\mathrm{L}(X, X)$, the latter implies that

$$\exp \circ \ln \circ h = h$$

and in particular that

$$\exp(\ln(A)) = (\exp \circ \ln \circ h)(1) = h(1) = A\ .$$

□

If $\mathbb{K} \in \{\mathbb{R}, \mathbb{C}\}$ and $(X, \|\ \|)$ a non-trivial $\mathbb{K}$-Banach space, then we define an extension $\widehat{\ln}$ of by

$$\widehat{\ln}(1 + D) := \sum_{k=1}^{n-1} \frac{(-1)^{k+1}}{k} D^k\ ,$$

for every nilpotent $D \in \mathrm{L}(X, X)$ of degree $n \in \mathbb{N}^*$.

Supplementary to Proposition 1.29 (ii), we have the following.

If $D \in \mathrm{L}(X, X)$ is nilpotent of degree $n \in \mathbb{N}^*$, then

$$\exp(\widehat{\ln}(1 + D)) = 1 + D\ . \tag{1.15}$$

For the proof, we note that if $D \in \mathrm{L}(X, X)$ is nilpotent of degree 1, then $D = 0$ and hence

$$\exp(\widehat{\ln}(1 + D)) = \exp(\widehat{\ln}(1)) = \exp(0) = 1 = 1 + D\ .$$

If $D \in \mathrm{L}(X, X)$ is nilpotent of degree $n \geqslant 2$, we proceed as follows. We define h by

$$h(z) := 1 + z\,D\ ,$$

for every $z \in \mathbb{K}$. Then, it follows for $z \in \mathbb{K}$ and $\delta \in \mathbb{K}^*$ that

$$\widehat{\ln}(h(z+\delta)) - \widehat{\ln}(h(z)) = \sum_{k=1}^{n-1} \frac{(-1)^{k+1}}{k}\,[(z+\delta)^k - z^k]D^k$$
$$= \sum_{k=1}^{n-1} \frac{(-1)^{k+1}}{k} \left[\sum_{l=1}^{k} \binom{k}{l} z^{k-l}\delta^l\right] D^k$$

and hence that

$$\widehat{\ln}(h(z+\delta)) - \widehat{\ln}(h(z)) - \delta \sum_{k=1}^{n-1} (-1)^{k-1}\, z^{k-1} D^k$$
$$= \delta^2 \sum_{k=2}^{n-1} \frac{(-1)^{k+1}}{k} \left[\sum_{l=2}^{k} \binom{k}{l} z^{k-l}\delta^{l-2}\right] D^k\ .$$

Therefore,

$$\left\| \frac{1}{\delta}\,[\,\widehat{\ln}(h(z+\delta)) - \widehat{\ln}(h(z))] - \sum_{k=1}^{n-1} (-1)^{k-1}\, z^{k-1} D^k \right\|_{op}$$
$$= \left\| \delta \sum_{k=2}^{n-1} \frac{(-1)^{k+1}}{k} \left[\sum_{l=2}^{k} \binom{k}{l} z^{k-l}\delta^{l-2}\right] D^k \right\|_{op}$$
$$\leqslant |\delta| \sum_{k=2}^{n-1} \left[\sum_{l=2}^{k} \binom{k}{l} |z|^{k-l}\, |\delta|^{l-2}\right] \frac{\|D\|_{op}^k}{k}$$

and hence

$$\lim_{\delta\to 0,\delta\neq 0} \frac{1}{\delta}\,[\,\widehat{\ln}(h(z+\delta)) - \widehat{\ln}(h(z))] = \sum_{k=1}^{n-1}(-1)^{k-1}\,z^{k-1}D^k$$

$$= \left[\sum_{k=0}^{n-2}(-zD)^k\right] D = (1+zD)^{-1}D = D\,[h(z)]^{-1} \ .$$

It follows that $\widehat{\ln} \circ h$ is differentiable with derivative

$$(\,\widehat{\ln} \circ h)'(z) = D\,[h(z)]^{-1} \ ,$$

for every $z \in \mathbb{K}$. Further, from Theorem 1.11 and the chain rule, along with some calculation, it follows that $\exp \circ \widehat{\ln} \circ h$ is differentiable, with derivative

$$(\exp \circ \widehat{\ln} \circ h)'(z) = D \circ [h(z)]^{-1} \circ (\exp \circ \widehat{\ln} \circ h)(z) \ ,$$

for every $z \in \mathbb{K}$). Hence, if $\omega \in \mathrm{L}(\mathrm{L}(X, X), \mathbb{K})$, then $\omega \circ (\exp \circ \widehat{\ln} \circ h - h)$ is differentiable and holomorphic, respectively, with a derivative that is constant of value 0. Hence, since $[\omega \circ (\exp \circ \widehat{\ln} \circ h - h)](0) = 0$, it follows that

$$\omega \circ \exp \circ \widehat{\ln} \circ h = \omega \circ h \ .$$

Since $\mathrm{L}(X, \mathbb{K})$ separates the points of $\mathrm{L}(X, X)$, the latter implies that

$$\exp \circ \widehat{\ln} \circ h = h$$

and in particular that

$$\exp(\,\widehat{\ln}(1+D)) = (\exp \circ \widehat{\ln} \circ h)(1) = h(1) = 1 + D \ .$$

The latter can be used to prove the following.

Proposition 1.30 (Surjectivity of the matrix exponential map III) *Let $n \in \mathbb{N}^*$. For every $G \in GL(n, \mathbb{C})$, there is an $A \in \mathrm{M}(n, \mathbb{C})$ such that $G = \exp(A)$.*

Proof For this purpose, let $G \in GL(n, \mathbb{C})$, $\lambda_1, \ldots, \lambda_m \in \mathbb{C}^*$, where $m \in \{1, \ldots, n\}$, be the pairwise different eigenvalues of G and $\hat{G} \in \mathbb{C}^n \to \mathbb{C}^n$ the linear map defined by $\hat{G}(z) := G \cdot z$, for every $z \in \mathbb{C}^n$. According to linear algebra, the generalized eigenspaces of $\hat{G}$ span $\mathbb{C}^n$, i.e.,

$$\ker(\hat{G} - \lambda_1)^n \oplus \ldots \oplus \ker(\hat{G} - \lambda_m)^n = \mathbb{C}^n \ ,$$

are left invariant by $\hat{G}$; the linear operators $S : \mathbb{C}^n \to \mathbb{C}^n$ and $D : \mathbb{C}^n \to \mathbb{C}^n$, defined by

$$S := \sum_{j=1}^{m} \lambda_j P_j \ , \quad D := \sum_{j=1}^{m} (\hat{G} - \lambda_j) P_j \ ,$$

where for every $j \in \{1, \ldots, m\}$, P_j denotes the projection onto $\ker(\hat{G} - \lambda_j)^n$, leave the generalized eigenspaces invariant and satisfy

$$\hat{G} = S + D \ , \ [S, D] = 0 \ , \ D^n = 0 \ .$$

We note that the projections $P_1, \ldots, P_n$ commute pairwise, since $P_j P_k = 0$, for different $j, k \in \{1, \ldots, m\}$. In particular, S is bijective and we have that

$$S = \exp\left(\sum_{j=1}^{n} \ln(\lambda_j) P_j \right) ,$$

where $\ln(\lambda_j) \in \mathbb{C}$ is such that $\exp(\ln(\lambda_j)) = \lambda_j$, for every $j \in \{1, \ldots, m\}$. Further, since S and D leave the generalized eigenspace invariant, we have that $[S, P_j] = [D, P_j] = 0$,

$$\begin{aligned} 0 &= -S^{-1}[S, P_j]S^{-1} = -S^{-1}SP_jS^{-1} + S^{-1}P_jSS^{-1} \\ &= -P_jS^{-1} + S^{-1}P_j = [S^{-1}, P_j] \ , \end{aligned}$$

for every $j \in \{1, \ldots, m\}$ and that

$$\begin{aligned} 0 &= -S^{-1}[S, D]S^{-1} = -S^{-1}SDS^{-1} + S^{-1}DSS^{-1} \\ &= -DS^{-1} + S^{-1}D = [S^{-1}, D] \ . \end{aligned}$$

Note that the latter implies that

$$(S^{-1}D)^n = (S^{-1})^n D^n = 0 \ .$$

Hence, it follows that

$$\exp(\widehat{\ln}(1 + S^{-1}D)) = 1 + S^{-1}D$$

and finally that

$$\begin{aligned} \hat{G} = S(1 + S^{-1}D) &= \exp\left(\sum_{j=1}^{n} \ln(\lambda_j) P_j \right) \exp(\widehat{\ln}(1 + S^{-1}D)) \\ &= \exp\left(\sum_{j=1}^{n} \ln(\lambda_j) P_j + \widehat{\ln}(1 + S^{-1}D) \right) . \end{aligned}$$

□

A Strongly Continuous Unitary Representation of SU(2)

2

The goal of this chapter is the construction of the spinor representation of SU(2), used in the description of Spin $\frac{1}{2}$ particles in quantum mechanics.

2.1 Basic Properties of SO(2)

In the following, we collect additional information about SO(2). In a first step, we derive parametrizations of SO(2). We note that from (1.3), it follows that a real $n \times n$-matrix is orthogonal if and only if its column vectors form an orthonormal basis of $\mathbb{R}^n$, equipped with the Euclidean scalar product. In particular, if

$$\begin{pmatrix} R_{11} & R_{12} \\ R_{21} & R_{22} \end{pmatrix} \in \mathrm{SO}(2) ,$$

then

$$\begin{aligned} &R_{11}R_{12} + R_{21}R_{22} = 0 , \quad R_{11}R_{22} - R_{12}R_{21} = 1 , \\ &R_{11}^2 + R_{21}^2 = R_{12}^2 + R_{22}^2 = 1 , \end{aligned} \tag{2.1}$$

and hence

$$\begin{aligned} R_{11}(R_{22} - R_{11}) - R_{21}(R_{12} + R_{21}) = 0 , \\ (R_{11} - R_{22})R_{22} - R_{12}(R_{21} + R_{12}) = 0 , \end{aligned}$$

or equivalently

H. R. Beyer, *Quantum Spin and Representations of the Poincaré Group, Part I*,
Synthesis Lectures on Engineering, Science, and Technology,
https://doi.org/10.1007/978-3-031-84140-8_2

$$R_{11}(R_{22}-R_{11})-R_{21}(R_{12}+R_{21})=0\ ,$$
$$-R_{22}(R_{22}-R_{11})-R_{12}(R_{12}+R_{21})=0\ ,$$

implying that

$$-(R_{22}-R_{11})^2-(R_{21}+R_{12})^2=0\ .$$

Hence, it follows that

$$R_{22}=R_{11}\ ,\ \ R_{21}=-R_{12}\ .$$

Substituting the latter into the first 2 equations gives

$$R_{11}R_{12}-R_{12}R_{11}=0\ ,\ \ R_{11}^2+R_{12}^2=1\ .$$

Hence, we have that

$$\begin{pmatrix} R_{11} & R_{12} \\ R_{21} & R_{22} \end{pmatrix} = \begin{pmatrix} R_{11} & R_{12} \\ -R_{12} & R_{11} \end{pmatrix}$$

as well as that

$$R_{11}^2+R_{12}^2=1\ .$$

On the other hand, if R_{11}, R_{12} are real numbers such that $R_{11}^2+R_{12}^2=1$, then the determinant of the matrix

$$\begin{pmatrix} R_{11} & R_{12} \\ -R_{12} & R_{11} \end{pmatrix}$$

is equal to 1, the column vectors have norm 1 and are orthogonal.

Hence, it follows that

$$\mathrm{SO}(2)=\left\{\begin{pmatrix} R_{11} & R_{12} \\ -R_{12} & R_{11} \end{pmatrix} : R_{11}, R_{12}\in\mathbb{R}\wedge R_{11}^2+R_{12}^2=1\right\}\ . \tag{2.2}$$

We note that if $R_{11}, R_{12}\in\mathbb{R}$ are such that $R_{11}^2+R_{12}^2=1$, then

$$\begin{pmatrix} R_{11} & R_{12} \\ -R_{12} & R_{11} \end{pmatrix}^{-1} = \begin{pmatrix} R_{11} & -R_{12} \\ R_{12} & R_{11} \end{pmatrix} = \begin{pmatrix} R_{11} & R_{12} \\ -R_{12} & R_{11} \end{pmatrix}^t\ .$$

Proposition 2.1 (A parametrization of SO(2))

(i) For every $R\in\mathrm{SO}(2)$*, there is a uniquely determined* $\varphi\in(-\pi,\pi]$ *such that*

$$R=\begin{pmatrix} \cos(\varphi) & -\sin(\varphi) \\ \sin(\varphi) & \cos(\varphi) \end{pmatrix}\ .$$

(ii) For every $\varphi \in \mathbb{R}$,

$$\exp\left(\varphi \begin{pmatrix} 0 & -1 \\ 1 & 0 \end{pmatrix}\right) = \begin{pmatrix} \cos(\varphi) & -\sin(\varphi) \\ \sin(\varphi) & \cos(\varphi) \end{pmatrix} \in \mathrm{SO}(2) \ .$$

Proof 'Part (i)': For the proof, let

$$R = \begin{pmatrix} a & b \\ -b & a \end{pmatrix} ,$$

where $a, b \in \mathbb{R}$ are such that $a^2 + b^2 = 1$, be some element of SO(2). Then $a \in [-1, 1]$. Hence $\alpha := \arccos(a) \in [0, \pi]$ satisfies $\cos(\alpha) = \cos(-\alpha) = a$. Further, since

$$1 = a^2 + b^2 = \cos^2(\alpha) + b^2 = \cos^2(\alpha) + \sin^2(\alpha) \ ,$$

it follows that $b \in \{-\sin(\alpha), -\sin(-\alpha)\}$. Hence, there is $\varphi \in [-\pi, \pi]$ such that $\cos(\varphi) = a$ and at the same time such that $-\sin(\varphi) = b$. In particular, if $\varphi = -\pi$, then $a = \cos(-\pi) = \cos(\pi) = -1$ and $b = -\sin(-\pi) = \sin(\pi) = 0$ and hence $a = \cos(\pi)$ and $b = -\sin(\pi)$. Therefore, it follows that there is $\varphi \in (-\pi, \pi]$ such that $\cos(\varphi) = a$ and at the same time such that $-\sin(\varphi) = b$. If $\varphi' \in (-\pi, \pi]$ is such that $\cos(\varphi) = \cos(\varphi') = a$, $-\sin(\varphi) = -\sin(\varphi') = b$, then $e^{-i\varphi} = e^{-i\varphi'}$ and hence $e^{i(\varphi'-\varphi)} = 1$. Since $\varphi, \varphi' \in (-\pi, \pi]$, we have that $-\pi < \varphi' \leqslant \pi$ and $-\pi \leqslant -\varphi < \pi$ and hence that $-2\pi < \varphi' - \varphi < 2\pi$. With the aid of the latter, it follows that $\varphi' = \varphi$.

'Part (ii)': Further, by using that

$$A^2 = -E \ ,$$

where

$$A := \begin{pmatrix} 0 & -1 \\ 1 & 0 \end{pmatrix} ,$$

and E is the 2×2 unit matrix, it follows that

$$A^{2k} = (-1)^k E \ , \ A^{2k+1} = (-1)^k A \ ,$$

for $k \in \mathbb{N}$, and hence for $\varphi \in \mathbb{R}$ that

$$\begin{aligned}
\exp(\varphi A) &= \sum_{k=0}^{\infty} \frac{\varphi^k}{k!} A^k = \sum_{k=0}^{\infty} \frac{\varphi^{2k}}{(2k)!} A^{2k} + \sum_{k=1}^{\infty} \frac{\varphi^{2k+1}}{(2k+1)!} A^{2k+1} \\
&= \left[\sum_{k=0}^{\infty} (-1)^k \frac{\varphi^{2k}}{(2k)!}\right] E + \left[\sum_{k=1}^{\infty} (-1)^k \frac{\varphi^{2k+1}}{(2k+1)!}\right] A \\
&= \cos(\varphi) E + \sin(\varphi) A = \begin{pmatrix} \cos(\varphi) & -\sin(\varphi) \\ \sin(\varphi) & \cos(\varphi) \end{pmatrix} \in \mathrm{SO}(2) \ .
\end{aligned}$$

□

As a consequence, we arrive at a direct proof of the fact that

$$\exp(T_E \mathrm{SO}(2)) = \mathrm{SO}(2) \ .$$

For a different proof of the same result, for the more general case of $\mathrm{SU}(n)$, $n \in \mathbb{N}^*$, see Proposition 1.22.

Corollary 2.2 (Surjectivity of the exponential map) *We have that*

$$\exp(\{A \in \mathrm{M}(2, \mathbb{R}) : A \textit{ is skew-symmetric } \}) = \mathrm{SO}(2) \ .$$

Proof According to Proposition 2.1, we have that

$$\exp(\{A \in \mathrm{M}(2, \mathbb{R}) : A \text{ is skew-symmetric } \}) \supset \mathrm{SO}(2) \ .$$

If $A \in \mathrm{M}(2, \mathbb{R})$ is skew-symmetric, we conclude as follows. If $A = 0$, then $\exp(A) = 1 \in \mathrm{SO}(2)$. If $A \neq 0$, then, there is $\varphi \neq 0$ such that

$$A = \varphi \begin{pmatrix} 0 & -1 \\ 1 & 0 \end{pmatrix}$$

and hence, according to Proposition 2.1,

$$\exp(A) \in \mathrm{SO}(2) \ .$$

Hence it follows that

$$\exp(\{A \in \mathrm{M}(2, \mathbb{R}) : A \text{ is skew-symmetric } \}) \subset \mathrm{SO}(2) \ .$$

□

Remark 2.3 We note that,

$$|Ax|^2 = |{}^t(-x_2, x_1)|^2 = x_1^2 + x_2^2 = |x|^2 \ ,$$

for every $x = {}^t(x_1, x_2) \in \mathbb{R}^2$, where

$$A := \begin{pmatrix} 0 & -1 \\ 1 & 0 \end{pmatrix} \ ,$$

and $\mathbb{R}^2$ is equipped with the Euclidean scalar product. Hence $\|A\|_{\mathrm{op}} = 1$, where $\| \ \|_{\mathrm{op}}$ denotes the induced operator norm on $\mathrm{M}(2, \mathbb{R})$.

The representation (2.2) leads to the isomorphism $f : S^1 \to \mathrm{SO}(2)$ of topological groups, where $S^1 \subset \mathbb{C}$ is equipped with the induced topology and the operations of complex multiplication and inversion and SO(2), given by

$$f(z) := \begin{pmatrix} x & y \\ -y & x \end{pmatrix} , \tag{2.3}$$

for every $z = (x, y) \in S^1$. Since complex multiplication is commutative, this implies in particular that SO(2) is commutative. The details of the corresponding proof are left to the reader.

Exercise 2.1 Show that $S^1 \subset \mathbb{C}$ equipped with the induced topology and the operations of complex multiplication and inversion is a topological group and that by (2.3), for every $z = (x, y) \in S^1$, there is defined an isomorphism $f : S^1 \to \mathrm{SO}(2)$ of topological groups.

2.2 Basic Properties of SO(3)

In the following, we collect additional information about SO(3). In a first step, we arrive at parametrizations of SO(3) similar to that of Proposition 2.1 (ii).

Theorem 2.4 (A parametrization of SO(3), I) *For every* $R \in \mathrm{SO}(3)$, *there are* $\varphi \in [0, \pi]$ *as well as a skew-symmetric* $A \in \mathrm{M}(3, \mathbb{R})$ *satisfying* $\|A\|_{\mathrm{op}} = 1$ *and such that*

$$R = \exp(\varphi A) ,$$

where $\mathbb{R}^3$ *is equipped with the Euclidean scalar product and* $\| \; \|_{\mathrm{op}}$ *denotes the induced operator norm on* $\mathrm{M}(3, \mathbb{R})$.

Proof For the proof, let $R \in \mathrm{SO}(3)$. If $R = 1$, we have that $\exp(0.A) = 1$, for every skew-symmetric $A \in \mathrm{M}(3, \mathbb{R})$ such hat $\|A\|_{\mathrm{op}} = 1$. In the following, we assume that $R \neq 1$. We consider SO(3) as subset of $\mathrm{M}(3, \mathbb{C})$ and $\mathbb{C}^3$ as equipped with the canonical scalar product $\langle \, | \, \rangle$. Since R is normal, there is an orthonormal basis

$$e_1 = \begin{pmatrix} e_{11} \\ e_{12} \\ e_{13} \end{pmatrix} , \; e_2 = \begin{pmatrix} e_{21} \\ e_{22} \\ e_{23} \end{pmatrix} , \; e_3 = \begin{pmatrix} e_{31} \\ e_{32} \\ e_{33} \end{pmatrix}$$

of eigenvectors of R. In particular, let $\lambda_1, \lambda_2, \lambda_3$ be the corresponding eigenvalues such that

$$R\, e_k = \lambda_k e_k$$

for every $k \in \{1, 2, 3\}$. Since the characteristic polynomial, that is associated with R, has real coefficents and is of 3rd order, it follows from the mean value theorem that at least one of the eigenvalues is real. Without loss of generality, we can assume that λ_1 is real. Since the coefficients of the characteristic polynomial are real, there are 2 cases. Either, λ_2 and λ_3 are both real, possibly identical, or λ_2 and λ_3 are both non-real and conjugate complex. In the first case, without loss of generality, e_1, e_2, e_3 can be assumed to have real coefficients, and we define $V \in \mathrm{O}(3)$ by

$$V := \begin{pmatrix} e_{11} & e_{21} & e_{31} \\ e_{12} & e_{22} & e_{32} \\ e_{13} & e_{23} & e_{33} \end{pmatrix} .$$

Then

$$R\,V = \begin{pmatrix} \lambda_1 e_{11} & \lambda_2 e_{21} & \lambda_3 e_{31} \\ \lambda_1 e_{12} & \lambda_2 e_{22} & \lambda_3 e_{32} \\ \lambda_1 e_{13} & \lambda_2 e_{23} & \lambda_3 e_{33} \end{pmatrix} = V \cdot \begin{pmatrix} \lambda_1 & 0 & 0 \\ 0 & \lambda_2 & 0 \\ 0 & 0 & \lambda_3 \end{pmatrix}$$

and hence

$$V^{-1} R\,V = \begin{pmatrix} \lambda_1 & 0 & 0 \\ 0 & \lambda_2 & 0 \\ 0 & 0 & \lambda_3 \end{pmatrix} .$$

Further, since

$$R^* R = E \ ,$$

where E denotes the 3×3 unit matrix,

$$1 = \langle e_k | e_k \rangle = \langle R e_k | R e_k \rangle = \langle \lambda_k e_k | \lambda_k e_k \rangle = \lambda_k^2 \, \langle e_k | e_k \rangle = \lambda_k^2 \ ,$$

for every $k \in \{1, 2, 3\}$. Hence, we have that $\lambda_k \in \{-1, 1\}$, for every $k \in \{1, 2, 3\}$, and since

$$1 = \det(R) = \det(V^{-1} R\,V) = \lambda_1 \lambda_2 \lambda_3 \ ,$$

we can assume without loss of generality that $\lambda_1 = 1$, $\lambda_2 = \lambda_3 = -1$. Otherwise, it would follow that $\lambda_1 = \lambda_2 = \lambda_3 = 1$ and hence that $V^{-1} R\,V = 1$ and subsequently that $R = V E\,V^{-1} = V V^{-1} = 1$, a case we already considered. Summarizing the previous, we have that

$$V^{-1} R\,V = \begin{pmatrix} 1 & 0 & 0 \\ 0 & -1 & 0 \\ 0 & 0 & -1 \end{pmatrix} . \tag{2.4}$$

The second case below, leads to (2.6), which, for the case $\varphi = \pi$, is analogous to (2.4), only the matrices V differ. Therefore, the reasoning after (2.6) applies also to this first case. In the second case, we conclude as follows. Since the coefficients of the characteristic polynomial are real, it follows that λ_3 is the complex conjugate of λ_2. In addition, we can assume without loss of generality that the coefficients of e_1 are real as well as that

$$e_3 = \begin{pmatrix} e_{31} \\ e_{32} \\ e_{33} \end{pmatrix} = \begin{pmatrix} (e_{21})^* \\ (e_{22})^* \\ (e_{23})^* \end{pmatrix} . \tag{2.5}$$

Further, since

$$R^* R = E ,$$

we conclude that

$$1 = \langle e_i | e_i \rangle = \langle R e_i | R e_i \rangle = |\lambda_i|^2 ,$$

for every $i \in \{1, 2, 3\}$, and hence that $\lambda_1, \lambda_2, \lambda_3 \in S^1$. In particular, $\lambda_1 \in \{-1,\ 1\}$, and, without loss of generality, we can assume that there is $\varphi \in [0, \pi]$ such that

$$\lambda_2 = e^{-i\varphi} , \ \lambda_3 = e^{i\varphi} .$$

We define

$$v_1 := e_1 , \ v_2 := \frac{1}{\sqrt{2}} (e_2 + e_3) , \ v_3 := \frac{-i}{\sqrt{2}} (e_2 - e_3) .$$

Obviously, it follows that v_1, v_2, v_3 is an orthonormal basis of $\mathbb{C}^3$ and that

$$e_2 = \frac{1}{\sqrt{2}} (v_2 + i v_3) , \ e_3 = \frac{1}{\sqrt{2}} (v_2 - i v_3) .$$

In particular, (2.5) implies that the components of v_1, v_2, v_3 are real. Further,

$$\begin{aligned} R v_2 &= \frac{1}{\sqrt{2}} (e^{-i\varphi} e_2 + e^{i\varphi} e_3) = \frac{1}{2} \left[e^{-i\varphi} (v_2 + i v_3) + e^{i\varphi} (v_2 - i v_3) \right] \\ &= \frac{1}{2} \left(e^{i\varphi} + e^{-i\varphi} \right) v_2 - \frac{i}{2} \left(e^{i\varphi} - e^{-i\varphi} \right) v_3 = \cos(\varphi)\, v_2 + \sin(\varphi)\, v_3 , \\ R v_3 &= \frac{-i}{\sqrt{2}} (e^{-i\varphi} e_2 - e^{i\varphi} e_3) = \frac{-i}{2} \left[e^{-i\varphi} (v_2 + i v_3) - e^{i\varphi} (v_2 - i v_3) \right] \\ &= \frac{i}{2} \left(e^{i\varphi} - e^{-i\varphi} \right) v_2 + \frac{1}{2} \left(e^{i\varphi} + e^{-i\varphi} \right) v_3 = - \sin(\varphi)\, v_2 + \cos(\varphi)\, v_3 . \end{aligned}$$

We define $V \in \mathrm{O}(3)$ by

$$V := \begin{pmatrix} v_{11} & v_{21} & v_{31} \\ v_{12} & v_{22} & v_{32} \\ v_{13} & v_{23} & v_{33} \end{pmatrix} .$$

Then

$$
\begin{aligned}
RV &= \begin{pmatrix} \lambda_1\, v_{11} & \cos(\varphi)\, v_{21} + \sin(\varphi)\, v_{31} & -\sin(\varphi)\, v_{21} + \cos(\varphi)\, v_{31} \\ \lambda_1\, v_{12} & \cos(\varphi)\, v_{22} + \sin(\varphi)\, v_{32} & -\sin(\varphi)\, v_{22} + \cos(\varphi)\, v_{32} \\ \lambda_1\, v_{13} & \cos(\varphi)\, v_{23} + \sin(\varphi)\, v_{33} & -\sin(\varphi)\, v_{23} + \cos(\varphi)\, v_{33} \end{pmatrix} \\
&= \begin{pmatrix} v_{11} & v_{21} & v_{31} \\ v_{12} & v_{22} & v_{32} \\ v_{13} & v_{23} & v_{33} \end{pmatrix} \cdot \begin{pmatrix} \lambda_1 & 0 & 0 \\ 0 & \cos(\varphi) & -\sin(\varphi) \\ 0 & \sin(\varphi) & \cos(\varphi) \end{pmatrix} \\
&= V \begin{pmatrix} \lambda_1 & 0 & 0 \\ 0 & \cos(\varphi) & -\sin(\varphi) \\ 0 & \sin(\varphi) & \cos(\varphi) \end{pmatrix}
\end{aligned}
$$

and hence

$$
V^{-1} R\, V = \begin{pmatrix} \lambda_1 & 0 & 0 \\ 0 & \cos(\varphi) & -\sin(\varphi) \\ 0 & \sin(\varphi) & \cos(\varphi) \end{pmatrix} . \tag{2.6}
$$

In particular, the latter implies that

$$
\lambda_1 = \det(V^{-1} R\, V) = \det(R) = 1 .
$$

Further, by using that

$$
\begin{aligned}
A^2 &= - \begin{pmatrix} 0 & 0 & 0 \\ 0 & 1 & 0 \\ 0 & 0 & 1 \end{pmatrix} , \quad \begin{pmatrix} 0 & 0 & 0 \\ 0 & 1 & 0 \\ 0 & 0 & 1 \end{pmatrix} \cdot \begin{pmatrix} 0 & 0 & 0 \\ 0 & 1 & 0 \\ 0 & 0 & 1 \end{pmatrix} = \begin{pmatrix} 0 & 0 & 0 \\ 0 & 1 & 0 \\ 0 & 0 & 1 \end{pmatrix} , \\
A^3 &= -A , \quad \begin{pmatrix} 0 & 0 & 0 \\ 0 & 1 & 0 \\ 0 & 0 & 1 \end{pmatrix} \cdot A = A ,
\end{aligned}
$$

where

$$
A := \begin{pmatrix} 0 & 0 & 0 \\ 0 & 0 & -1 \\ 0 & 1 & 0 \end{pmatrix} ,
$$

it follows that

$$
A^{2k} = (-1)^k \begin{pmatrix} 0 & 0 & 0 \\ 0 & 1 & 0 \\ 0 & 0 & 1 \end{pmatrix} , \quad A^{2k+1} = (-1)^k A ,
$$

for $k \in \mathbb{N}^*$, and hence that

$$\begin{aligned}
\exp(\varphi A) &= \sum_{k=0}^{\infty} \frac{\varphi^k}{k!} A^k = E + \sum_{k=1}^{\infty} \frac{\varphi^{2k}}{(2k)!} A^{2k} + \sum_{k=1}^{\infty} \frac{\varphi^{2k+1}}{(2k+1)!} A^{2k+1} \\
&= E + \left[\sum_{k=1}^{\infty} (-1)^k \frac{\varphi^{2k}}{(2k)!} \right] \begin{pmatrix} 0 & 0 & 0 \\ 0 & 1 & 0 \\ 0 & 0 & 1 \end{pmatrix} + \left[\sum_{k=1}^{\infty} (-1)^k \frac{\varphi^{2k+1}}{(2k+1)!} \right] A \\
&= E + [\cos(\varphi) - 1] \begin{pmatrix} 0 & 0 & 0 \\ 0 & 1 & 0 \\ 0 & 0 & 1 \end{pmatrix} + \sin(\varphi)\, A \\
&= E + \begin{pmatrix} 0 & 0 & 0 \\ 0 & \cos(\varphi) - 1 & 0 \\ 0 & 0 & \cos(\varphi) - 1 \end{pmatrix} + \sin(\varphi) \begin{pmatrix} 0 & 0 & 0 \\ 0 & 0 & -1 \\ 0 & 1 & 0 \end{pmatrix} \\
&= \begin{pmatrix} 1 & 0 & 0 \\ 0 & \cos(\varphi) & -\sin(\varphi) \\ 0 & \sin(\varphi) & \cos(\varphi) \end{pmatrix} = V^{-1} R\, V\ .
\end{aligned}$$

From the latter, it follows that

$$R = V \exp(\varphi A) V^{-1} = \exp(\varphi\, VA\, V^{-1})\ .$$

Since V is a real matrix and A is a skew-symmetric real matrix, it follows that

$$(VA\, V^{-1})^* = (VA\, V^*)^* = V A^* V^* = -VA\, V^* = -VA\, V^{-1}\ ,$$

i.e., that $VA\, V^{-1}$ is a skew-symmetric real matrix. Further, since

$$|Ax|^2 = |{}^t(0, -x_3, x_2)|^2 = x_2^2 + x_3^2 \leqslant |x|^2\ , \quad |A\, {}^t(0, 1, 0)| = 1\ ,$$

for every $x = {}^t(x_1, x_2, x_3) \in \mathbb{R}^3$, we have that $\|A\|_{\text{op}} = 1$ and hence also that $\|VA\, V^{-1}\|_{\text{op}} = 1$. □

Proposition 2.5 (A parametrization of SO(3), II) *Let* $n \in S^2$, $\varphi \in \mathbb{R}$,

$$A := \begin{pmatrix} 0 & -n_3 & n_2 \\ n_3 & 0 & -n_1 \\ -n_2 & n_1 & 0 \end{pmatrix}\ ,$$

$\langle\,|\,\rangle$ *be the Euclidean scalar product on* $\mathbb{R}^3$ *and* $\|\ \|_{\text{op}}$ *the associated operator norm on* $\mathrm{M}(3, \mathbb{R})$.

(i) Then $\|A\|_{\text{op}} = 1$, $\exp(\varphi A) \in \mathrm{SO}(3)$ *and*

$$\exp(\varphi A)\, x = \langle x|n\rangle\, n + \cos(\varphi)\, [\, x - \langle x|n\rangle\, n\,] + \sin(\varphi)\, (n \times x)\ ,$$

for every $x \in \mathbb{R}^3$.

(ii) If $e \in S^2$ *is orthogonal to* n*, then* $e, n \times e, n$ *is a positively oriented orthonormal basis of* $\mathbb{R}^3$*, and the representation matrix* $M_{\exp(\varphi A)}$ *of* $\exp(\varphi A)$ *with respect to this basis is given by*

$$M_{\exp(\varphi A)} = \begin{pmatrix} \cos(\varphi) & \sin(\varphi) & 0 \\ -\sin(\varphi) & \cos(\varphi) & 0 \\ 0 & 0 & 1 \end{pmatrix} .$$

Proof In the following, $\langle | \rangle$ denotes the Euclidean scalar product for $\mathbb{R}^3$ and $\| \, \|_{\text{op}}$ the associated operator norm on $\text{M}(3, \mathbb{R})$. If $n \in S^2$, $\varphi \in \mathbb{R}$ and

$$A := \begin{pmatrix} 0 & -n_3 & n_2 \\ n_3 & 0 & -n_1 \\ -n_2 & n_1 & 0 \end{pmatrix} ,$$

then

$$Ax = \begin{pmatrix} 0 & -n_3 & n_2 \\ n_3 & 0 & -n_1 \\ -n_2 & n_1 & 0 \end{pmatrix} \cdot \begin{pmatrix} x_1 \\ x_2 \\ x_3 \end{pmatrix} = \begin{pmatrix} n_2x_3 - n_3x_2 \\ n_3x_1 - n_1x_3 \\ n_1x_2 - n_2x_1 \end{pmatrix} = n \times x ,$$

for every $x \in \mathbb{R}^3$. As a consequence,

$$|Ax| = \sqrt{|n|^2 - \langle n|x \rangle^2} ,$$

for every $x \in S^2$, and

$$\|A\|_{\text{op}} = \sup_{x \in S^2} |Ax| = |n| = 1 .$$

Further,

$$A^2 = \begin{pmatrix} -(n_2^2 + n_3^2) & n_1n_2 & n_1n_3 \\ n_1n_2 & -(n_1^2 + n_3^2) & n_2n_3 \\ n_1n_3 & n_2n_3 & -(n_1^2 + n_2^2) \end{pmatrix} , \quad A^3 = -A .$$

Hence

$$A^{2k+1} = (-1)^k A \text{ and } A^{2k} = (-1)^{k-1} A^2 , \tag{2.7}$$

for $k \in \mathbb{N}$ and $k \in \mathbb{N}^*$, respectively. As a consequence,

$$\begin{aligned} R := \exp(\varphi A) &= E + \sum_{k=0}^{\infty} \frac{\varphi^{2k+1}}{(2k+1)!} A^{2k+1} + \sum_{k=1}^{\infty} \frac{\varphi^{2k}}{(2k)!} A^{2k} \\ &= E + \left[\sum_{k=0}^{\infty} \frac{(-1)^k \varphi^{2k+1}}{(2k+1)!} \right] A - \left[\sum_{k=1}^{\infty} \frac{(-1)^k \varphi^{2k}}{(2k)!} \right] A^2 \\ &= E + \sin(\varphi)\, A + [1 - \cos(\varphi)]\, A^2 \ . \end{aligned} \tag{2.8}$$

Hence it follows for $x \in \mathbb{R}^3$ that

$$\begin{aligned} Rx &= x + \sin(\varphi) A x + [1 - \cos(\varphi)]\, A^2 x \\ &= x + \sin(\varphi)\, n \times x + [1 - \cos(\varphi)]\, n \times (n \times x) \\ &= x + \sin(\varphi)\, n \times x + [1 - \cos(\varphi)] \cdot [\langle x | n \rangle\, n - x] \\ &= \langle x | n \rangle\, n + \cos(\varphi)\, [\, x - \langle x | n \rangle\, n\,] + \sin(\varphi)\, (n \times x) \ . \end{aligned}$$

Further, Since $n \in S^2$, we have that

$$\begin{aligned} |Rx|^2 &= |\, \langle x | n \rangle\, |^2 + \cos^2(\varphi)\, |x - \langle x | n \rangle\, n|^2 + \sin^2(\varphi)\, |n \times x|^2 \\ &= |\, \langle x | n \rangle\, |^2 + \cos^2(\varphi)\ \langle x - \langle x | n \rangle\, n | x - \langle x | n \rangle\, n \rangle + \sin^2(\varphi)\, |n \times x|^2 \\ &= |\, \langle x | n \rangle\, |^2 + \cos^2(\varphi)\ (\, |x|^2 - |\, \langle x | n \rangle\, |^2) + \sin^2(\varphi)\ (\, |x|^2 - |\, \langle x | n \rangle\, |^2) \\ &= |\, \langle x | n \rangle\, |^2 + |x|^2 - |\, \langle x | n \rangle\, |^2 = |x|^2 \ , \end{aligned}$$

for every $x \in \mathbb{R}^3$ and hence that

$$\begin{aligned} \langle Rx | Ry \rangle &= \frac{1}{2}\, (|Rx + Ry|^2 - |Rx|^2 - |Ry|^2) \\ &= \frac{1}{2}\, (|R(x + y)|^2 - |Rx|^2 - |Ry|^2) \\ &= \frac{1}{2}\, (|x + y|^2 - |x|^2 - |y|^2) = \langle x | y \rangle \ , \end{aligned}$$

for all $x, y \in \mathbb{R}^3$. Therefore $R \in \mathrm{O}(3)$. If $x \in \mathbb{R}^3$, we note that $n, x - \langle x | n \rangle\, n, n \times x$ are orthogonal. In addition, we have that

$$\begin{aligned} &Rn = n \ , \\ &R(x - \langle x | n \rangle\, n) \\ &= \langle x - \langle x | n \rangle\, n | n \rangle\, n + \cos(\varphi)\, [\, x - \langle x | n \rangle\, n - \langle x - \langle x | n \rangle\, n | n \rangle\, n\,] \\ &\quad + \sin(\varphi)\, (n \times (x - \langle x | n \rangle\, n)) \\ &= \cos(\varphi)\, [\, x - \langle x | n \rangle\, n] + \sin(\varphi)\, (n \times x) \ , \\ &R(n \times x) \\ &= \langle n \times x | n \rangle\, n + \cos(\varphi)\, [\, n \times x - \langle n \times x | n \rangle\, n\,] + \sin(\varphi)\, (n \times (n \times x)) \\ &= \cos(\varphi)(n \times x) + \sin(\varphi)\, [n \times (n \times x)] \\ &= -\sin(\varphi)\, [x - \langle x | n \rangle\, n] + \cos(\varphi)(n \times x) \end{aligned}$$

as well as that

$$|x - \langle x|n\rangle\, n|^2 = \langle x - \langle x|n\rangle\, n | x - \langle x|n\rangle\, n\rangle = |x|^2 - 2\,\langle x|n\rangle^2 + \langle x|n\rangle^2$$
$$= |x|^2 - \langle x|n\rangle^2 \ ,$$
$$|n \times x|^2 = \langle n \times x | n \times x\rangle = \langle n|n\rangle\,\langle x|x\rangle - \langle n|x\rangle\,\langle x|n\rangle = |x|^2 - \langle x|n\rangle^2 \ ,$$

i.e., $x - \langle x|n\rangle\, n$ and $n \times x$ are of equal length. Hence, if $e \in S^2$ is orthogonal to n, then $e, n \times e, n$ is an orthonormal basis of $\mathbb{R}^3$, and the representation matrix M_R of R with respect to this basis is given by

$$M_R = \begin{pmatrix} \cos(\varphi) & \sin(\varphi) & 0 \\ -\sin(\varphi) & \cos(\varphi) & 0 \\ 0 & 0 & 1 \end{pmatrix} .$$

In particular, there is $U \in \mathrm{GL}(3, \mathbb{R})$ such that $M_R = U\,R\,U^{-1}$. Hence,

$$1 = \det(M_R) = \det(U\,R\,U^{-1}) = \det(U)\det(R)\det(U^{-1})$$
$$= \det(U)\det(U^{-1})\det(R) = \det(UU^{-1}) \cdot \det(R)$$
$$= \det(E)\det(R) = \det(R)$$

and therefore $R \in \mathrm{SO}(3)$. Since

$$\begin{vmatrix} e_1 & (n \times e)_1 & n_1 \\ e_2 & (n \times e)_2 & n_2 \\ e_3 & (n \times e)_3 & n_3 \end{vmatrix} = \langle e|(n \times e) \times n\rangle = \langle e|\,\langle n|n\rangle\, e - \langle e|n\rangle\, n\rangle = \langle e|e\rangle = 1 \ ,$$

the orthonormal basis $e, n \times e, n$ is positively oriented. □

As a consequence, we arrive at a direct proof of the fact that

$$\exp(T_E\mathrm{SO}(3)) = \mathrm{SO}(3) \ .$$

For a different proof of the same result, for the more general case of $\mathrm{SU}(n)$, $n \in \mathbb{N}^*$, see Proposition 1.22.

Corollary 2.6 (Surjectivity of the exponential map) *We have that*

$$\exp(\{A \in \mathrm{M}(3, \mathbb{R}) : A \textit{ is skew-symmetric }\}) = \mathrm{SO}(3) \ .$$

Proof According to Theorem 2.4, we have that

$$\exp(\{A \in \mathrm{M}(3, \mathbb{R}) : A \text{ is skew-symmetric }\}) \supset \mathrm{SO}(3) \ .$$

If $A \in \mathrm{M}(3, \mathbb{R})$ is skew-symmetric, we conclude as follows. If $A = 0$, then $\exp(A) = 1 \in \mathrm{SO}(3)$. If $A \neq 0$, then, according to Proposition 2.5,

$$\exp(A) = \exp[\, \|A\|_{\mathrm{op}}\, (\|A\|_{\mathrm{op}}^{-1} A)\,] \in \mathrm{SO}(3) \ .$$

Hence it follows that

$$\exp(\{A \in \mathrm{M}(3, \mathbb{R}) : A \text{ is skew-symmetric }\}) \subset \mathrm{SO}(3) \ .$$

□

The results from Proposition 2.5 allow the derivation of a result concerning the injectivity of the exponential map.

Proposition 2.7 (Injectivity of the exponential map) *The map* $u : U \to \mathrm{SO}(3)$, *defined by* $u(A) := \exp(A)$, *for every*

$$A \in U := \{A \in \mathrm{M}(3, \mathbb{R}) : A \textit{ is skew-symmetric and } \|A\|_{\mathrm{op}} < \pi\} \ ,$$

is injective.

Proof In a first step, we are going to show that the map $h : (0, \pi) \times S^2 \to \mathrm{SO}(3)$, defined by

$$h(\varphi, n) := \exp\left(\varphi \begin{pmatrix} 0 & -n_3 & n_2 \\ n_3 & 0 & -n_1 \\ -n_2 & n_1 & 0 \end{pmatrix}\right) ,$$

for every $(\varphi, n) \in (0, \pi) \times S^2$, is injective. If $\varphi \in (0, \pi)$, $n \in S^2$ and

$$A = \begin{pmatrix} 0 & -n_3 & n_2 \\ n_3 & 0 & -n_1 \\ -n_2 & n_1 & 0 \end{pmatrix} ,$$

then it follows from Proposition 2.5 (ii) that $\exp(\varphi A)$ has the simple eigenvalues 1, $e^{i\varphi} \in S^1 \cap (\mathbb{R} \times (0, \infty))$ and $e^{-i\varphi} \in S^1 \cap (\mathbb{R} \times (-\infty, 0))$. For later use, we note that, as a consequence, $\exp(\varphi A) \neq 1$. Hence, if $\psi \in (0, \pi)$, $\bar{n} \in S^2$ are such that

$$\exp(\psi \bar{A}) = \exp(\varphi A) \ ,$$

where

$$\bar{A} = \begin{pmatrix} 0 & -\bar{n}_3 & \bar{n}_2 \\ \bar{n}_3 & 0 & -\bar{n}_1 \\ -\bar{n}_2 & \bar{n}_1 & 0 \end{pmatrix} ,$$

then $e^{i\psi} = e^{i\varphi}$ and therefore $\psi = \varphi$. Since,

$$n = \exp(\varphi A)n = \exp(\psi \bar{A})n \ ,$$

n is an eigenvector of $\exp(\psi \bar{A})$ to the eigenvalue 1 and hence $\bar{n} = \pm n$. Further, if $\bar{n} = -n$, it follows from Proposition 2.5 (i) that

$$\begin{aligned}
&[\exp(\psi \bar{A}) - \exp(\varphi A)]\, x \\
&= \langle x|n\rangle\, n + \cos(\varphi)\, [\, x - \langle x|n\rangle\, n\,] - \sin(\varphi)\, (n \times x) \\
&\quad - \{\langle x|n\rangle\, n + \cos(\varphi)\, [\, x - \langle x|n\rangle\, n\,] + \sin(\varphi)\, (n \times x)\} \\
&= -2\sin(\varphi)\, (n \times x) \ ,
\end{aligned}$$

for every $x \in \mathbb{R}^3$ and hence the contradiction that $n \times x = 0$, for every $x \in \mathbb{R}^3$. ↯ Therefore, $\bar{n} = n$ and consequently $\bar{A} = A$. In the next step, we are going to show that the map $u : U \to \mathrm{SO}(3)$, where

$$U := \{A \in \mathrm{M}(3, \mathbb{R}) : A \text{ is skew-symmetric and } 0 < \|A\|_{\mathrm{op}} < \pi\} \ ,$$

defined by $u(A) := \exp(A)$, for every $A \in U$, is injective. If $A \in U$, then $\varphi := \|A\|_{\mathrm{op}} \in (0, \pi)$, $\|\varphi^{-1} A\|_{\mathrm{op}} = 1$. Since $\varphi^{-1} A$ is skew-symmetric, there is $n \in \mathbb{R}^3$ such that

$$\varphi^{-1} A = \begin{pmatrix} 0 & -n_3 & n_2 \\ n_3 & 0 & -n_1 \\ -n_2 & n_1 & 0 \end{pmatrix} .$$

Then

$$\varphi^{-1} A x = \begin{pmatrix} 0 & -n_3 & n_2 \\ n_3 & 0 & -n_1 \\ -n_2 & n_1 & 0 \end{pmatrix} \cdot \begin{pmatrix} x_1 \\ x_2 \\ x_3 \end{pmatrix} = \begin{pmatrix} n_2 x_3 - n_3 x_2 \\ n_3 x_1 - n_1 x_3 \\ n_1 x_2 - n_2 x_1 \end{pmatrix} = n \times x \ ,$$

for every $x \in \mathbb{R}^3$. As a consequence,

$$|\varphi^{-1} A x| = \sqrt{|n|^2 - \langle n|x\rangle^2} \ ,$$

for every $x \in S^2$, and

$$1 = \|\varphi^{-1} A\|_{\mathrm{op}} = \sup_{x \in S^2} |\varphi^{-1} A x| = |n| \ ,$$

Therefore $n \in S^2$ and

$$\exp(A) = \exp(\varphi\, \varphi^{-1} A) = h(\varphi, n) \ .$$

Hence, if $\bar{A} \in U$ such that $\exp(\bar{A}) = \exp(A)$, it follows from the injectivity of h that $\bar{\varphi} = \varphi$ and $\bar{n} = n$, where $\bar{\varphi} := \|\bar{A}\|_{\mathrm{op}}$ and $\bar{n} \in \mathbb{R}^3$ is such that

$$\bar{\varphi}^{-1}\bar{A} = \begin{pmatrix} 0 & -\bar{n}_3 & \bar{n}_2 \\ \bar{n}_3 & 0 & -\bar{n}_1 \\ -\bar{n}_2 & \bar{n}_1 & 0 \end{pmatrix} .$$

Therefore

$$\bar{A} = \bar{\varphi}\,\bar{\varphi}^{-1}\bar{A} = \varphi\,\varphi^{-1}A = A \; .$$

Since, as noted above $1 \notin \mathrm{Ran}(u)$, the injectivity of u follows. □

Exercise 2.2 Using (2.8) and (2.7), show that $R^* \cdot R = E$ and conclude that $R \in \mathrm{O}(3)$.

We note that R from Proposition 2.5 leaves the points on $\mathbb{R}.n$ fixed, i.e., $R\,\lambda n = \lambda n$, for every $\lambda \in \mathbb{R}$. Further, for $x \in \mathbb{R}^3 \setminus (\mathbb{R}.n)$, $\langle x|n\rangle\, n$ is the orthogonal projection of x onto $\mathbb{R}.n$, where $\langle\,|\,\rangle$ denotes the Euclidean scalar product on $\mathbb{R}^3$, and $x - \langle x|n\rangle\, n$ is the vector that is orthogonal to $\mathbb{R}.n$, starting from $\mathbb{R}.n$ and ending at the tip of x. Further, $n \times x$ is orthogonal to x and n and hence also orthogonal to $x - \langle x|n\rangle\, n$. In particular,

$$\begin{aligned} |x - \langle x|n\rangle\, n|^2 &= \langle x - \langle x|n\rangle\, n | x - \langle x|n\rangle\, n\rangle = |x|^2 - 2\,\langle x|n\rangle^2 + \langle x|n\rangle^2 \\ &= |x|^2 - \langle x|n\rangle^2 > 0 \;, \\ |n \times x|^2 &= \langle n \times x | n \times x\rangle = \langle n|n\rangle\,\langle x|x\rangle - \langle n|x\rangle\,\langle x|n\rangle = |x|^2 - \langle x|n\rangle^2 \;, \end{aligned}$$

i.e., $x - \langle x|n\rangle\, n$ and $n \times x$ are of equal length. Further,

$$\begin{aligned} \langle x - \langle x|n\rangle\, n | (n \times x) \times n\rangle &= \langle x - \langle x|n\rangle\, n |\, \langle n|n\rangle\, x - \langle x|n\rangle\, n\rangle \\ &= |x - \langle x|n\rangle\, n|^2 \;, \end{aligned}$$

i.e.,

☞ If $x \notin \mathbb{R}.n$, then $x - \langle x|n\rangle\, n$, $n \times x$ and n form a right-handed orthogonal basis. Hence, if $0 \leqslant \varphi \leqslant \pi$, $\exp(\varphi A)$ from Corollary 2.5 describes a counterclockwise rotation by φ about the axis $\mathbb{R}.n$, see Fig. 2.1, whereas if $-\pi < \varphi < 0$,$\exp(\varphi A)$ describes a clockwise rotation by φ about the axis $\mathbb{R}.n$.

In the following, we derive a decomposition of rotations into rotations associated with proper Euler angles, (Fig. 2.2).

Proposition 2.8 (A decomposition of rotations using proper Euler angles)

(i) For every $R \in \mathrm{SO}(3)$, there are $\psi, \theta \in (-\pi, \pi]$ and $\varphi \in [-\pi, 0]$ such that

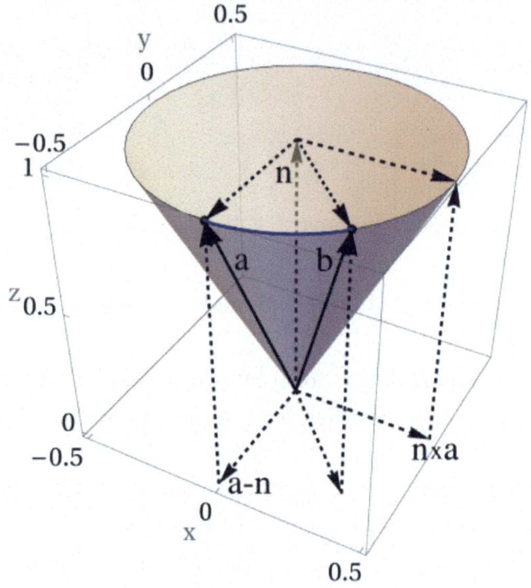

Fig. 2.1 Counterclockwise rotation of $\mathbf{a} = {}^t(0, -1/2, 1)$, about the z-axis by the angle $\varphi = \pi/4$, into $\mathbf{b} = \exp(\varphi A)\,\mathbf{a}$

$$R = \begin{pmatrix} \cos(\psi) & -\sin(\psi) & 0 \\ \sin(\psi) & \cos(\psi) & 0 \\ 0 & 0 & 1 \end{pmatrix} \cdot \begin{pmatrix} 1 & 0 & 0 \\ 0 & \cos(\varphi) & -\sin(\varphi) \\ 0 & \sin(\varphi) & \cos(\varphi) \end{pmatrix}$$
$$\cdot \begin{pmatrix} \cos(\theta) & -\sin(\theta) & 0 \\ \sin(\theta) & \cos(\theta) & 0 \\ 0 & 0 & 1 \end{pmatrix} . \tag{2.9}$$

(ii) *For every* $R \in \mathrm{SO}(3)$ *such that* $R_{13}^2 + R_{23}^2 \neq 0$, *there are uniquely determined* $\psi, \theta \in (-\pi, \pi]$ *and* $\varphi \in (-\pi, 0)$ *such that (2.9) is satisfied.*

Proof "Part (i)": For the proof, let

$$R = \begin{pmatrix} R_{11} & R_{12} & R_{13} \\ R_{21} & R_{22} & R_{23} \\ R_{31} & R_{32} & R_{33} \end{pmatrix}$$

be some element of SO(3). Then

$$\begin{pmatrix} \cos(\psi) & \sin(\psi) & 0 \\ -\sin(\psi) & \cos(\psi) & 0 \\ 0 & 0 & 1 \end{pmatrix} R = \begin{pmatrix} V_{11} & V_{12} & V_{13} \\ V_{21} & V_{22} & V_{23} \\ V_{31} & V_{32} & V_{33} \end{pmatrix} \in \mathrm{SO}(3) ,$$

where

$$\begin{pmatrix} V_{11} & V_{12} & V_{13} \\ V_{21} & V_{22} & V_{23} \\ V_{31} & V_{32} & V_{33} \end{pmatrix} = \begin{pmatrix} \cos(\psi)\,R_{11} + \sin(\psi)\,R_{21} \\ -\sin(\psi)\,R_{11} + \cos(\psi)\,R_{21} \\ R_{31} \end{pmatrix.}$$

$$\begin{matrix} \cos(\psi)\,R_{12} + \sin(\psi)\,R_{22} & \cos(\psi)\,R_{13} + \sin(\psi)\,R_{23} \\ -\sin(\psi)\,R_{12} + \cos(\psi)\,R_{22} & -\sin(\psi)\,R_{13} + \cos(\psi)\,R_{23} \\ R_{32} & R_{33} \end{matrix}\Bigg) ,$$

and $\psi \in (-\pi, \pi]$. In particular,

$$V_{13} = \cos(\psi)\,R_{13} + \sin(\psi)\,R_{23} = \mathrm{Im}(\exp(i\psi)(R_{23} + i\,R_{13})) ,$$
$$V_{23} = -\sin(\psi)\,R_{13} + \cos(\psi)\,R_{23} = \mathrm{Re}(\exp(i\psi)(R_{23} + i\,R_{13})) .$$

We consider cases.

Case (1) If $R_{13}^2 + R_{23}^2 = 0$, then $V_{13} = V_{23} = 0$, for every $\psi \in (\pi, \pi]$. We note that it follows that $V_{23}^2 + V_{33}^2 \neq 0$.

Case (2) If $R_{13}^2 + R_{23}^2 \neq 0$, then $V_{13} = 0$ if

$$\exp(i\psi) = \cos(\psi) + i\sin(\psi) = \frac{R_{23} - i\,R_{13}}{|R_{23} - i\,R_{13}|} .$$

We note that the implies that

$$V_{23} = \mathrm{Re}\left(\frac{R_{23} - i\,R_{13}}{|R_{23} - i\,R_{13}|}\,(R_{23} + i\,R_{13})\right) = |R_{23} - i\,R_{13}| > 0 .$$

Hence there is $\psi \in (-\pi, \pi]$ such that $V_{13} = 0$. We note that it follows that $V_{23}^2 + V_{33}^2 \neq 0$.

Further, for $\psi \in (\pi, \pi]$ from both cases,

$$\begin{pmatrix} 1 & 0 & 0 \\ 0 & \cos(\varphi) & \sin(\varphi) \\ 0 & -\sin(\varphi) & \cos(\varphi) \end{pmatrix} \cdot \begin{pmatrix} V_{11} & V_{12} & 0 \\ V_{21} & V_{22} & V_{23} \\ V_{31} & V_{32} & V_{33} \end{pmatrix}$$
$$= \begin{pmatrix} W_{11} & W_{12} & 0 \\ W_{21} & W_{22} & W_{23} \\ W_{31} & W_{32} & W_{33} \end{pmatrix} \in \mathrm{SO}(3) ,$$

where

$$\begin{pmatrix} W_{11} & W_{12} & 0 \\ W_{21} & W_{22} & W_{23} \\ W_{31} & W_{32} & W_{33} \end{pmatrix} = \begin{pmatrix} V_{11} \\ \cos(\varphi)\,V_{21} + \sin(\varphi)\,V_{31} \\ -\sin(\varphi)\,V_{21} + \cos(\varphi)\,V_{31} \end{pmatrix.}$$

$$\begin{matrix} V_{12} & 0 \\ \cos(\varphi)\,V_{22} + \sin(\varphi)\,V_{32} & \cos(\varphi)\,V_{23} + \sin(\varphi)\,V_{33} \\ -\sin(\varphi)\,V_{22} + \cos(\varphi)\,V_{32} & -\sin(\varphi)\,V_{23} + \cos(\varphi)\,V_{33} \end{matrix}\Bigg) ,$$

and $\varphi \in (-\pi, \pi]$. In particular,

$$W_{23} = \cos(\varphi)\, V_{23} + \sin(\varphi)\, V_{33} = \text{Im}(\exp(i\varphi)(V_{33} + i\,V_{23}))\ ,$$
$$W_{33} = -\sin(\varphi)\, V_{23} + \cos(\varphi)\, V_{33} = \text{Re}(\exp(i\varphi)(V_{33} + i\,V_{23}))\ ,$$

and hence $W_{23} = 0$, if

$$\exp(i\varphi) = \cos(\varphi) + i\,\sin(\varphi) = \frac{V_{33} - i\,V_{23}}{|V_{33} - i\,V_{23}|}\ .$$

As a consequence, for both cases, there is $\varphi \in [-\pi, 0]$ such that $W_{23} = 0$ and the same time such that

$$W_{33} = \text{Re}\left(\frac{V_{33} - i\,V_{23}}{|V_{33} - i\,V_{23}|}\,(V_{33} + i\,V_{23})\right) = |V_{33} - i\,V_{23}| > 0\ .$$

For such φ, it follows that

$$\begin{pmatrix} W_{11} & W_{12} & 0 \\ W_{21} & W_{22} & W_{23} \\ W_{31} & W_{32} & W_{33} \end{pmatrix} = \begin{pmatrix} W_{11} & W_{12} & 0 \\ W_{21} & W_{22} & 0 \\ W_{31} & W_{32} & W_{33} \end{pmatrix} = \begin{pmatrix} W_{11} & W_{12} & 0 \\ W_{21} & W_{22} & 0 \\ 0 & 0 & 1 \end{pmatrix}\ ,$$

where we use that the column vectors of every element of SO(3) form an orthonormal basis of $\mathbb{R}^3$, the latter equipped with the Euclidean scalar product. In particular, it follows that

$$\begin{pmatrix} W_{11} & W_{12} \\ W_{21} & W_{22} \end{pmatrix} \in \text{SO}(2)$$

and hence from Proposition 2.1 the existence of $\theta \in (-\pi, \pi]$ such that

$$\begin{pmatrix} W_{11} & W_{12} \\ W_{21} & W_{22} \end{pmatrix} = \begin{pmatrix} \cos(\theta) & -\sin(\theta) \\ \sin(\theta) & \cos(\theta) \end{pmatrix}\ .$$

Hence, we have that

$$\begin{pmatrix} 1 & 0 & 0 \\ 0 & \cos(\varphi) & \sin(\varphi) \\ 0 & -\sin(\varphi) & \cos(\varphi) \end{pmatrix} \cdot \begin{pmatrix} \cos(\psi) & \sin(\psi) & 0 \\ -\sin(\psi) & \cos(\psi) & 0 \\ 0 & 0 & 1 \end{pmatrix} R$$
$$= \begin{pmatrix} \cos(\theta) & -\sin(\theta) & 0 \\ \sin(\theta) & \cos(\theta) & 0 \\ 0 & 0 & 1 \end{pmatrix}\ ,$$

finally, it follows the representation (2.9).

"Part (ii)": If $R \in \text{SO}(3)$ is such that $R_{13}^2 + R_{23}^2 \neq 0$ and $\psi, \theta \in (-\pi, \pi]$, $\varphi \in [-\pi, 0]$ are such that (2.9) is valid, then $\varphi \notin \{-\pi, 0\}$ because if $\varphi = 0$, then

$$R = \begin{pmatrix} \cos(\psi+\theta) & -\sin(\psi+\theta) & 0 \\ \sin(\psi+\theta) & \cos(\psi+\theta) & 0 \\ 0 & 0 & 1 \end{pmatrix}$$

and if $\varphi = -\pi$, then

$$\begin{aligned} R &= \begin{pmatrix} \cos(\psi) & -\sin(\psi) & 0 \\ \sin(\psi) & \cos(\psi) & 0 \\ 0 & 0 & 1 \end{pmatrix} \cdot \begin{pmatrix} 1 & 0 & 0 \\ 0 & -1 & 0 \\ 0 & 0 & -1 \end{pmatrix} \cdot \begin{pmatrix} \cos(\theta) & -\sin(\theta) & 0 \\ \sin(\theta) & \cos(\theta) & 0 \\ 0 & 0 & 1 \end{pmatrix} \\ &= \begin{pmatrix} \cos(\psi) & -\sin(\psi) & 0 \\ \sin(\psi) & \cos(\psi) & 0 \\ 0 & 0 & 1 \end{pmatrix} \cdot \begin{pmatrix} \cos(\theta) & -\sin(\theta) & 0 \\ -\sin(\theta) & -\cos(\theta) & 0 \\ 0 & 0 & -1 \end{pmatrix} \\ &= \begin{pmatrix} \cos(\psi-\theta) & \sin(\psi-\theta) & 0 \\ \sin(\psi-\theta) & -\cos(\psi-\theta) & 0 \\ 0 & 0 & -1 \end{pmatrix} . \end{aligned}$$

Hence, in both cases $R_{13} = R_{23} = 0$. As a consequence, for every $R \in \mathrm{SO}(3)$ such that $R_{13}^2 + R_{23}^2 \neq 0$, there are $\psi, \theta \in (-\pi, \pi]$, $\varphi \in (-\pi, 0)$ such that (2.9) is valid. Further, for $\psi, \theta \in (-\pi, \pi]$, $\varphi \in (-\pi, 0)$, it follows that

$$\begin{aligned} R &= \begin{pmatrix} \cos(\psi) & -\sin(\psi) & 0 \\ \sin(\psi) & \cos(\psi) & 0 \\ 0 & 0 & 1 \end{pmatrix} \cdot \begin{pmatrix} 1 & 0 & 0 \\ 0 & \cos(\varphi) & -\sin(\varphi) \\ 0 & \sin(\varphi) & \cos(\varphi) \end{pmatrix} \\ &\quad \cdot \begin{pmatrix} \cos(\theta) & -\sin(\theta) & 0 \\ \sin(\theta) & \cos(\theta) & 0 \\ 0 & 0 & 1 \end{pmatrix} \\ &= \begin{pmatrix} \cos(\psi) & -\sin(\psi)\cos(\varphi) & \sin(\psi)\sin(\varphi) \\ \sin(\psi) & \cos(\psi)\cos(\varphi) & -\cos(\psi)\sin(\varphi) \\ 0 & \sin(\varphi) & \cos(\varphi) \end{pmatrix} \\ &\quad \cdot \begin{pmatrix} \cos(\theta) & -\sin(\theta) & 0 \\ \sin(\theta) & \cos(\theta) & 0 \\ 0 & 0 & 1 \end{pmatrix} \end{aligned}$$

and hence that

$$R_{13}^2 + R_{23}^2 = \sin^2(\psi)\sin^2(\varphi) + \cos^2(\psi)\sin^2(\varphi) = \sin^2(\varphi) \neq 0 .$$

Concerning the uniqueness of such a representation, if $\psi_1, \theta_1, \psi_2, \theta_2 \in (-\pi, \pi]$ and $\varphi_1, \varphi_2 \in [-\pi, 0]$ are such that

$$
\begin{pmatrix} \cos(\psi_1) & -\sin(\psi_1) & 0 \\ \sin(\psi_1) & \cos(\psi_1) & 0 \\ 0 & 0 & 1 \end{pmatrix} \cdot \begin{pmatrix} 1 & 0 & 0 \\ 0 & \cos(\varphi_1) & -\sin(\varphi_1) \\ 0 & \sin(\varphi_1) & \cos(\varphi_1) \end{pmatrix}
\cdot \begin{pmatrix} \cos(\theta_1) & -\sin(\theta_1) & 0 \\ \sin(\theta_1) & \cos(\theta_1) & 0 \\ 0 & 0 & 1 \end{pmatrix}
$$
$$
= \begin{pmatrix} \cos(\psi_2) & -\sin(\psi_2) & 0 \\ \sin(\psi_2) & \cos(\psi_2) & 0 \\ 0 & 0 & 1 \end{pmatrix} \cdot \begin{pmatrix} 1 & 0 & 0 \\ 0 & \cos(\varphi_2) & -\sin(\varphi_2) \\ 0 & \sin(\varphi_2) & \cos(\varphi_2) \end{pmatrix}
\cdot \begin{pmatrix} \cos(\theta_2) & -\sin(\theta_2) & 0 \\ \sin(\theta_2) & \cos(\theta_2) & 0 \\ 0 & 0 & 1 \end{pmatrix},
$$

we note that the previous equation is equivalent to

$$
\begin{pmatrix} \cos(\psi_1-\psi_2) & -\sin(\psi_1-\psi_2) & 0 \\ \sin(\psi_1-\psi_2) & \cos(\psi_1-\psi_2) & 0 \\ 0 & 0 & 1 \end{pmatrix} \cdot \begin{pmatrix} 1 & 0 & 0 \\ 0 & \cos(\varphi_1) & -\sin(\varphi_1) \\ 0 & \sin(\varphi_1) & \cos(\varphi_1) \end{pmatrix}
$$
$$
= \begin{pmatrix} 1 & 0 & 0 \\ 0 & \cos(\varphi_2) & -\sin(\varphi_2) \\ 0 & \sin(\varphi_2) & \cos(\varphi_2) \end{pmatrix} \cdot \begin{pmatrix} \cos(\theta_2-\theta_1) & -\sin(\theta_2-\theta_1) & 0 \\ \sin(\theta_2-\theta_1) & \cos(\theta_2-\theta_1) & 0 \\ 0 & 0 & 1 \end{pmatrix}
$$

and hence equivalent to

$$
\begin{pmatrix} \cos(\psi_1-\psi_2) & -\sin(\psi_1-\psi_2)\cos(\varphi_1) & \sin(\psi_1-\psi_2)\sin(\varphi_1) \\ \sin(\psi_1-\psi_2) & \cos(\psi_1-\psi_2)\cos(\varphi_1) & -\cos(\psi_1-\psi_2)\sin(\varphi_1) \\ 0 & \sin(\varphi_1) & \cos(\varphi_1) \end{pmatrix}
$$
$$
= \begin{pmatrix} \cos(\theta_2-\theta_1) & -\sin(\theta_2-\theta_1) & 0 \\ \cos(\varphi_2)\sin(\theta_2-\theta_1) & \cos(\varphi_2)\cos(\theta_2-\theta_1) & -\sin(\varphi_2) \\ \sin(\varphi_2)\sin(\theta_2-\theta_1) & \sin(\varphi_2)\cos(\theta_2-\theta_1) & \cos(\varphi_2) \end{pmatrix}.
$$

Hence $\cos(\varphi_1) = \cos(\varphi_2)$ and $\varphi_1 = \varphi_2$. In addition,

$$
\sin(\psi_1-\psi_2)\sin(\varphi_1) = 0\,, \quad -\cos(\psi_1-\psi_2)\sin(\varphi_1) = -\sin(\varphi_1)
$$
$$
\sin(\varphi_1)\sin(\theta_2-\theta_1) = 0\,, \quad \sin(\varphi_1)\cos(\theta_2-\theta_1) = \sin(\varphi_1)\,.
$$

Hence if $\varphi_1 \in (-\pi, 0)$, then

$$
\sin(\psi_1-\psi_2) = 0\,, \quad \cos(\psi_1-\psi_2) = 1
$$
$$
\sin(\theta_2-\theta_1) = 0\,, \quad \cos(\theta_2-\theta_1) = 1\,.
$$

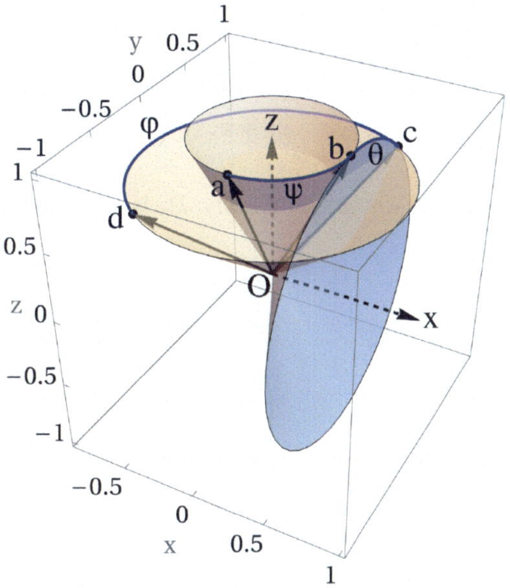

Fig. 2.2 Decomposition of the application of a rotation $R \in SO(3)$ to the vector $\mathbf{a} = {}^t(0, -1/2, 1)$, into a rotation of $\mathbf{a}$ around the z-axis about the angle $\theta = \pi/2$, subsequent rotation of the result, the vector $\mathbf{b} = {}^t(1/2, 0, 1)$, around the x-axis about the angle $\varphi = -\pi/4$, and final rotation of the result, the vector $\mathbf{c} = {}^t(1/2, 1/\sqrt{2}, 1/\sqrt{2})$, around the z-axis about the angle $\psi = \pi$ so that $\mathbf{d} = R\mathbf{a} = {}^t(-1/2, -1/\sqrt{2}, 1/\sqrt{2})$. The angles ψ, φ and θ are Euler angles corresponding to R

Since $\psi_1, \theta_1, \psi_2, \theta_2 \in (-\pi, \pi]$, we have that $\psi_1 - \psi_2, \theta_1 - \theta_2 \in (-2\pi, 2\pi)$ and hence that $\psi_1 = \psi_2$ as well as that $\theta_1 = \theta_2$. □

The following result is often used, e.g., in the calculation of Fourier transforms of spherically symmetric functions.

Proposition 2.9 *For every $x \in \mathbb{R}^3$, there is $R \in SO(3)$ such that*

$$Rx = \begin{pmatrix} 0 \\ 0 \\ |x| \end{pmatrix} ,$$

where $|x|$ denotes the Euclidean norm of x.

Proof For this purpose, let $x \in \mathbb{R}^3$. We consider 3 cases.
1. If $x = {}^t(0, 0, x_3)$ is such that $x_3 \geqslant 0$, then the unit matrix satisfies the requirements for R.
2. If $x = {}^t(0, 0, x_3)$ is such that $x_3 < 0$, then

$$R := \begin{pmatrix} 1 & 0 & 0 \\ 0 & -1 & 0 \\ 0 & 0 & -1 \end{pmatrix}$$

satisfies the requirements for R.

3. If $x = {}^t(x_1, x_2, x_3) \in \mathbb{R}^3$ is such that $x_1^2 + x_2^2 \neq 0$, then we define $D_x : \mathbb{R}^3 \to \mathbb{R}^3$ by

$$D_x(y) := \frac{x_2 y_1 - x_1 y_2}{x_1^2 + x_2^2} \begin{pmatrix} x_2 \\ -x_1 \\ 0 \end{pmatrix} + \frac{x_1 y_1 + x_2 y_2}{(x_1^2 + x_2^2)\,|x|} \left[x \times \begin{pmatrix} x_2 \\ -x_1 \\ 0 \end{pmatrix} \right] + \frac{y_3}{|x|}\, x \ ,$$

for every $y \in \mathbb{R}^3$. Obviously, D_x is linear and satisfies

$$D_x({}^t(0,0,1)) = \frac{1}{|x|}\, x \ , \ \ D_x({}^t(x_2, -x_1, 0)) = {}^t(x_2, -x_1, 0) \ .$$

By using that

$$\begin{pmatrix} x_2 \\ -x_1 \\ 0 \end{pmatrix} \ , \ x \times \begin{pmatrix} x_2 \\ -x_1 \\ 0 \end{pmatrix} = \begin{pmatrix} x_1 x_3 \\ x_2 x_3 \\ -(x_1^2 + x_2^2) \end{pmatrix} \ , \ x$$

are pairwise orthogonal with norms

$$\sqrt{x_1^2 + x_2^2} \ , \ |x| \sqrt{x_1^2 + x_2^2} \ , \ |x| \ ,$$

with respect to the Euclidean scalar product $\langle \, | \, \rangle$ on $\mathbb{R}^3$, it follows that

$$\langle D_x(y) | D_x(y) \rangle = \frac{(x_2 y_1 - x_1 y_2)^2 + (x_1 y_1 + x_2 y_2)^2}{x_1^2 + x_2^2} + y_3^2 = \langle y | y \rangle \ ,$$

for all $y \in \mathbb{R}^3$, and hence that

$$\begin{aligned} \langle D_x(y) | D_x(z) \rangle &= \frac{1}{2} \left(|D_x(y) + D_x(z)|^2 - |D_x(y)|^2 - |D_x(z)|^2 \right) \\ &= \frac{1}{2} \left(|D_x(y + z)|^2 - |D_x(y)|^2 - |D_x(z)|^2 \right) \\ &= \frac{1}{2} \left(|y + z|^2 - |y|^2 - |z|^2 \right) = \langle y | z \rangle \ , \end{aligned}$$

for all $y, z \in \mathbb{R}^3$. Therefore D_x is orthogonal. Further, $\det(D_x')$ is given by the scalar product of

$$\frac{1}{x_1^2 + x_2^2} \left[x_2 \begin{pmatrix} x_2 \\ -x_1 \\ 0 \end{pmatrix} + \frac{x_1}{|x|} \left(x \times \begin{pmatrix} x_2 \\ -x_1 \\ 0 \end{pmatrix} \right) \right]$$

and

$$
\begin{aligned}
&\frac{1}{x_1^2+x_2^2}\left[-x_1\begin{pmatrix}x_2\\-x_1\\0\end{pmatrix}+\frac{x_2}{|x|}\left(x\times\begin{pmatrix}x_2\\-x_1\\0\end{pmatrix}\right)\right]\times\frac{1}{|x|}\,x\\
&=\frac{1}{(x_1^2+x_2^2)|x|}\left\{x_1\left[x\times\begin{pmatrix}x_2\\-x_1\\0\end{pmatrix}\right]-\frac{x_2}{|x|}\left[x\times\left(x\times\begin{pmatrix}x_2\\-x_1\\0\end{pmatrix}\right)\right]\right\}\\
&=\frac{1}{(x_1^2+x_2^2)|x|}\left\{x_1\left[x\times\begin{pmatrix}x_2\\-x_1\\0\end{pmatrix}\right]+\frac{x_2}{|x|}\,|x|^2\begin{pmatrix}x_2\\-x_1\\0\end{pmatrix}\right\}\\
&=\frac{1}{x_1^2+x_2^2}\left\{x_2\begin{pmatrix}x_2\\-x_1\\0\end{pmatrix}+\frac{x_1}{|x|}\left[x\times\begin{pmatrix}x_2\\-x_1\\0\end{pmatrix}\right]\right\}.
\end{aligned}
$$

Hence

$$
\det(D_x'(y))=\frac{1}{(x_1^2+x_2^2)^2}\left[x_2^2(x_1^2+x_2^2)+x_1^2(x_1^2+x_2^2)\right]=1\ ,
$$

for every $y\in\mathbb{R}^3$. As a consequence, the representation matrix D of D_x with respect to the canonical basis of $\mathbb{R}^3$ is in SO(3) and such that

$$
D\begin{pmatrix}0\\0\\1\end{pmatrix}=\frac{1}{|x|}\,x\ .
$$

Hence $R := D^{-1}$ satisfies the requirements. □

2.3 Construction of a Double Cover of SO(3)

In the following, we are going to construct a double cover $\Phi_1 : \mathrm{SU}(2)\to\mathrm{SO}(3)$ of SO(3) that later on is used in the definition of the spinor representation of SU(2). In a first step, we derive a parametrization of SU(2). If

$$
\begin{pmatrix}U_{11} & U_{12}\\ U_{21} & U_{22}\end{pmatrix}\in\mathrm{SU}(2)\ ,
$$

then

$$
\begin{aligned}
&U_{11}^*U_{12}+U_{21}^*U_{22}=0\ ,\ U_{11}U_{22}-U_{12}U_{21}=1\ ,\\
&U_{11}^*U_{11}+U_{21}^*U_{21}=U_{12}^*U_{12}+U_{22}^*U_{22}=1\ ,
\end{aligned}
\tag{2.10}
$$

and hence

$$U_{11}(U_{22} - U_{11}^*) - U_{21}(U_{12} + U_{21}^*) = 0 \ ,$$
$$(U_{11} - U_{22}^*)U_{22} - U_{12}(U_{21} + U_{12}^*) = 0 \ ,$$

or equivalently

$$U_{11}(U_{22} - U_{11}^*) - U_{21}(U_{12} + U_{21}^*) = 0 \ ,$$
$$-U_{22}^*(U_{22} - U_{11}^*) - U_{12}^*(U_{12} + U_{21}^*) = 0 \ ,$$

implying that

$$(U_{11} - U_{22}^*)(U_{22} - U_{11}^*) - (U_{21} + U_{12}^*)(U_{12} + U_{21}^*) = 0$$

or equivalently that

$$|U_{22} - U_{11}^*|^2 + |U_{21} + U_{12}^*|^2 = 1 \ .$$

Hence, it follows that

$$U_{22} = U_{11}^* \ , \ U_{21} = -U_{12}^*$$

Substituting the latter into the first 2 equations gives

$$U_{11}^* U_{12} - U_{12} U_{11}^* = 0 \ , \ U_{11} U_{11}^* + U_{12} U_{12}^* = 1 \ .$$

Hence, we have that

$$\begin{pmatrix} U_{11} & U_{12} \\ U_{21} & U_{22} \end{pmatrix} = \begin{pmatrix} U_{11} & U_{12} \\ -U_{12}^* & U_{11}^* \end{pmatrix} \ .$$

as well as that

$$U_{11}^* U_{12} + U_{12} U_{11}^* = 0 \ , \ |U_{11}|^2 + |U_{12}|^2 = 1 \ .$$

On the other hand, if U_{11}, U_{12} are complex numbers such that $|U_{11}|^2 + |U_{12}|^2 = 1$, then the determinant of the matrix

$$\begin{pmatrix} U_{11} & U_{12} \\ -U_{12}^* & U_{11}^* \end{pmatrix}$$

is equal to 1, the column vectors have norm 1 and are orthogonal, since

$$U_{11}^* \begin{pmatrix} U_{11} \\ -U_{12}^* \end{pmatrix} + U_{12}^* \begin{pmatrix} U_{12} \\ U_{11}^* \end{pmatrix} = \begin{pmatrix} 0 \\ 0 \end{pmatrix} \ .$$

Hence, it follows that

$$\mathrm{SU}(2) = \left\{ \begin{pmatrix} U_{11} & U_{12} \\ -U_{12}^* & U_{11}^* \end{pmatrix} : U_{11}, U_{12} \in \mathbb{C} \wedge |U_{11}|^2 + |U_{12}|^2 = 1 \right\} \ .$$

We note that if $U_{11}, U_{12} \in \mathbb{C}$ are such that $|U_{11}|^2 + |U_{12}|^2 = 1$, then

$$\begin{pmatrix} U_{11} & U_{12} \\ -U_{12}^* & U_{11}^* \end{pmatrix}^{-1} = \begin{pmatrix} U_{11}^* & -U_{12} \\ U_{12}^* & U_{11} \end{pmatrix} = \begin{pmatrix} U_{11} & U_{12} \\ -U_{12}^* & U_{11}^* \end{pmatrix}^* .$$

In the next step, we define an auxiliary real subspace V_3 of $\mathrm{M}(2, \mathbb{C})$ with inner product $\langle\ ,\ \rangle$ such that $(V_3, \langle\ ,\ \rangle)$ is isomorphic to Euclidean 3-space, via the map σ.

Definition 2.10 We define the auxiliary real subspace V_3 of $\mathrm{M}(2, \mathbb{C})$ of all traceless anti-Hermitian matrices,

$$V_3 := \{A \in iH(2, \mathbb{C}) : \mathrm{Tr}(A) = 0\} ,$$

where $H(2, \mathbb{C})$ denotes the real subspace of $\mathrm{M}(2, \mathbb{C})$ consisting of all Hermitian matrices, and $\sigma : \mathbb{R}^3 \to V_3$ by

$$\sigma(x) := i \begin{pmatrix} x^3 & x^1 - ix^2 \\ x^1 + ix^2 & -x^3 \end{pmatrix} = i \sum_{k=1}^{3} x^k \sigma_k , \tag{2.11}$$

for every $x \in \mathbb{R}^3$, where

$$\sigma_1 := \begin{pmatrix} 0 & 1 \\ 1 & 0 \end{pmatrix} , \quad \sigma_2 := \begin{pmatrix} 0 & -i \\ i & 0 \end{pmatrix} , \quad \sigma_3 := \begin{pmatrix} 1 & 0 \\ 0 & -1 \end{pmatrix}$$

are the Pauli spin matrices. In addition, we define $\langle\ ,\ \rangle : V_3 \times V_3 \to \mathbb{R}$ by

$$\begin{aligned} \langle A, B\rangle &:= \frac{1}{2} \left[\det(A + B) - \det(A) - \det(B) \right] \\ &= \frac{1}{2} \left(a_{12} b_{12}^* + a_{12}^* b_{12} - 2 a_{11} b_{11} \right) = -\frac{1}{2} \mathrm{Tr}(AB) , \end{aligned}$$

for all

$$A = \begin{pmatrix} a_{11} & a_{12} \\ -a_{12}^* & -a_{11} \end{pmatrix} , \quad B = \begin{pmatrix} b_{11} & b_{12} \\ -b_{12}^* & -b_{11} \end{pmatrix} \in V_3 .$$

Exercise 2.3 Show the following multiplication table, where the first factor of the products is from the first column and the second factor is from the first row (Table 2.1).

Lemma 2.11 *(i)* $\langle\ ,\ \rangle$ *defines a scalar product on* V_3.
(ii) $\sigma : \mathbb{R}^3 \to V_3$ *is an isomorphism of real vector spaces. In addition,*

$$\langle \sigma(x), \sigma(y) \rangle = \langle x | y \rangle ,$$

for all $x, y \in \mathbb{R}^3$, *where* $\langle\,|\,\rangle$ *denotes the Euclidean scalar product on* $\mathbb{R}^3$.
(iii) We have that

Table 2.1 Multiplication table of the Pauli spin matrices

$\cdot$	σ_1	σ_2	σ_3
σ_1	E	$i\sigma_3$	$-i\sigma_2$
σ_2	$-i\sigma_3$	E	$i\sigma_1$
σ_3	$i\sigma_2$	$-i\sigma_1$	E

$$\sigma(x \times y) = -\frac{1}{2}\left[\sigma(x), \sigma(y)\right], \quad \frac{1}{4}\text{Tr}\left(\sigma(x)[\sigma(y), \sigma(z)]\right) = x \cdot (y \times z) ,$$

for all $x, y, z \in \mathbb{R}^3$.

Proof '(ii)': For $x, y \in \mathbb{R}^3$ and $\lambda \in \mathbb{R}$, it follows that

$$\sigma(x+y) = i\begin{pmatrix} (x+y)^3 & (x+y)^1 - i(x+y)^2 \\ (x+y)^1 + i(x+y)^2 & -(x+y)^3 \end{pmatrix}$$
$$= i\begin{pmatrix} x^3 + y^3 & x^1 + y^1 - ix^2 - iy^2 \\ x^1 + y^1 + ix^2 + iy^2 & -x^3 - y^3 \end{pmatrix} = \sigma(x) + \sigma(y) ,$$

and

$$\sigma(\lambda x) = i\begin{pmatrix} (\lambda x)^3 & (\lambda x)^1 - i(\lambda x)^2 \\ (\lambda x)^1 + i(\lambda x)^2 & -(\lambda x)^3 \end{pmatrix} = \lambda\,\sigma(x) .$$

In particular, $\sigma(x) = 0$ implies that

$$x^3 = -x^3 = x^1 - ix^2 = x^1 + ix^2 = 0$$

and hence that $x = 0$. As a consequence, σ is a monomorphism. On the other hand, if

$$\begin{pmatrix} a_{11} & a_{12} \\ -a_{12}^* & -a_{11} \end{pmatrix}$$

is an element of $iH(2, \mathbb{C})$, then a_{11} is purely imaginary and

$$\sigma\left(\text{Im}(a_{12}), \text{Re}(a_{12}), -ia_{11}\right) = i\begin{pmatrix} -ia_{11} & \text{Im}(a_{12}) - i\text{Re}(a_{12}) \\ \text{Im}(a_{12}) + i\text{Re}(a_{12}) & ia_{11} \end{pmatrix}$$
$$= \begin{pmatrix} a_{11} & a_{12} \\ -a_{12}^* & -a_{11} \end{pmatrix} = A .$$

As a consequence, σ is also surjective. Further, it follows for $x, y \in \mathbb{R}^3$ that

$$\begin{aligned}\langle \sigma(x), \sigma(y) \rangle &= \frac{1}{2} \left[(x^2 + ix^1)(y^2 + iy^1)^* + (x^2 + ix^1)^*(y^2 + iy^1) - 2ix^3iy^3 \right] \\ &= \frac{1}{2} \left[(x^2 + ix^1)(y^2 - iy^1) + (x^2 - ix^1)(y^2 + iy^1) + 2x^3y^3 \right] \\ &= \frac{1}{2} \left[2(x^2y^2 + x^1y^1) + 2x^3y^3 \right] = \langle x | y \rangle \ .\end{aligned}$$

Since $\sigma : \mathbb{R}^3 \to V_3$ is an isomorphism, also the validity of (i) follows. '(iii)': Further, for $x, y, z \in \mathbb{R}^3$, it follows that

$$\begin{aligned}[\sigma(x), \sigma(y)] &= \left[i \sum_{k=1}^{3} x^k \sigma_k, i \sum_{l=1}^{3} y^l \sigma_l \right] = - \sum_{k=1}^{3} \sum_{l=1}^{3} x^k y^l \, [\sigma_k, \sigma_l] \\ &= -2i \sum_{j=1}^{3} \left[\sum_{k=1}^{3} \sum_{l=1}^{3} \varepsilon_{klj} \, x^k y^l \right] \sigma_j = -2i \sum_{j=1}^{3} \left[\sum_{k=1}^{3} \sum_{l=1}^{3} \varepsilon_{jkl} \, x^k y^l \right] \sigma_j \\ &= -2i \sum_{j=1}^{3} (x \times y)_j \sigma_j = -2\sigma(x \times y) \ ,\end{aligned}$$

where ε denotes the Levi-Civita symbol in 3 dimensions, i.e., $\varepsilon_{jkl} = 1$ if (j, k, l) is a cyclic permutation of $(1, 2, 3)$, $\varepsilon_{jkl} = -1$ if (j, k, l) is an anti-cyclic permutation of $(1, 2, 3)$ and $\varepsilon_{jkl} = 0$, otherwise, where $(j, k, l) \in \{1, 2, 3\}^3$, as well as that

$$\begin{aligned}x \cdot (y \times z) &= \langle \sigma(x), \sigma(y \times z) \rangle = -\frac{1}{2} \mathrm{Tr}(\sigma(x)\sigma(y \times z)) \\ &= \frac{1}{4} \mathrm{Tr}(\sigma(x)[\sigma(y), \sigma(z)]) \ .\end{aligned}$$

□

Remark 2.12 We note that, since σ is an isomorphism that preserves scalar products, and since $\sigma(e_j) = i\sigma_j$, for every $j \in \{1, 2, 3\}$, it follows that $i\sigma_1, i\sigma_2, i\sigma_3$ is an orthonormal basis of $(V_3, \langle \ , \ \rangle)$.

We define for every $G \in \mathrm{SU}(2)$ a corresponding linear transformation $R_G : V_3 \to V_3$ by

$$R_G(A) := GA\,G^* = GA\,G^{-1} \ , \tag{2.12}$$

for every $A \in iH(2, \mathbb{C})$ with trace 0. We note that R_G is well-defined, since if $A \in iH(2, \mathbb{C})$ with trace 0, then

$$[-iGAG^*]^* = [G(-iA)G^*]^* = G(-iA)G^* = -iGAG^* \ ,$$

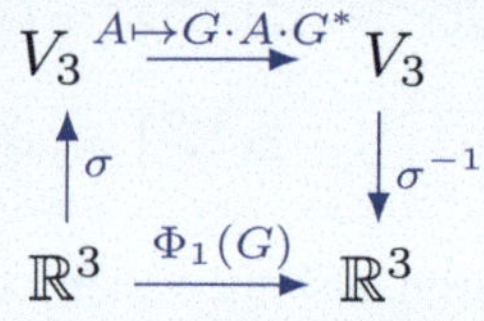

Fig. 2.3 The map $\Phi_1 : \mathrm{SU}(2) \to \mathrm{SO}(3)$, defined by $[\Phi_1(G)](x) := \sigma^{-1}(G \cdot \sigma(x) \cdot G^*)$, for every $G \in \mathrm{SU}(2)$ and $x \in \mathbb{R}^3$, is a double covering of SO(3)

and hence $GAG^* \in iH(2, \mathbb{C})$ and

$$\mathrm{Tr}(GA\,G^*) = \mathrm{Tr}(GA\,G^{-1}) = \mathrm{Tr}(A) = 0 \ .$$

The double cover of SO(3) that later on is used in the definition of the spinor representation of SU(2) is given in the following theorem (Fig. 2.3 and Fig. 2.4).

Theorem 2.13 *By*

$$\Phi_1(G) := \sigma^{-1} \circ R_G \circ \sigma \tag{2.13}$$

for every $G \in \mathrm{SU}(2)$, *there is given an epimorphism* $\Phi_1 : \mathrm{SU}(2) \to \mathrm{SO}(3)$, *where we identify the elements of* $\mathrm{L}(\mathbb{R}^3)$ *with their corresponding representation matrices, with respect to the canonical basis* e_1, e_2, e_3 *of* $\mathbb{R}^3$. *In particular,*

$$\Phi_1(G_1) = \Phi_1(G_2)$$

for $G_1, G_2 \in \mathrm{SU}(2)$ *if and only if* $G_2 = \pm G_1$.

Proof If $G_1, G_2 \in \mathrm{SU}(2)$ and $A \in iH(2, \mathbb{C})$ with trace 0, then

$$\begin{aligned} R_{G_1G_2}(A) &= G_1G_2A\,(G_1G_2)^{-1} = G_1G_2A\,G_2^{-1}G_1^{-1} = G_1R_{G_2}(A)G_1^{-1} \\ &= R_{G_1}(R_{G_2}(A)) = (R_{G_1} \circ R_{G_2})(A) \ , \end{aligned}$$

and it follows that $R_{G_1G_2} = R_{G_1} \circ R_{G_2}$. Hence, we have that

$$\begin{aligned} \Phi_1(G_1G_2) &= \sigma^{-1} \circ R_{G_1G_2} \circ \sigma = \sigma^{-1} \circ R_{G_1} \circ R_{G_2} \circ \sigma \\ &= \sigma^{-1} \circ R_{G_1} \circ \sigma \circ \sigma^{-1} \circ R_{G_2} \circ \sigma = \Phi_1(G_1) \circ \Phi_1(G_2) \ . \end{aligned}$$

Further, for $G \in \mathrm{SU}(2)$, we have that

$$\langle \Phi_1(G)(x)|\Phi_1(G)(y)\rangle = \langle \sigma(\Phi_1(G)(x)), \sigma(\Phi_1(G)(y))\rangle$$
$$= \left\langle i\sum_{k=1}^{3} x^k G\sigma_k G^*, i\sum_{l=1}^{3} y^l G\sigma_l G^*\right\rangle = \sum_{k=1}^{3}\sum_{l=1}^{3} x^k y^k \left\langle iG\sigma_k G^*, iG\sigma_l G^*\right\rangle$$
$$= -\frac{1}{2}\sum_{k=1}^{3}\sum_{l=1}^{3} x^k y^k \mathrm{Tr}(iG\sigma_k G^* iG\sigma_l G^*) = \frac{1}{2}\sum_{k=1}^{3}\sum_{l=1}^{3} x^k y^k \mathrm{Tr}(G\sigma_k\sigma_l G^*)$$
$$= \frac{1}{2}\sum_{k=1}^{3}\sum_{l=1}^{3} x^k y^k \mathrm{Tr}(G\sigma_k\sigma_l G^*) = \frac{1}{2}\sum_{k=1}^{3}\sum_{l=1}^{3} x^k y^k \mathrm{Tr}(\sigma_k\sigma_l)$$
$$= \sum_{k=1}^{3}\sum_{l=1}^{3} \delta_{kl} x^k y^k = \langle x|y\rangle \ ,$$

for all $x, y \in \mathbb{R}^3$. Further, we conclude that

$$\Phi_1(G)(e_1)\cdot(\Phi_1(G)(e_2)\times\Phi_1(G)(e_3))$$
$$= \frac{1}{4}\mathrm{Tr}(\sigma(\Phi_1(G)(e_1))[\sigma(\Phi_1(G)(e_2)), \sigma(\Phi_1(G)(e_3))])$$
$$= \frac{1}{4}\mathrm{Tr}(iG\sigma_1 G^*[iG\sigma_2 G^*, iG\sigma_3 G^*]) = -\frac{i}{4}\mathrm{Tr}(G\sigma_1 G^*[G\sigma_2 G^*, G\sigma_3 G^*])$$
$$-\frac{i}{4}\mathrm{Tr}(G\sigma_1 G^*[G\sigma_2 G^* G\sigma_3 G^* - G\sigma_3 G^* G\sigma_2 G^*])$$
$$= -\frac{i}{4}\mathrm{Tr}(G\sigma_1 G^*[G\sigma_2\sigma_3 G^* - G\sigma_3\sigma_2 G^*]) = -\frac{i}{4}\mathrm{Tr}(G\sigma_1[\sigma_2, \sigma_3]G^*)$$
$$= \frac{1}{2}\mathrm{Tr}(G\sigma_1^2 G^*) = 1$$

and hence that

$$\det(\Phi_1(G)) = 1 \ .$$

Therefore, it follows that $\Phi_1(G) \in \mathrm{SO}(3)$. Also, we infer that

$$\sigma(\Phi_1(G)(x)) = R_G(\sigma(x)) = G\sigma(x)G^* = \sum_{k=1}^{3}\left\langle i\sigma_k, G\sigma(x)G^*\right\rangle i\sigma_k$$
$$= \sum_{k=1}^{3}\left\langle i\sigma_k, Gi\sum_{l=1}^{3} x^l\sigma_l G^*\right\rangle i\sigma_k = i\sum_{k=1}^{3}\left[\sum_{l=1}^{3} x^l \left\langle i\sigma_k, iG\sigma_l G^*\right\rangle\right]\sigma_k$$
$$= -\frac{i}{2}\sum_{k=1}^{3}\left[\sum_{l=1}^{3} x^l\,\mathrm{Tr}(i\sigma_k iG\sigma_l G^*)\right]\sigma_k = \frac{i}{2}\sum_{k=1}^{3}\left[\sum_{l=1}^{3} x^l\,\mathrm{Tr}(\sigma_k G\sigma_l G^*)\right]\sigma_k$$

and hence that

$$\Phi_1(G)(x) = \frac{1}{2}\sum_{k=1}^{3}\left[\sum_{l=1}^{3}\mathrm{Tr}(\sigma_k G \sigma_l G^*)\, x^l\right] e_k\ ,$$

for every $x = {}^t(x^1, x^2, x^3) \in \mathbb{R}^3$. As a consequence,

$$\begin{aligned}\Phi_1(G) &= \left(\frac{1}{2}\mathrm{Tr}(\sigma_k G \sigma_l G^*)\right)_{kl\in\{1,2,3\}} \\ &= \begin{pmatrix} \mathrm{Re}(g_{11}^2 - g_{12}^2) & \mathrm{Im}(g_{11}^2 + g_{12}^2) & -2\,\mathrm{Re}(g_{11}g_{12}) \\ -\mathrm{Im}(g_{11}^2 - g_{12}^2) & \mathrm{Re}(g_{11}^2 + g_{12}^2) & 2\,\mathrm{Im}(g_{11}g_{12}) \\ 2\,\mathrm{Re}(g_{11}g_{12}^*) & 2\,\mathrm{Im}(g_{11}g_{12}^*) & |g_{11}|^2 - |g_{12}|^2 \end{pmatrix}\ .\end{aligned}$$

In particular, if

$$G = \begin{pmatrix} e^{-i\psi/2} & 0 \\ 0 & e^{i\psi/2} \end{pmatrix}\ ,$$

for $\psi \in \mathbb{R}$, then

$$\Phi_1(G) = \begin{pmatrix} \cos(\psi) & -\sin(\psi) & 0 \\ \sin(\psi) & \cos(\psi) & 0 \\ 0 & 0 & 1 \end{pmatrix}\ ,$$

and if

$$G = \begin{pmatrix} \cos(\theta/2) & -i\sin(\theta/2) \\ -i\sin(\theta/2) & \cos(\theta/2) \end{pmatrix}\ ,$$

for $\theta \in \mathbb{R}$, then

$$\begin{aligned}\Phi_1(G) &= \begin{pmatrix} 1 & 0 & 0 \\ 0 & \cos^2(\theta/2) - \sin^2(\theta/2) & -2\sin(\theta/2)\cos(\theta/2) \\ 0 & 2\sin(\theta/2)\cos(\theta/2) & \cos^2(\theta/2) - \sin^2(\theta/2) \end{pmatrix} \\ &= \begin{pmatrix} 1 & 0 & 0 \\ 0 & \cos(\theta) & -\sin(\theta) \\ 0 & \sin(\theta) & \cos(\theta) \end{pmatrix}\ .\end{aligned}$$

Hence, it follows from Lemma 2.8 that Φ_1 is surjective. In addition, we note that if

$$G = \begin{pmatrix} \cos(\theta/2) & \sin(\theta/2) \\ -\sin(\theta/2) & \cos(\theta/2) \end{pmatrix}\ ,$$

for $\theta \in \mathbb{R}$, then

$$\Phi_1(G) = \begin{pmatrix} \cos^2(\theta/2) - \sin^2(\theta/2) & 0 & -2\sin(\theta/2)\cos(\theta/2) \\ 0 & 1 & 0 \\ 2\sin(\theta/2)\cos(\theta/2) & 0 & \cos^2(\theta/2) - \sin^2(\theta/2) \end{pmatrix}$$
$$= \begin{pmatrix} \cos(\theta) & 0 & -\sin(\theta) \\ 0 & 1 & 0 \\ \sin(\theta) & 0 & \cos(\theta) \end{pmatrix} .$$

Further, if

$$G = \begin{pmatrix} g_{11} & g_{12} \\ -g_{12}^* & g_{11}^* \end{pmatrix} \in \mathrm{SU}(2)$$

is such that $\Phi_1(G) = E$, then

$$\begin{pmatrix} \mathrm{Re}(g_{11}^2 - g_{12}^2) & \mathrm{Im}(g_{11}^2 + g_{12}^2) & -2\,\mathrm{Re}(g_{11}g_{12}) \\ -\,\mathrm{Im}(g_{11}^2 - g_{12}^2) & \mathrm{Re}(g_{11}^2 + g_{12}^2) & 2\,\mathrm{Im}(g_{11}g_{12}) \\ 2\,\mathrm{Re}(g_{11}g_{12}^*) & 2\,\mathrm{Im}(g_{11}g_{12}^*) & |g_{11}|^2 - |g_{12}|^2 \end{pmatrix} = \begin{pmatrix} 1 & 0 & 0 \\ 0 & 1 & 0 \\ 0 & 0 & 1 \end{pmatrix} .$$

Then $g_{11}g_{12} = 0$. If $g_{11} = 0$, we arrive at the contradiction that $-|g_{12}|^2 = 1$. ↯ If $g_{12} = 0$, then

$$\begin{pmatrix} \mathrm{Re}(g_{11}^2) & \mathrm{Im}(g_{11}^2) & 0 \\ -\,\mathrm{Im}(g_{11}^2) & \mathrm{Re}(g_{11}^2) & 0 \\ 0 & 0 & |g_{11}|^2 \end{pmatrix} = \begin{pmatrix} 1 & 0 & 0 \\ 0 & 1 & 0 \\ 0 & 0 & 1 \end{pmatrix} .$$

and hence $g_{11}^2 = 1$, implying that $g_{11} \in \{-1, 1\}$. As consequence, $G = \pm E$. Finally, obviously, $E, -E \in \mathrm{SU}(2)$ and $M_E = M_{-E} = E$. □

From the proof of Theorem 2.13, we obtain the following corollaries.

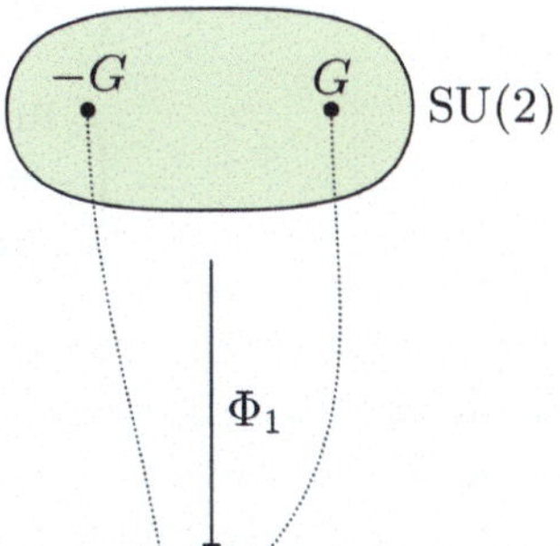

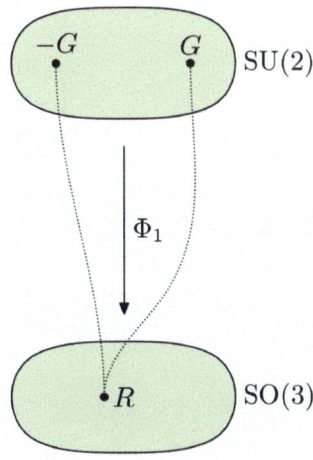

Fig. 2.4 Depiction of the action of Φ_1

Corollary 2.14 (Particular images of Φ_1) *For $\psi, \theta, \varphi \in \mathbb{R}$, we have that*

$$\Phi_1\left(\begin{pmatrix} e^{i\psi/2} & 0 \\ 0 & e^{-i\psi/2} \end{pmatrix}\right) = \begin{pmatrix} \cos(\psi) & \sin(\psi) & 0 \\ -\sin(\psi) & \cos(\psi) & 0 \\ 0 & 0 & 1 \end{pmatrix},$$

$$\Phi_1\left(\begin{pmatrix} \cos(\theta/2) & i\sin(\theta/2) \\ i\sin(\theta/2) & \cos(\theta/2) \end{pmatrix}\right) = \begin{pmatrix} 1 & 0 & 0 \\ 0 & \cos(\theta) & \sin(\theta) \\ 0 & -\sin(\theta) & \cos(\theta) \end{pmatrix},$$

$$\Phi_1\left(\begin{pmatrix} \cos(\varphi/2) & \sin(\varphi/2) \\ -\sin(\varphi/2) & \cos(\varphi/2) \end{pmatrix}\right) = \begin{pmatrix} \cos(\varphi) & 0 & -\sin(\varphi) \\ 0 & 1 & 0 \\ \sin(\varphi) & 0 & \cos(\varphi) \end{pmatrix}.$$

Further, for $G \in \mathrm{SU}(2)$, we have that

$$\Phi_1(G) = \begin{pmatrix} \mathrm{Re}(g_{11}^2 - g_{12}^2) & \mathrm{Im}(g_{11}^2 + g_{12}^2) & -2\,\mathrm{Re}(g_{11}g_{12}) \\ -\mathrm{Im}(g_{11}^2 - g_{12}^2) & \mathrm{Re}(g_{11}^2 + g_{12}^2) & 2\,\mathrm{Im}(g_{11}g_{12}) \\ 2\,\mathrm{Re}(g_{11}g_{12}^*) & 2\,\mathrm{Im}(g_{11}g_{12}^*) & |g_{11}|^2 - |g_{12}|^2 \end{pmatrix}.$$

Corollary 2.15 (Continuity of Φ_1) *Let $G \in \mathrm{SU}(2)$. If $G_1, G_2, \ldots$ is a sequence in* SU(2) *that converges component-wise to $G \in \mathrm{SU}(2)$, then the sequence in* SO(3),

$$\Phi_1(G_1), \Phi_1(G_2), \ldots ,$$

converges component-wise to $\Phi_1(G)$.

Proof According to the proof of Theorem 2.13, for $G \in \mathrm{SU}(2)$, the corresponding $\Phi_1(G)$ is given by

$$\begin{aligned} \Phi_1(G) &= \left(\frac{1}{2}\mathrm{Tr}(\sigma_k G \sigma_l G^*)\right)_{kl\in\{1,2,3\}} \\ &= \begin{pmatrix} \mathrm{Re}(g_{11}^2 - g_{12}^2) & \mathrm{Im}(g_{11}^2 + g_{12}^2) & -2\,\mathrm{Re}(g_{11}g_{12}) \\ -\mathrm{Im}(g_{11}^2 - g_{12}^2) & \mathrm{Re}(g_{11}^2 + g_{12}^2) & 2\,\mathrm{Im}(g_{11}g_{12}) \\ 2\,\mathrm{Re}(g_{11}g_{12}^*) & 2\,\mathrm{Im}(g_{11}g_{12}^*) & |g_{11}|^2 - |g_{12}|^2 \end{pmatrix}. \end{aligned}$$

Hence, if $G_1, G_2, \ldots$ is a sequence in SU(2) that converges component-wise to $G \in \mathrm{SU}(2)$, then the corresponding sequence $\Phi_1(G_1), \Phi_2(G_2), \ldots$ in SO(3) converges component-wise to $\Phi_1(G)$. □

Exercise 2.4 For $G \in \mathrm{SU}(2)$, show that

$$\sum_{k=1}^{3} [\Phi_1(G)]_{kl}\, \sigma_k = G \cdot \sigma_l \cdot G^* \ , \tag{2.14}$$

for all $l \in \{1, 2, 3\}$.

Exercise 2.5 Show for every $G \in \mathrm{SU}(2)$ that

$$\Phi_1(G^*) = [\Phi_1(G)]^t \ , \tag{2.15}$$

and as consequence that, in addition to (2.14), for every $G \in \mathrm{SU}(2, \mathbb{C})$, we have that

$$\sum_{k=1}^{3} [\Phi_1(G)]_{lk}\, \sigma_k = G^* \cdot \sigma_l \cdot G \ , \tag{2.16}$$

for all $l \in \{1, 2, 3\}$.

The following proposition gives one-parameter subgroups of SU(2) that are mapped by Φ_1 into one-parameter subgroups of rotations.

Proposition 2.16 *Let $n \in S^2$, $\varphi \in \mathbb{R}$ and*

$$\begin{aligned} G(\varphi) &:= \begin{pmatrix} \cos(\frac{\varphi}{2}) - in_3 \sin(\frac{\varphi}{2}) & -(n_2 + in_1) \sin(\frac{\varphi}{2}) \\ (n_2 - in_1) \sin(\frac{\varphi}{2}) & \cos(\frac{\varphi}{2}) + in_3 \sin(\frac{\varphi}{2}) \end{pmatrix} \\ &= \cos\left(\frac{\varphi}{2}\right) E - i \sin\left(\frac{\varphi}{2}\right) \sum_{k=1}^{3} n_k \sigma_k \in \mathrm{SU}(2) \ . \end{aligned}$$

Then

$$G(\varphi) = \exp\left(-\frac{i\varphi}{2} \sum_{k=1}^{3} n_k \sigma_k \right) \ , \tag{2.17}$$

and $\Phi_1(G(\varphi))$ is given by

$$\Phi_1(G(\varphi))(x) = \exp(\varphi A) \cdot x \ ,$$

for every $x \in \mathbb{R}^3$, where

$$A = \begin{pmatrix} 0 & -n_3 & n_2 \\ n_3 & 0 & -n_1 \\ -n_2 & n_1 & 0 \end{pmatrix} \ .$$

Proof If

$$g_{11}(\varphi) = \cos(\varphi/2) - in_3\sin(\varphi/2)\ ,\ \ g_{12}(\varphi) = -(n_2 + in_1)\sin(\varphi/2)\ ,$$

it follows that

$$\begin{aligned}
&g_{11}^2(\varphi) - g_{12}^2(\varphi)\\
&= \cos^2(\varphi/2) - n_3^2\sin^2(\varphi/2) - 2in_3\sin(\varphi/2)\cos(\varphi/2)\\
&\quad - (n_2^2 - n_1^2 + 2in_1n_2)\sin^2(\varphi/2)\\
&= \cos^2(\varphi/2) - in_3\sin(\varphi) - (n_2^2 + n_3^2 - n_1^2 + 2in_1n_2)\sin^2(\varphi/2)\\
&= 1 - in_3\sin(\varphi) - (n_2^2 + n_3^2 - n_1^2 + 1 + 2in_1n_2)\sin^2(\varphi/2)\\
&= 1 - in_3\sin(\varphi) - 2(n_2^2 + n_3^2 + in_1n_2)\sin^2(\varphi/2)\\
&= 1 - in_3\sin(\varphi) - (n_2^2 + n_3^2 + in_1n_2)[1 - \cos(\varphi)]\\
&= 1 - (n_2^2 + n_3^2)\cdot[1 - \cos(\varphi)] + i\{-n_1n_2\cdot[1 - \cos(\varphi)] - n_3\sin(\varphi)\}\ ,\\
&g_{11}^2(\varphi) + g_{12}^2(\varphi)\\
&= \cos^2(\varphi/2) - n_3^2\sin^2(\varphi/2) - 2in_3\sin(\varphi/2)\cos(\varphi/2)\\
&\quad + (n_2^2 - n_1^2 + 2in_1n_2)\sin^2(\varphi/2)\\
&= \cos^2(\varphi/2) - in_3\sin(\varphi) + (n_2^2 - n_3^2 - n_1^2 + 2in_1n_2)\sin^2(\varphi/2)\\
&= 1 - in_3\sin(\varphi) + (n_2^2 - n_3^2 - n_1^2 - 1 + 2in_1n_2)\sin^2(\varphi/2)\\
&= 1 - in_3\sin(\varphi) - 2(n_1^2 + n_3^2 - in_1n_2)\sin^2(\varphi/2)\\
&= 1 - in_3\sin(\varphi) - (n_1^2 + n_3^2 - in_1n_2)\cdot[1 - \cos(\varphi)]\\
&= 1 - (n_1^2 + n_3^2)\cdot[1 - \cos(\varphi)] - i\{n_3\sin(\varphi) - n_1n_2\cdot[1 - \cos(\varphi)]\}\ ,\\
&g_{11}(\varphi)g_{12}(\varphi) = (-n_2 - in_1)\sin(\varphi/2)[\cos(\varphi/2) - in_3\sin(\varphi/2)]\\
&= (-n_2 - in_1)[\sin(\varphi/2)\cos(\varphi/2) - in_3\sin^2(\varphi/2)]\\
&= \frac{1}{2}(-n_2 - in_1)\{\sin(\varphi) - in_3[1 - \cos(\varphi)]\}\\
&= \frac{1}{2}\{-n_2\sin(\varphi) - n_1n_3[1 - \cos(\varphi)]\} - i\,\frac{1}{2}\{n_1\sin(\varphi) - n_2n_3[1 - \cos(\varphi)]\}\ ,\\
&g_{11}(\varphi)g_{12}(\varphi)^* = (-n_2 + in_1)\sin(\varphi/2)[\cos(\varphi/2) - in_3\sin(\varphi/2)]\\
&= (-n_2 + in_1)[\sin(\varphi/2)\cos(\varphi/2) - in_3\sin^2(\varphi/2)]\\
&= \frac{1}{2}(-n_2 + in_1)\{\sin(\varphi) - in_3[1 - \cos(\varphi)]\}\\
&= \frac{1}{2}\{-n_2\sin(\varphi) + n_1n_3[1 - \cos(\varphi)]\} + \frac{i}{2}\{n_1\sin(\varphi) + n_2n_3[1 - \cos(\varphi)]\}\ ,\\
&|g_{11}(\varphi)|^2 - |g_{12}(\varphi)|^2 = \cos^2(\varphi/2) + n_3^2\sin^2(\varphi/2) - (n_1^2 + n_2^2)\sin^2(\varphi/2)\\
&= \cos^2(\varphi/2) + [n_3^2 - (n_1^2 + n_2^2)]\sin^2(\varphi/2)\\
&= 1 + [n_3^2 - (n_1^2 + n_2^2) - 1]\sin^2(\varphi/2)\\
&= 1 - 2(n_1^2 + n_2^2)\sin^2(\varphi/2) = 1 - (n_1^2 + n_2^2)[1 - \cos(\varphi)]\ .
\end{aligned}$$

Hence, it follows from the proofs of Theorems 2.13, 2.4 and Proposition 2.5 that

$$
\begin{aligned}
&\Phi_1(G(\varphi)) \\
&= \begin{pmatrix} \mathrm{Re}(g_{11}^2(\varphi) - g_{12}^2(\varphi)) & \mathrm{Im}(g_{11}^2(\varphi) + g_{12}^2(\varphi)) & -2\,\mathrm{Re}(g_{11}(\varphi)g_{12}(\varphi)) \\ -\mathrm{Im}(g_{11}^2(\varphi) - g_{12}^2(\varphi)) & \mathrm{Re}(g_{11}^2(\varphi) + g_{12}^2(\varphi)) & 2\,\mathrm{Im}(g_{11}(\varphi)g_{12}(\varphi)) \\ 2\,\mathrm{Re}(g_{11}(\varphi)g_{12}^*(\varphi)) & 2\,\mathrm{Im}(g_{11}(\varphi)g_{12}^*(\varphi)) & |g_{11}(\varphi)|^2 - |g_{12}(\varphi)|^2 \end{pmatrix} \\
&= \begin{pmatrix} 1-(n_2^2+n_3^2)\cdot[1-\cos(\varphi)] & -\{n_3\sin(\varphi) - n_1 n_2\cdot[1-\cos(\varphi)] \\ -\{-n_1 n_2\cdot[1-\cos(\varphi)] - n_3\sin(\varphi)\} & 1-(n_1^2+n_3^2)\cdot[1-\cos(\varphi)] \\ -n_2\sin(\varphi) + n_1 n_3[1-\cos(\varphi)] & n_1\sin(\varphi) + n_2 n_3[1-\cos(\varphi)] \end{pmatrix.} \\
&\qquad \begin{matrix} -\{-n_2\sin(\varphi) - n_1 n_3[1-\cos(\varphi)]\} \\ -\{n_1\sin(\varphi) - n_2 n_3[1-\cos(\varphi)]\} \\ 1-(n_1^2+n_2^2)[1-\cos(\varphi)] \end{matrix}\Bigg) \\
&= \begin{pmatrix} 1 & 0 & 0 \\ 0 & 1 & 0 \\ 0 & 0 & 1 \end{pmatrix} + \sin(\varphi)\begin{pmatrix} 0 & -n_3 & n_2 \\ n_3 & 0 & -n_1 \\ -n_2 & n_1 & 0 \end{pmatrix} \\
&\quad + [1-\cos(\varphi)]\begin{pmatrix} -(n_2^2+n_3^2) & n_1 n_2 & n_1 n_3 \\ n_1 n_2 & -(n_1^2+n_3^2) & n_2 n_3 \\ n_1 n_3 & n_2 n_3 & -(n_1^2+n_2^2) \end{pmatrix} \\
&= E + \sin(\varphi)\,A + [1-\cos(\varphi)]\,A^2 = \exp(\varphi A)\ ,
\end{aligned}
$$

where

$$
A = \begin{pmatrix} 0 & -n_3 & n_2 \\ n_3 & 0 & -n_1 \\ -n_2 & n_1 & 0 \end{pmatrix}\ ,
$$

$$
A^2 = \begin{pmatrix} -(n_2^2+n_3^2) & n_1 n_2 & n_1 n_3 \\ n_1 n_2 & -(n_1^2+n_3^2) & n_2 n_3 \\ n_1 n_3 & n_2 n_3 & -(n_1^2+n_2^2) \end{pmatrix}\ .
$$

Further, we note that $G : \mathbb{R} \to \mathrm{SU}(2)$, defined by

$$
G(\varphi) := \cos(\varphi/2)E - i\sin(\varphi/2)\sum_{k=1}^{3} n_k \sigma_k\ , \tag{2.18}
$$

for every $\varphi \in \mathbb{R}$, where $n \in S^2$, is a one-parameter unitary group, since

$$
\begin{aligned}
&G(\varphi_1)G(\varphi_2) = \\
&= \left[\cos(\varphi_1/2)E - i\sin(\varphi_1/2)\sum_{k=1}^{3} n_k\sigma_k\right] \\
&\quad \cdot \left[\cos(\varphi_2/2)E - i\sin(\varphi_2/2)\sum_{l=1}^{3} n_k\sigma_l\right] \\
&= \cos(\varphi_1/2)\cos(\varphi_2/2)E \\
&\quad - i[\sin(\varphi_1/2)\cos(\varphi_2/2) + \cos(\varphi_1/2)\sin(\varphi_2/2)]\sum_{k=1}^{3} n_k\sigma_k \\
&\quad - \sin(\varphi_1/2)\sin(\varphi_2/2)\sum_{k,l=1}^{3} n_k n_l \sigma_k \sigma_l \\
&= \cos(\varphi_1/2)\cos(\varphi_2/2)E - i\sin((\varphi_1+\varphi_2)/2)\sum_{k=1}^{3} n_k\sigma_k \\
&\quad - \sin(\varphi_1/2)\sin(\varphi_2/2)\sum_{k,l=1}^{3} n_k n_l \left(\delta_{kl}E + i\sum_{m=1}^{3}\varepsilon_{klm}\sigma_m\right) \\
&= (\cos(\varphi_1/2)\cos(\varphi_2/2) - \sin(\varphi_1/2)\sin(\varphi_2/2))E \\
&\quad - i\sin((\varphi_1+\varphi_2)/2)\sum_{k=1}^{3} n_k\sigma_k \\
&\quad - i\sin(\varphi_1/2)\sin(\varphi_2/2)\sum_{m=1}^{3}\left(\sum_{k,l=1}^{3}\varepsilon_{mkl}n_k n_l\right)\sigma_m \\
&= \cos((\varphi_1+\varphi_2)/2)E - i\sin((\varphi_1+\varphi_2)/2)\sum_{k=1}^{3} n_k\sigma_k \\
&\quad - i\sin(\varphi_1/2)\sin(\varphi_2/2)\sum_{m=1}^{3}(n\times n)_m\sigma_m \\
&= G(\varphi_1+\varphi_2)\ ,
\end{aligned}
$$

for all $\varphi_1, \varphi_2 \in \mathbb{R}$. In addition, G is differentiable, and the generator of G is given by

$$
\frac{1}{i}\,G'(0) = \frac{1}{i}\left(-\frac{i}{2}\sum_{k=1}^{3} n_k\sigma_k\right) = -\frac{1}{2}\sum_{k=1}^{3} n_k\sigma_k\ .
$$

Hence, it follows from the spectral theorem for linear self-adjoint operators in complex Hilbert spaces that

$$G(\varphi) = \exp\left(-\frac{i\varphi}{2}\sum_{k=1}^{3} n_k\sigma_k\right),$$

for every $\varphi \in \mathbb{R}$. □

Finally, the following proposition gives a further parametrization of SU(2), with the help of the exponential map.

Proposition 2.17 (A parametrization of SU(2)) *For every $G \in \mathrm{SU}(2)$, there are $n \in S^2$ and $\varphi \in [0, \pi] \cup [2\pi, 3\pi]$ such that*

$$G = \exp\left(-\frac{i\varphi}{2}\sum_{k=1}^{3} n_k\sigma_k\right).$$

Proof If $G \in \mathrm{SU}(2)$, then the representation matrix of $\Phi_1(G)$, with respect to the canonical basis e_1, e_2, e_3 of $\mathbb{R}^3$, is an element of SO(3) and hence, according to Proposition 2.5, there are $n \in S^2$ and $\varphi \in [0, \pi]$ such that

$$\Phi_1(G)(x) = \exp(\varphi A)\cdot x = \Phi_1\left(\exp\left(-\frac{i\varphi}{2}\sum_{k=1}^{3} n_k\sigma_k\right)\right)(x),$$

for every $x \in \mathbb{R}^3$, where (2.18), (2.17) have been used and

$$A := \begin{pmatrix} 0 & -n_3 & n_2 \\ n_3 & 0 & -n_1 \\ -n_2 & n_1 & 0 \end{pmatrix}.$$

Hence, it follows that

$$\Phi_1(G) = \Phi_1\left(\exp\left(-\frac{i\varphi}{2}\sum_{k=1}^{3} n_k\sigma_k\right)\right)$$

and from Theorem 2.13 that

$$G = \pm\exp\left(-\frac{i\varphi}{2}\sum_{k=1}^{3} n_k\sigma_k\right).$$

If

$$G = -\exp\left(-\frac{i\varphi}{2}\sum_{k=1}^{3} n_k\sigma_k\right),$$

we proceed as follows. Since

$$\exp\left(-\frac{i\,(2\pi)}{2}\sum_{k=1}^{3} n_k\sigma_k\right) = \cos(\pi)E - i\sin(\pi)\sum_{k=1}^{3} n_k\sigma_k = -1 \ ,$$

we have that

$$G = \exp\left(-\frac{i(\varphi+2\pi)}{2}\sum_{k=1}^{3} n_k\sigma_k\right) ,$$

where we used Theorem 1.8 (ii). □

2.4 SU(2)-Spinors

The fundamental representation of SU(2) governs the transformation of the components of spinor fields, defined in the next section.

For $G \in \mathrm{SU}(2)$, we define

$$D^{\frac{1}{2}}(G) := (\mathbb{C}^2 \to \mathbb{C}^2, \psi \mapsto G\psi) \ .$$

From the linearity of matrix multiplication, it follows that $D^{\frac{1}{2}}(G)$ is linear and, assuming that $\mathbb{C}^2$ is equipped with the canonical scalar product $\langle\,|\,\rangle$, $D^{\frac{1}{2}}(G)$ is unitary. For the proof, let $a, b \in \mathbb{C}$ be such that $|a|^2 + |b|^2 = 1$ and

$$G = \begin{pmatrix} a & b \\ -b^* & a^* \end{pmatrix} .$$

Then it follows that

$$\begin{aligned}
&\langle G\psi_1 | G\psi_2\rangle \\
&= \left\langle {}^t(a\psi_{11} + b\psi_{12}, -b^*\psi_{11} + a^*\psi_{12}) | {}^t(a\psi_{21} + b\psi_{22}, -b^*\psi_{21} + a^*\psi_{22})\right\rangle \\
&= (a\psi_{11} + b\psi_{12})^*(a\psi_{21} + b\psi_{22}) + (-b^*\psi_{11} + a^*\psi_{12})^*(-b^*\psi_{21} + a^*\psi_{22}) \\
&= (a^*\psi_{11}^* + b^*\psi_{12}^*)(a\psi_{21} + b\psi_{22}) + (-b\psi_{11}^* + a\psi_{12}^*)(-b^*\psi_{21} + a^*\psi_{22}) \\
&= |a|^2\psi_{11}^*\psi_{21} + b^*a\psi_{12}^*\psi_{21} + a^*b\psi_{11}^*\psi_{22} + |b|^2\psi_{12}^*\psi_{22} \\
&\quad + |b|^2\psi_{11}^*\psi_{21} - ab^*\psi_{12}^*\psi_{21} - ba^*\psi_{11}^*\psi_{22} + |a|^2\psi_{12}^*\psi_{22} \\
&= (|a|^2 + |b|^2)(\psi_{11}^*\psi_{21} + \psi_{12}^*\psi_{22}) = \langle\psi_1|\psi_2\rangle \ ,
\end{aligned}$$

for all $\psi_1 = {}^t(\psi_{11}, \psi_{12}), \psi_2 = {}^t(\psi_{21}, \psi_{22}) \in \mathbb{C}^2$. Hence, $D^{\frac{1}{2}}(G)$ is isometric. Further, since $GG^*\psi = \psi$, for every $\psi \in \mathbb{C}^2$, $D^{\frac{1}{2}}(G)$ is surjective and hence an unitary linear operator.

Further,

$$D^{\frac{1}{2}} := \begin{pmatrix} \mathrm{SU}(2) & \to & \mathrm{L}(\mathbb{C}^2) \\ G & \mapsto & D^{\frac{1}{2}}(G) \end{pmatrix} ,$$

is a unitary representation of SU(2), the so called fundamental representation of SU(2),

since for $G_1, G_2 \in \mathrm{SU}(2)$, we have that

$$\begin{aligned} D^{\frac{1}{2}}(G_1 G_2)\psi &= (G_1 \cdot G_2) \cdot \psi = G_1 \cdot (G_2 \cdot \psi) = D^{\frac{1}{2}}(G_1) D^{\frac{1}{2}}(G_2)\psi , \\ D^{\frac{1}{2}}(E)\psi &= E\psi = \psi , \end{aligned}$$

for every $\psi \in \mathbb{C}^2$, where E denotes the 2×2 unit matrix, and hence that

$$D^{\frac{1}{2}}(G_1 G_2) = D^{\frac{1}{2}}(G_1) \cdot D^{\frac{1}{2}}(G_2) , \quad D^{\frac{1}{2}}(E) = \mathrm{id}_{\mathbb{C}^2} .$$

In particular, we note that

$$D^{\frac{1}{2}}(-G) = -D^{\frac{1}{2}}(G) ,$$

for all $G \in \mathrm{SU}(2)$ and that

the representation $D^{\frac{1}{2}}$ is irreducible, i.e., there is no non-trivial proper subspace of $\mathbb{C}^2$ that is left invariant by every $D^{\frac{1}{2}}(G)$, $G \in \mathrm{SU}(2)$.

For the proof, we note that every non-trivial proper subspace of $\mathbb{C}^2$ is of the form $\mathbb{C}.\psi$, where $\psi = {}^t(\psi_1, \psi_2) \in \mathbb{C}^2$ is such that $|\psi_1|^2 + |\psi_2|^2 = 1$. Further, for $c, d \in \mathbb{C}$ satisfying $|c|^2 + |d|^2 = 1$, it follows that

$$G = \begin{pmatrix} \psi_1^* c + \psi_2 d^* & \psi_2^* c - \psi_1 d^* \\ -(\psi_2 c^* - \psi_1^* d) & \psi_1 c^* + \psi_2^* d \end{pmatrix} \in \mathrm{SU}(2) ,$$

where we note that

$$\begin{aligned} &(\psi_1^* c + \psi_2 d^*) \cdot (\psi_1 c^* + \psi_2^* d) + (\psi_2^* c - \psi_1 d^*) \cdot (\psi_2 c^* - \psi_1^* d) \\ &= |\psi_1|^2 \, |c|^2 + |\psi_2|^2 \, |d|^2 + \psi_2 \psi_1 d^* c^* + \psi_1^* \psi_2^* c \, d \\ &\quad + |\psi_2|^2 \, |c|^2 + |\psi_1|^2 \, |d|^2 - \psi_1 \psi_2 d^* c^* - \psi_2^* \psi_1^* c \, d \\ &= (|\psi_1|^2 + |\psi_2|^2) \cdot (|c|^2 + |d|^2) = 1 . \end{aligned}$$

In addition, we have that

$$G\psi = \begin{pmatrix} \psi_1^* c + \psi_2 d^* & \psi_2^* c - \psi_1 d^* \\ -(\psi_2 c^* - \psi_1^* d) & \psi_1 c^* + \psi_2^* d \end{pmatrix} \cdot \begin{pmatrix} \psi_1 \\ \psi_2 \end{pmatrix}$$
$$= \begin{pmatrix} |\psi_1|^2 c + \psi_1 \psi_2 d^* + |\psi_2|^2 c - \psi_1 \psi_2 d^* \\ -\psi_2 \psi_1 c^* + |\psi_1|^2 d + \psi_1 \psi_2 c^* + |\psi_2|^2 d \end{pmatrix} = \begin{pmatrix} c \\ d \end{pmatrix} .$$

In particular, if

$$\begin{pmatrix} c \\ d \end{pmatrix} = \begin{pmatrix} -\psi_2^* \\ \psi_1^* \end{pmatrix} ,$$

then $G\psi$ is orthogonal to ψ, such that $D^{\frac{1}{2}}(\mathbb{C}.\psi) \not\subset \mathbb{C}.\psi$, i.e., $\mathbb{C}.\psi$ is not left invariant by every $D^{\frac{1}{2}}(G)$, $G \in \mathrm{SU}(2)$.

The elements of the representation space $\mathbb{C}^2$ of SU(2) are SU(2) two-component spinors. Under a rotation $R \in \mathrm{SO}(3)$, a two-component spinor

$$\begin{pmatrix} \psi_1 \\ \psi_2 \end{pmatrix}$$

is transformed into

$$\pm G \cdot \begin{pmatrix} \psi_1 \\ \psi_2 \end{pmatrix} ,$$

where $G \in \Phi_1^{-1}(R)$. Since the pure states of quantum mechanics are rays in Hilbert space, the sign ambiguity in this transformation law is not observable.

2.5 A Strongly Continuous Unitary Representation of SU(2)

In quantum mechanics, particles with spin $\frac{1}{2}$ are described by spin $\frac{1}{2}$ fields. For the motivation of the transformation properties of such fields under rotations, we consider in the following the transformation properties of vector fields from differential topology.

For this purpose, let M and N be differentiable (C^∞-) manifolds, $f : M \to N$ be differentiable and $p \in M$. Then the corresponding induced map, from the tangent space $T_p M$ of M in p to the tangent space $T_{f(p)} N$ of N in $f(p)$, $f_{*p} : T_p M \to T_{f(p)} N$ is given by

$$(f_{*p} v) \cdot \varphi := v \cdot (\varphi \circ f) ,$$

for every $v \in T_p M$ and every C^∞-function φ, defined in an open neighborhood of $f(p)$. In particular, if u is a member of the atlas on M, with domain containing p, and $\bar{u}$ is a member of the atlas on N, with domain containing $f(p)$, then it follows for $l \in \{1, \ldots, n\}$ that

$$
\begin{aligned}
\left(f_{*p}\left.\frac{\partial}{\partial u_l}\right|_p\right)\cdot\varphi &:= \left.\frac{\partial}{\partial u_l}\right|_p\cdot(\varphi\circ f) = \frac{\partial\,(\varphi\circ f\circ u^{-1})}{\partial x_l}(u(p)) \\
&= \frac{\partial\,(\varphi\circ\tilde{u}^{-1}\circ\tilde{u}\circ f\circ u^{-1})}{\partial x_l}(u(p)) \\
&= \sum_{k=1}^{n}\frac{\partial\,(\varphi\circ\tilde{u}^{-1})}{\partial x_k}(\tilde{u}(f(p))\cdot\frac{\partial\,(\tilde{u}_k\circ f\circ u^{-1})}{\partial x_l}(u(p)) \\
&= \sum_{k=1}^{n}\frac{\partial\,(\tilde{u}_k\circ f\circ u^{-1})}{\partial x_l}(u(p))\left(\left.\frac{\partial}{\partial\tilde{u}_k}\right|_{f(p)}\cdot\varphi\right)
\end{aligned}
$$

and hence that

$$
f_{*p}\left.\frac{\partial}{\partial u_l}\right|_p = \sum_{k=1}^{n}\frac{\partial\,(\tilde{u}_k\circ f\circ u^{-1})}{\partial x_l}(u(p))\left.\frac{\partial}{\partial\tilde{u}_k}\right|_{f(p)}.
$$

As a consequence, we have that

$$
f_{*p}v = f_{*p}\sum_{l=1}^{n}v_l\left.\frac{\partial}{\partial u_l}\right|_p = \sum_{k=1}^{n}\left[\sum_{l=1}^{n}\frac{\partial\,(\tilde{u}_k\circ f\circ u^{-1})}{\partial x_l}(u(p))\,v_l\right]\left.\frac{\partial}{\partial\tilde{u}_k}\right|_{f(p)},
$$

for very $v\in T_pM$. Further, if X is a differentiable vector field, defined on an open neighborhood U of p, then

$$
f_{*p}X(p) = \sum_{k=1}^{n}\left[\sum_{l=1}^{n}\frac{\partial\,(\tilde{u}_k\circ f\circ u^{-1})}{\partial x_l}(u(p))\,X_l(p)\right]\left.\frac{\partial}{\partial\tilde{u}_k}\right|_{f(p)},
$$

where $X_1,\ldots,X_n$ are the coefficient functions in the expansion of X, in terms of the coordinate vector fields $\frac{\partial}{\partial u_1},\ldots,\frac{\partial}{\partial u_n}$. In particular, if $M=N=\mathbb{R}^n$, $f:\mathbb{R}^n\to\mathbb{R}^n$ is given by

$$
f(x) := R\cdot x = {}^t\!\left(\sum_{m=1}^{n}R_{1m}x_m,\ldots,\sum_{k=1}^{n}R_{nm}x_m\right),
$$

for every $x={}^t(x_1,\ldots,x_n)\in\mathbb{R}^n$, where $R\in\mathrm{SO}(n)$, $u=\tilde{u}=\mathrm{id}_{\mathbb{R}^n}$, X is defined on the whole of $\mathbb{R}^n$, then

$$
f_{*p}X(p) = \sum_{k=1}^{n}\left[\sum_{l=1}^{n}R_{kl}X_l(p)\right]\left.\frac{\partial}{\partial x_k}\right|_{f(p)},
$$

and hence the value at p of the rotated vector field X_R, resulting from X through the rotation R, is given by

$$X_R(p) = f_{*f^{-1}(p)}X(f^{-1}p) = \sum_{k=1}^{n}\left[\sum_{l=1}^{n} R_{kl}X_l(f^{-1}(p))\right]\frac{\partial}{\partial x_k}\bigg|_p ,$$

for every $p \in \mathbb{R}^n$.

As a consequence, the coefficient functions $X_{R1}, \dots, X_{Rn}$ in the expansion of the rotated vector field, in terms of the coordinate vector fields $\frac{\partial}{\partial x_1}, \dots, \frac{\partial}{\partial x_n}$ are given for every $k \in \{1, \dots, n\}$ by

$$X_{Rk}(x) = \sum_{l=1}^{n} R_{kl}X_l(R^{-1} \cdot x) , \tag{2.19}$$

for every $x = {}^t(x_1, \dots, x_n) \in \mathbb{R}^n$.

Another derivation of the transformation law (2.19) proceeds as follows. If $X : \mathbb{R}^n \to \mathbb{R}^n$ and $x \in \mathbb{R}^n$, then $X(x) = \gamma'(0)$, where the parametrized curve $\gamma : \mathbb{R} \to \mathbb{R}^n$ is defined by

$$\gamma(\tau) := x + \tau.{}^t(X_1(x), \dots, X_n(x)) ,$$

for every $\tau \in \mathbb{R}$. We note that $\gamma(0) = x$. If $R \in \mathrm{SO}(n)$, then the parametrized curve γ is mapped by R into the rotated parametrized curve $\gamma_R : \mathbb{R} \to \mathbb{R}^n$ given by

$$\gamma_R(\tau) := R \cdot \gamma(\tau) ,$$

for every $\tau \in \mathbb{R}$. We note that the point

$$\gamma_R(0) = R \cdot x$$

is the rotation x_R of x. Then the tangent vector $X(x)$ is mapped into a tangent vector $X_R(x_R)$ at the point x_R given by

$$X_R(x_R) = R \cdot {}^t(X_1(x), \dots, X_n(x)) ,$$

compare Fig. 2.5 for the case $n = 2$. Hence, the tangent vector $X_R(x)$ at the point x is given by

$$X_R(x) = R \cdot {}^t(X_1(R^{-1} \cdot x), \dots, X_n(R^{-1} \cdot x)) .$$

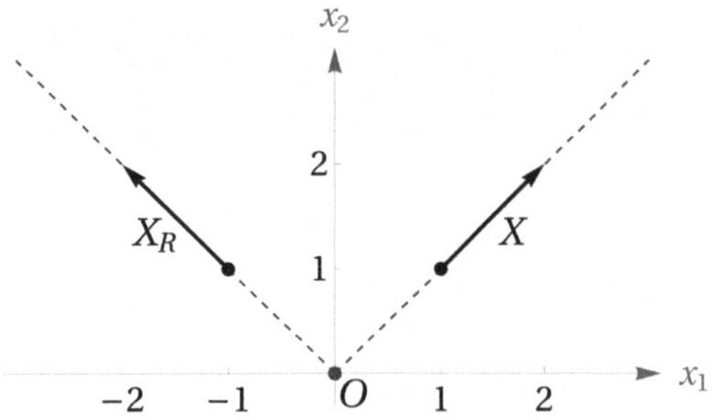

Fig. 2.5 Rotated tangent vector X_R resulting from the tangent vector X by a rotation $R \in \mathrm{SO}(2)$ about the origin of $90°$ degrees in counterclockwise direction. In this case, $x = (1, 1)$ and $x_R = (-1, 1)$. Compare text

For the case $n = 3$, $R \in \mathrm{SO}(3)$ and a spinor field $X = {}^t(X_1, X_2)$, we are going to assume the analogous transformation law,

$$X_R(x) = G \cdot \begin{pmatrix} X_1(R^{-1} \cdot x) \\ X_2(R^{-1} \cdot x) \end{pmatrix} = G \cdot X([\Phi_1(G)]^{-1}(x)) \ , \tag{2.20}$$

for every $x = (x_1, x_2, x_3) \in \mathbb{R}^3$, where $G \in \Phi_1^{-1}(R)$.

As said before, since the pure states of quantum mechanics are rays in Hilbert space, the sign ambiguity in this transformation law is not observable.

In the following, we continue the investigation of spinor fields. With the help of (2.20), we arrive at a strongly continuous unitary spinor representation $U_{1/2}$ of SU(2) on $(L^2_{\mathbb{C}}(\mathbb{R}^3))^2$ as follows.

In this connection, we use the fact from quantum mechanics, see, e.g., [7, Sect. 2.7.1], that the map

$$U : \mathrm{O}(3) \to L(L^2_{\mathbb{C}}(\mathbb{R}^3), L^2_{\mathbb{C}}(\mathbb{R}^3)) \ ,$$

defined by

$$U(M)f := f \circ T_M^{-1} = f \circ T_{M^*} \ ,$$

for every $M \in \mathrm{O}(3)$, where $T_M \in L(\mathbb{R}^3, \mathbb{R}^3)$ is defined by

$$T_M u := M \cdot u \ ,$$

for every $u \in \mathbb{R}^3$, is a unitary representation of O(3). In addition, U is strongly continuous, i.e., if $M_1, M_2, \ldots$ is a sequence in O(3) that converges component-wise to $M \in \mathrm{O}(3)$, then

$$\lim_{\nu\to\infty} \|[U(M_\nu) - U(M)]f\|_2 = 0 ,$$

for every $f \in L^2_{\mathbb{C}}(\mathbb{R}^3)$.

For $G \in \mathrm{SU}(2)$, we define $U_{1/2}(G)$ by

$$\begin{aligned} U_{1/2}(G)\begin{pmatrix} f_1 \\ f_2 \end{pmatrix} &:= G \cdot \begin{pmatrix} U(\Phi_1(G))f_1 \\ U(\Phi_1(G))f_2 \end{pmatrix} \\ &= \begin{pmatrix} G_{11}U(\Phi_1(G))f_1 + G_{12}U(\Phi_1(G))f_2 \\ G_{21}U(\Phi_1(G))f_1 + G_{22}U(\Phi_1(G))f_2 \end{pmatrix} , \end{aligned} \tag{2.21}$$

for every ${}^t(f_1, f_2) \in (L^2_{\mathbb{C}}(\mathbb{R}^3))^2$, where as usual we identify $L(\mathbb{R}^3, \mathbb{R}^3)$ and $\mathrm{M}(3, \mathbb{R})$. The map $U_{1/2}(G)$ is obviously linear and satisfies

$$\begin{aligned} &\langle U_{1/2}(G)\,{}^t(f_1, f_2) | U_{1/2}(G)\,{}^t(g_1, g_2)\rangle \\ &= \langle G_{11}U(\Phi_1(G))f_1 + G_{12}U(\Phi_1(G))f_2 | G_{11}U(\Phi_1(G))g_1 + G_{12}U(\Phi_1(G))g_2\rangle_2 \\ &\quad + \langle G_{21}U(\Phi_1(G))f_1 + G_{22}U(\Phi_1(G))f_2 | G_{21}U(\Phi_1(G))g_1 + G_{22}U(\Phi_1(G))g_2\rangle_2 \\ &= |G_{11}|^2 \langle f_1|g_1\rangle_2 + G_{11}^* G_{12} \langle f_1|g_2\rangle_2 + G_{12}^* G_{11} \langle f_2|g_1\rangle_2 + |G_{12}|^2 \langle f_2|g_2\rangle_2 \\ &\quad + |G_{21}|^2 \langle f_1|g_1\rangle_2 + G_{21}^* G_{22} \langle f_1|g_2\rangle_2 + G_{22}^* G_{21} \langle f_2|g_1\rangle_2 + |G_{22}|^2 \langle f_2|g_2\rangle_2 \\ &= (|G_{11}|^2 + |G_{21}|^2) \langle f_1|g_1\rangle_2 + (G_{11}^* G_{12} + G_{21}^* G_{22}) \langle f_1|g_2\rangle_2 \\ &\quad + (G_{11}G_{12}^* + G_{21}G_{22}^*) \langle f_2|g_1\rangle_2 + (|G_{12}|^2 + |G_{22}|^2) \langle f_2|g_2\rangle_2 \\ &= \langle f_1|g_1\rangle_2 + \langle f_2|g_2\rangle_2 = \langle {}^t(f_1, f_2) | {}^t(g_1, g_2)\rangle , \end{aligned}$$

for all ${}^t(f_1, f_2), {}^t(g_1, g_2) \in (L^2_{\mathbb{C}}(\mathbb{R}^3))^2$, where $\langle\,|\,\rangle_2$ denotes the scalar product for $L^2_{\mathbb{C}}(\mathbb{R}^3)$ and $\langle\,|\,\rangle$ denotes the scalar product for $(L^2_{\mathbb{C}}(\mathbb{R}^3))^2$. Hence, $U_{1/2}(G)$ is isometric. In particular, it follows that by (2.21), there is defined a map

$$U_{1/2} : \mathrm{SU}(2) \to L((L^2_{\mathbb{C}}(\mathbb{R}^3))^2, (L^2_{\mathbb{C}}(\mathbb{R}^3))^2) .$$

Further, for $G, H \in \mathrm{SU}(2)$ and ${}^t(f_1, f_2) \in (L^2_{\mathbb{C}}(\mathbb{R}^3))^2$, we have that

$$
U_{1/2}(G)\,U_{1/2}(H)\begin{pmatrix} f_1 \\ f_2 \end{pmatrix} = U_{1/2}(G)\begin{pmatrix} H_{11}U(\Phi_1(H))f_1 + H_{12}U(\Phi_1(H))f_2 \\ H_{21}U(\Phi_1(H))f_1 + H_{22}U(\Phi_1(H))f_2 \end{pmatrix}
$$

$$
= G\cdot\begin{pmatrix} U(\Phi_1(G))[H_{11}U(\Phi_1(H))f_1 + H_{12}U(\Phi_1(H))f_2] \\ U(\Phi_1(G))[H_{21}U(\Phi_1(H))f_1 + H_{22}U(\Phi_1(H))f_2] \end{pmatrix}
$$

$$
= G\cdot\begin{pmatrix} H_{11}U(\Phi_1(G\cdot H))f_1 + H_{12}U(\Phi_1(G\cdot H))f_2 \\ H_{21}U(\Phi_1(G\cdot H))f_1 + H_{22}U(\Phi_1(G\cdot H))f_2 \end{pmatrix}
$$

$$
= G\cdot H\cdot\begin{pmatrix} U(\Phi_1(G\cdot H))f_1 \\ U(\Phi_1(G\cdot H))f_2 \end{pmatrix} = U_{1/2}(G\cdot H)\begin{pmatrix} f_1 \\ f_2 \end{pmatrix}
$$

and hence that

$$
U_{1/2}(G\cdot H) = U_{1/2}(G)\,U_{1/2}(H)\ .
$$

Since $U_{1/2}(E)$ is given by the identical map on $(L^2_{\mathbb{C}}(\mathbb{R}^3))^2$, it follows that $U_{1/2}(G)$ is unitary for every $G \in \mathrm{SU}(2)$. As consequence, $U_{1/2}$ is an unitary representation of SU(2). Further, if $G, H \in \mathrm{SU}(2)$ and ${}^t(f_1, f_2) \in (L^2_{\mathbb{C}}(\mathbb{R}^3))^2$, we have that

$$
U_{1/2}(G)f - U_{1/2}(H)f = G\cdot\begin{pmatrix} U(\Phi_1(G))f_1 \\ U(\Phi_1(G))f_2 \end{pmatrix} - H\cdot\begin{pmatrix} U(\Phi_1(H))f_1 \\ U(\Phi_1(H))f_2 \end{pmatrix}
$$

$$
= (G-H)\cdot\begin{pmatrix} U(\Phi_1(G))f_1 \\ U(\Phi_1(G))f_2 \end{pmatrix} + H\cdot\begin{pmatrix} [U(\Phi_1(G)) - U(\Phi_1(H))]f_1 \\ [U(\Phi_1(G)) - U(\Phi_1(H))]f_2 \end{pmatrix}
$$

$$
= \begin{pmatrix} (G_{11}-H_{11})U(\Phi_1(G))f_1 + (G_{12}-H_{12})U(\Phi_1(G))f_2 \\ (G_{21}-H_{21})U(\Phi_1(G))f_1 + (G_{22}-H_{22})U(\Phi_1(G))f_2 \end{pmatrix}
$$

$$
+ \begin{pmatrix} H_{11}[U(\Phi_1(G)) - U(\Phi_1(H))]f_1 + H_{12}[U(\Phi_1(G)) - U(\Phi_1(H))]f_2 \\ H_{21}[U(\Phi_1(G)) - U(\Phi_1(H))]f_1 + H_{22}[U(\Phi_1(G)) - U(\Phi_1(H))]f_2 \end{pmatrix}
$$

and hence that

$$
\frac{1}{4}\,\|U_{1/2}(G)f - U_{1/2}(H)f\|^2
$$

$$
\leqslant \frac{1}{2}\left\|\begin{pmatrix} (G_{11}-H_{11})U(\Phi_1(G))f_1 + (G_{12}-H_{12})U(\Phi_1(G))f_2 \\ (G_{21}-H_{21})U(\Phi_1(G))f_1 + (G_{22}-H_{22})U(\Phi_1(G))f_2 \end{pmatrix}\right\|^2
$$

$$
+ \frac{1}{2}\left\|\begin{pmatrix} H_{11}[U(\Phi_1(G)) - U(\Phi_1(H))]f_1 + H_{12}[U(\Phi_1(G)) - U(\Phi_1(H))]f_2 \\ H_{21}[U(\Phi_1(G)) - U(\Phi_1(H))]f_1 + H_{22}[U(\Phi_1(G)) - U(\Phi_1(H))]f_2 \end{pmatrix}\right\|^2
$$

$$
= \frac{1}{2}\,\|(G_{11}-H_{11})U(\Phi_1(G))f_1 + (G_{12}-H_{12})U(\Phi_1(G))f_2\|_2^2
$$

$$
+ \frac{1}{2}\,\|(G_{21}-H_{21})U(\Phi_1(G))f_1 + (G_{22}-H_{22})U(\Phi_1(G))f_2\|_2^2
$$

$$
+ \frac{1}{2}\,\|H_{11}[U(\Phi_1(G)) - U(\Phi_1(H))]f_1 + H_{12}[U(\Phi_1(G)) - U(\Phi_1(H))]f_2\|_2^2
$$

$$
+ \frac{1}{2}\,\|H_{21}[U(\Phi_1(G)) - U(\Phi_1(H))]f_1 + H_{22}[U(\Phi_1(G)) - U(\Phi_1(H))]f_2\|_2^2
$$

$$\begin{aligned}
&\leqslant |G_{11} - H_{11}|^2 \cdot \|f_1\|^2 + |G_{12} - H_{12}|^2 \cdot \|f_2\|_2^2 \\
&\quad + |G_{21} - H_{21}|^2 \cdot \|f_1\|_2^2 + |G_{22} - H_{22}|^2 \cdot \|f_2\|_2^2 \\
&\quad + |H_{11}|^2 \cdot \|[U(\Phi_1(G)) - U(\Phi_1(H))]f_1\|_2^2 + |H_{12}|^2 \cdot \|[U(\Phi_1(G)) - U(\Phi_1(H))]f_2\|_2^2 \\
&\quad + |H_{21}|^2 \cdot \|[U(\Phi_1(G)) - U(\Phi_1(H))]f_1\|^2 + |H_{22}|^2 \cdot \|[U(\Phi_1(G)) - U(\Phi_1(H))]f_2\|_2^2 \, .
\end{aligned}$$

Since U is strongly continuous, it follows with help of Corollary 2.15 that $U_{1/2}$ is strongly continuous, i.e., if $G_1, G_2, \dots$ is a sequence in SU(2) that converges component-wise to $G \in \mathrm{SU}(2)$, then

$$\lim_{\nu \to \infty} \|[U_{1/2}(G_\nu) - U_{1/2}(G)]f\| = 0 \, ,$$

for every $f \in (L^2_{\mathbb{C}}(\mathbb{R}^3))^2$.

$$U_{1/2} : \mathrm{SU}(2) \to L((L^2_{\mathbb{C}}(\mathbb{R}^3))^2, (L^2_{\mathbb{C}}(\mathbb{R}^3))^2) \, ,$$

defined by

$$\begin{aligned}
U_{1/2}(G)\begin{pmatrix} f_1 \\ f_2 \end{pmatrix} &:= G \cdot \begin{pmatrix} U(\Phi_1(G))f_1 \\ U(\Phi_1(G))f_2 \end{pmatrix} \\
&= \begin{pmatrix} G_{11}U(\Phi_1(G))f_1 + G_{12}U(\Phi_1(G))f_2 \\ G_{21}U(\Phi_1(G))f_1 + G_{22}U(\Phi_1(G))f_2 \end{pmatrix} ,
\end{aligned}$$

for every ${}^t(f_1, f_2) \in (L^2_{\mathbb{C}}(\mathbb{R}^3))^2$, where as usual we identify $L(\mathbb{R}^3, \mathbb{R}^3)$ and $\mathrm{M}(3, \mathbb{R})$, is an unitary representation of SU(2). In addition, $U_{1/2}$ is strongly continuous, i.e., if $G_1, G_2, \dots$ is a sequence in SU(2) that converges component-wise to $G \in \mathrm{SU}(2)$, then

$$\lim_{\nu \to \infty} \|[U_{1/2}(G_\nu) - U_{1/2}(G)]f\| = 0 \, ,$$

for every $f \in (L^2_{\mathbb{C}}(\mathbb{R}^3))^2$.

2.5.1 Associated Generators and Pauli Interaction Hamiltonian

If G is a continuous one-parameter subgroup of SU(2), i.e., G is a map from $\mathbb{R}$ to SU(2) such that

$$G(0) = E \, , \quad G(s_1 + s_2) = G(s_1) \cdot G(s_2) \, ,$$

for all $s_1, s_2 \in \mathbb{R}$ and such that, for every sequence $s_1, s_2, \dots$ in $\mathbb{R}$ that is convergent to $s \in \mathbb{R}$, the corresponding sequence $G(s_1), G(s_2), \dots$ converges component-wise to $G(s)$, then $U_{1/2} \circ G$ is a strongly continuous one-parameter unitary group. According to Stone's theorem, there is a unique densely-defined, linear and self-adjoint operator A_G in

$$X := (L^2_{\mathbb{C}}(\mathbb{R}^3))^2$$

such that

$$\exp(isA_G) = (U_{1/2} \circ G)(s) \ ,$$

for every $s \in \mathbb{R}$ and, in particular, that $A_G : D(A_G) \to X$ is given by

$$D(A_G) = \{f \in X : \lim_{s \to 0, s \neq 0} \frac{1}{s} \left[(U_{1/2} \circ G)(s) - \mathrm{id}_X\right] f \text{ exists}\}$$

and for every $f \in D(A_G)$

$$A_G f = \frac{1}{i} \lim_{s \to 0, s \neq 0} \frac{1}{s} \left[(U_{1/2} \circ G)(s) - \mathrm{id}_X\right] f \ .$$

For $n \in S^2$, we define $G_n : \mathbb{R} \to \mathrm{SU}(2)$ by

$$G_n(\varphi) := \exp\left(\frac{i\varphi}{2} \sum_{k=1}^{3} n_k \sigma_k \right) \in \mathrm{SU}(2) \ ,$$

for every $\varphi \in \mathbb{R}$.[1] According to Proposition 2.16, for $\varphi \in \mathbb{R}$, $\Phi_1(G_n(\varphi))$ is given by

$$\Phi_1(G_n(\varphi))(x) = \exp(\varphi A_n) \cdot x \ ,$$

for every $x \in \mathbb{R}^3$, where

$$A_n = \begin{pmatrix} 0 & n_3 & -n_2 \\ -n_3 & 0 & n_1 \\ n_2 & -n_1 & 0 \end{pmatrix} .$$

We note that, as a consequence of the continuity of the exponential function, G_n is continuous and that

$$(U_{1/2} \circ G_n)(\varphi) = G_n(\varphi) \cdot \begin{pmatrix} U(\Phi_1(G_n(\varphi))) f_1 \\ U(\Phi_1(G_n(\varphi))) f_2 \end{pmatrix} = G_n(\varphi) \cdot \begin{pmatrix} U(\exp(\varphi A_n)) f_1 \\ U(\exp(\varphi A_n)) f_2 \end{pmatrix} .$$

Since, according to quantum mechanics, we have for every $j \in \{1, 2, 3\}$ that

$$U(\exp(\varphi A_{e_j})) = f \circ T_{[\exp(\varphi A_{e_j})]^*} = f \circ T_{\exp(\varphi A_{-e_j})} = \exp(\frac{i\varphi}{\hbar} \hat{L}_j) \ ,$$

---

[1] Note the difference of sign compared to the definition of G in Proposition 2.16.

where e_1, e_2, e_3 is the canonical basis for $\mathbb{R}^3$, it follows that

$$(U_{1/2} \circ G_{e_j})(\varphi) = G_{e_j}(\varphi) \cdot \begin{pmatrix} \exp(\frac{i\varphi}{\hbar} \hat{L}_j) f_1 \\ \exp(\frac{i\varphi}{\hbar} \hat{L}_j) f_2 \end{pmatrix} , \tag{2.22}$$

for every $\varphi \in \mathbb{R}$. Further, for $f \in X$ and $\varphi \in \mathbb{R}^*$, it follows that

$$\begin{aligned}
&\frac{1}{i\varphi} \left[(U_{1/2} \circ G_{e_j})(\varphi) - \mathrm{id}_X \right] f = \frac{1}{i\varphi} \left[G_{e_j}(\varphi) \cdot \begin{pmatrix} \exp(\frac{i\varphi}{\hbar} \hat{L}_j) f_1 \\ \exp(\frac{i\varphi}{\hbar} \hat{L}_j) f_2 \end{pmatrix} - \begin{pmatrix} f_1 \\ f_2 \end{pmatrix} \right] \\
&= G_{e_j}(\varphi) \cdot \begin{pmatrix} (i\varphi)^{-1} [\exp(\frac{i\varphi}{\hbar} \hat{L}_j) - 1] f_1 \\ (i\varphi)^{-1} [\exp(\frac{i\varphi}{\hbar} \hat{L}_j) - 1] f_2 \end{pmatrix} + \frac{1}{i\varphi} [G_{e_j}(\varphi) - 1] \begin{pmatrix} f_1 \\ f_2 \end{pmatrix} \\
&= \begin{pmatrix} (i\varphi)^{-1} [\exp(\frac{i\varphi}{\hbar} \hat{L}_j) - 1] f_1 \\ (i\varphi)^{-1} [\exp(\frac{i\varphi}{\hbar} \hat{L}_j) - 1] f_2 \end{pmatrix} + \frac{1}{2} \sigma_j \begin{pmatrix} f_1 \\ f_2 \end{pmatrix} \\
&\quad + [G_{e_j}(\varphi) - 1] \cdot \begin{pmatrix} (i\varphi)^{-1} [\exp(\frac{i\varphi}{\hbar} \hat{L}_j) - 1] f_1 \\ (i\varphi)^{-1} [\exp(\frac{i\varphi}{\hbar} \hat{L}_j) - 1] f_2 \end{pmatrix} \\
&\quad + \left\{ \frac{1}{i\varphi} [G_{e_j}(\varphi) - 1] - \frac{1}{2} \sigma_j \right\} \begin{pmatrix} f_1 \\ f_2 \end{pmatrix} .
\end{aligned}$$

Since,

$$\begin{aligned}
&\|G_{e_j}(\varphi) - 1\|_{\mathrm{op}} = \| \exp(\frac{i\varphi}{2} \sigma_j) - 1\|_{\mathrm{op}} = \| \sum_{k=1}^{\infty} \frac{1}{k!} (\frac{i\varphi}{2} \sigma_j)^k \|_{\mathrm{op}} \\
&\leqslant \sum_{k=1}^{\infty} \frac{1}{k!} \left(\frac{|\varphi|}{2} \right)^k = \frac{|\varphi|}{2} \sum_{k=1}^{\infty} \frac{1}{k!} \left(\frac{|\varphi|}{2} \right)^{k-1} \leqslant \frac{|\varphi|}{2} \sum_{k=1}^{\infty} \frac{1}{(k-1)!} \left(\frac{|\varphi|}{2} \right)^{k-1} \\
&= \frac{|\varphi|}{2} \exp\left(\frac{|\varphi|}{2} \right) , \\
&\| \frac{1}{\varphi} [G_{e_j}(\varphi) - 1] - \frac{i}{2} \sigma_j \|_{\mathrm{op}} = \| \frac{1}{\varphi} [\exp(\frac{i\varphi}{2} \sigma_j) - 1] - \frac{i}{2} \sigma_j \|_{\mathrm{op}} \\
&= \| \frac{1}{\varphi} \left[\sum_{k=1}^{\infty} \frac{1}{k!} (\frac{i\varphi}{2} \sigma_j)^k \right] - \frac{i}{2} \sigma_j \|_{\mathrm{op}} = \| \frac{1}{\varphi} \left[\sum_{k=2}^{\infty} \frac{1}{k!} (\frac{i\varphi}{2} \sigma_j)^k \right] \|_{\mathrm{op}} \\
&\leqslant \frac{1}{|\varphi|} \sum_{k=2}^{\infty} \frac{1}{k!} \left(\frac{|\varphi|}{2} \right)^k = \frac{|\varphi|}{4} \sum_{k=2}^{\infty} \frac{1}{k!} \left(\frac{|\varphi|}{2} \right)^{k-2} \leqslant \frac{|\varphi|}{4} \sum_{k=2}^{\infty} \frac{1}{(k-2)!} \left(\frac{|\varphi|}{2} \right)^{k-2} \\
&= \frac{|\varphi|}{4} \exp\left(\frac{|\varphi|}{2} \right) ,
\end{aligned}$$

where we used that $\|\sigma_j\|_{\mathrm{op}} = 1$, since

$$|\sigma_1 \cdot \begin{pmatrix} z_1 \\ z_2 \end{pmatrix}| = |\begin{pmatrix} 0 & 1 \\ 1 & 0 \end{pmatrix} \cdot \begin{pmatrix} z_1 \\ z_2 \end{pmatrix}| = |\begin{pmatrix} z_2 \\ z_1 \end{pmatrix}| = |\begin{pmatrix} z_1 \\ z_2 \end{pmatrix}| ,$$

$$|\sigma_2 \cdot \begin{pmatrix} z_1 \\ z_2 \end{pmatrix}| = |\begin{pmatrix} 0 & -i \\ i & 0 \end{pmatrix} \cdot \begin{pmatrix} z_1 \\ z_2 \end{pmatrix}| = |\begin{pmatrix} -iz_2 \\ iz_1 \end{pmatrix}| = |\begin{pmatrix} z_1 \\ z_2 \end{pmatrix}| ,$$

$$|\sigma_3 \cdot \begin{pmatrix} z_1 \\ z_2 \end{pmatrix}| = |\begin{pmatrix} 1 & 0 \\ 0 & -1 \end{pmatrix} \cdot \begin{pmatrix} z_1 \\ z_2 \end{pmatrix}| = |\begin{pmatrix} z_1 \\ -z_2 \end{pmatrix}| = |\begin{pmatrix} z_1 \\ z_2 \end{pmatrix}| ,$$

for every ${}^t(z_1, z_2) \in \mathbb{C}^2$, where $|\ |$ denotes the canonical norm for $\mathbb{C}^2$. We note that the coordinate projections $p_1, p_2 : X \to L^2_{\mathbb{C}}(\mathbb{R}^3)$ as well as the inclusions $\iota_1, \iota_2 : L^2_{\mathbb{C}}(\mathbb{R}^3) \to X$, defined by $p_1 f := f_1$, $p_2 f := f$, for every $f \in X$ and $\iota_1 f := {}^t(f, 0)$ and $\iota_2 f := {}^t(0, f)$, for every $f \in L^2_{\mathbb{C}}(\mathbb{R}^3)$ are linear and continuous, since

$$\|p_k f\|_2 = \|f_k\|_2 \leqslant \|f\| , \quad \|\iota_k g\| = \|g\|_2 ,$$

for every $f \in X$, $g \in L^2_{\mathbb{C}}(\mathbb{R}^3)$ and $k \in \{1, 2\}$. As a consequence, an element $f \in X$ is part of the domain $D(A_{G_{e_j}})$ of $A_{G_{e_j}}$ if and only if $f \in D(\hat{L}_j) \times D(\hat{L}_j)$ and if $f \in D(A_{G_{e_j}})$, then

$$A_{G_{e_j}} f = \frac{1}{\hbar} \begin{pmatrix} \hat{L}_j f_1 \\ \hat{L}_j f_2 \end{pmatrix} + \frac{1}{2} \sigma_j \cdot \begin{pmatrix} f_1 \\ f_2 \end{pmatrix} .$$

Hence, we define the, densely-defined, linear and self-adjoint, j-th component J_j of the total angular momentum operator $\hat{J}_j : D(\hat{L}_j) \times D(\hat{L}_j) \to X$ by

$$\hat{J}_j f := \hbar A_{G_{e_j}} f = \begin{pmatrix} \hat{L}_j f_1 \\ \hat{L}_j f_2 \end{pmatrix} + \frac{\hbar}{2} \sigma_j \cdot \begin{pmatrix} f_1 \\ f_2 \end{pmatrix} ,$$

for every $f \in D(\hat{L}_j) \times D(\hat{L}_j)$ as well as the, bounded linear and self-adjoint, j-th component $\hat{S}_j : X \to X$ of intrinsic angular momentum by

$$\hat{S}_j f := \frac{\hbar}{2} \sigma_j \cdot \begin{pmatrix} f_1 \\ f_2 \end{pmatrix} ,$$

for every $f \in X$.

Further, it follows from (2.22) that

$$\exp(\frac{i\varphi}{\hbar} \hat{J}_j) f = G_{e_j}(\varphi) \cdot \begin{pmatrix} \exp(\frac{i\varphi}{\hbar} \hat{L}_j) f_1 \\ \exp(\frac{i\varphi}{\hbar} \hat{L}_j) f_2 \end{pmatrix} , \tag{2.23}$$

for every $f \in X$ and $\varphi \in \mathbb{R}$. Also, we note that $-\hbar/2$ and $\hbar/2$ are eigenvalues of $\hat{S}_j$ as well as that

$$\ker\left(\hat{S}_1 + \frac{\hbar}{2}\right) = L^2_{\mathbb{C}}(\mathbb{R}^3)\,.\begin{pmatrix}1\\-1\end{pmatrix}\ ,\ \ker\left(\hat{S}_1 - \frac{\hbar}{2}\right) = L^2_{\mathbb{C}}(\mathbb{R}^3)\,.\begin{pmatrix}1\\1\end{pmatrix}\ ,$$
$$\ker\left(\hat{S}_2 + \frac{\hbar}{2}\right) = L^2_{\mathbb{C}}(\mathbb{R}^3)\,.\begin{pmatrix}1\\-i\end{pmatrix}\ ,\ \ker\left(\hat{S}_2 - \frac{\hbar}{2}\right) = L^2_{\mathbb{C}}(\mathbb{R}^3)\,.\begin{pmatrix}1\\i\end{pmatrix}\ ,$$
$$\ker\left(\hat{S}_3 + \frac{\hbar}{2}\right) = L^2_{\mathbb{C}}(\mathbb{R}^3)\,.\begin{pmatrix}0\\1\end{pmatrix}\ ,\ \ker\left(\hat{S}_3 - \frac{\hbar}{2}\right) = L^2_{\mathbb{C}}(\mathbb{R}^3)\,.\begin{pmatrix}1\\0\end{pmatrix}\ .$$

Hence, we obtain the following result.

For $j \in \{1, 2, 3\}$, there is a Hilbert basis of X, consisting of eigenvectors of $\hat{S}_j$, corresponding to the eigenvalues $-\hbar/2$ and $\hbar/2$. Therefore, $\hat{S}_j$ has a pure point spectrum given by

$$\sigma(\hat{S}_j) = \left\{-\frac{\hbar}{2}, \frac{\hbar}{2}\right\}\ .$$

In addition, we have that

$$\begin{aligned}
\exp\left(\frac{i\varphi}{\hbar}\,\hat{S}_j\right) f &= \sum_{k=0}^{\infty} \frac{1}{k!}\left(\frac{i\varphi}{\hbar}\,\hat{S}_j\right)^k f = \sum_{k=0}^{\infty} \frac{1}{k!}\left(\frac{i\varphi}{2}\,\sigma_j\right)^k f \\
&= \sum_{k=0}^{\infty} \frac{1}{(2k)!}\left(\frac{i\varphi}{2}\right)^{2k} \sigma_j^{2k} f + \sum_{k=0}^{\infty} \frac{1}{(2k+1)!}\left(\frac{i\varphi}{2}\right)^{2k+1} \sigma_j^{2k+1} f \\
&= \sum_{k=0}^{\infty} \frac{1}{(2k)!}\left(\frac{i\varphi}{2}\right)^{2k} f + \sum_{k=0}^{\infty} \frac{1}{(2k+1)!}\left(\frac{i\varphi}{2}\right)^{2k+1} \sigma_j f \\
&= \cosh\left(\frac{i\varphi}{2}\right) f + \sinh\left(\frac{i\varphi}{2}\right)\sigma_j f = \cos\left(\frac{\varphi}{2}\right) f + i\sin\left(\frac{\varphi}{2}\right)\sigma_j f\ ,
\end{aligned}$$

for every $f \in X$ and $\varphi \in \mathbb{R}$, where we used that

$$\sigma_j^{2k} = E\ ,\ \sigma_j^{2k+1} = \sigma_j\ ,$$

for every $k \in \mathbb{N}$. Hence,

$$\exp\left(\frac{i\varphi}{\hbar}\,\hat{S}_j\right) f = \cos\left(\frac{\varphi}{2}\right) f + i\sin\left(\frac{\varphi}{2}\right)\sigma_j f\ ,$$

for every $f \in X$ and $\varphi \in \mathbb{R}$. As a consequence,

$$\exp(\frac{i(\varphi + 2k\pi)}{\hbar}\hat{S}_j)f = (-1)^k \exp(\frac{i\varphi}{\hbar}\hat{S}_j)f ,$$

for every $f \in X$, $\varphi \in \mathbb{R}$ and $k \in \mathbb{Z}$, i.e., an increase of the angle φ about $2k\pi$, $k \in \mathbb{Z}$, results in a multiplication by a phase factor and hence leads to the same quantum state.

The Hamilton operator $\hat{H}_{\text{Pauli}}$, describing the interaction of the magnetic momentum of a charged particle, with charge q and spin $1/2$, with a magnetic field $B = {}^t(B_1, B_2, B_3)$, where the components of the latter are assumed to be a.e. defined on the domain of the system in question and measurable, is given by

$$\hat{H}_{\text{Pauli}} = -\frac{q}{mc}\sum_{j=1}^{3} T_{B_j}\hat{S}_j ,$$

where T_{B_j} denotes the maximal multiplication operator with the j-th component B_j of the magnetic field B. For instance, if the components of B are a.e. bounded, then $\hat{H}_{\text{Pauli}}$ constitutes a bounded perturbation.

A Family of Representations of the Poincaré Group 3

The focus of this chapter is the construction of a family of representations $\hat{\hat{U}}_a$, $a \geqslant 0$, of the Poincaré group in 3 steps. For every $a \geqslant 0$, in the first step, we construct a strongly continuous unitary representation of the restricted Lorentz group, the connected component of the Lorentz group that contains the unit element, and analyze the generators of one-parameter subgroups associated with rotations and Lorentz boosts. In the next step, we extend this representation to a strongly continuous unitary representation of the restricted Poincaré group. Here, we connect the generators of this representation, associated with one-parameter subgroups of translations, to the relativistic quantum mechanical description of the free motion of a scalar particle in Minkowski space. In the final step, we extend the latter representation to a representation of the whole Poincaré group, where the elements of the connected components $\mathscr{P}_{+}^{\uparrow}$ and $\mathscr{P}_{-}^{\uparrow}$ of $\mathscr{P}$ are represented by unitary linear operators, whereas the elements of the connected components $\mathscr{P}_{+}^{\downarrow}$ and $\mathscr{P}_{-}^{\downarrow}$ of $\mathscr{P}$ are represented by anti-unitary and anti-linear operators.

3.1 Basic Properties of the Lorentz Group

In the following, we provide basic information about the Lorentz group. In a first step, we define the underlying set $\mathscr{L}$ of the Lorentz group as the subset of $\mathrm{M}(4, \mathbb{R})$ consisting of Lorentz transformations and show that, equipped with the operation of matrix multiplication, $\mathscr{L}$ is subgroup of $GL(4, \mathbb{R})$. Here, the reader needs to take into account that in this text, we identify elements of $\mathrm{L}(\mathbb{K}^n, \mathbb{K}^n)$ with their matrix representation, with respect to the canonical basis of $\mathbb{K}^n$, where $\mathbb{K} \in \{\mathbb{R}, \mathbb{C}\}$ and n is a non-zero natural number.

H. R. Beyer, *Quantum Spin and Representations of the Poincaré Group, Part I*,
Synthesis Lectures on Engineering, Science, and Technology,
https://doi.org/10.1007/978-3-031-84140-8_3

Definition 3.1 (*Lorentz transformations*)

(i) We define the Lorentz product, a symmetric non-degenerate bilinear form, $\cdot : \mathbb{R}^4 \times \mathbb{R}^4 \to \mathbb{R}$ by

$$\begin{aligned} \cdot\,(x, y) &:= x \cdot y := x_0 y_0 - x_1 y_1 - x_2 y_2 - x_3 y_3 \\ &= x_0 y_0 - \left\langle {}^t(x_1, x_2, x_3) | {}^t(y_1, y_2, y_3) \right\rangle \ , \end{aligned}$$

for all $x, y \in \mathbb{R}^4$, where $\langle\, | \,\rangle$ denotes the Euclidean scalar product on $\mathbb{R}^3$.

(ii) A matrix $\Lambda \in \mathrm{M}(4, \mathbb{R})$ will be called a Lorentz transformation if

$$(\Lambda x) \cdot (\Lambda x) = x \cdot x \ , \tag{3.1}$$

for all $x \in \mathbb{R}^4$.

(iii) We define

$$\mathcal{L} := \{\Lambda \in \mathrm{M}(4, \mathbb{R}) : \Lambda \text{ is a Lorentz transformation}\} \ .$$

We note that, since the Lorentz inner product defines a symmetric bilinear form on $\mathbb{R}^4$, it follows that

$$(x + y) \cdot (x + y) = x \cdot x + y \cdot x + x \cdot y + y \cdot y = x \cdot x + 2x \cdot y + y \cdot y \ ,$$

for all $x, y \in \mathbb{R}^4$ and hence that the Lorentz inner product satisfies the following polarization identity

$$x \cdot y = \frac{1}{2} [(x + y) \cdot (x + y) - x \cdot x - y \cdot y] \ ,$$

for all $x, y \in \mathbb{R}^4$. Hence it follows for a Lorentz transformation $\Lambda \in \mathrm{M}(4, \mathbb{R})$ that

$$\begin{aligned} (\Lambda x) \cdot (\Lambda y) &= \frac{1}{2} [(\Lambda x + \Lambda y) \cdot (\Lambda x + \Lambda y) - (\Lambda x) \cdot (\Lambda x) - (\Lambda y) \cdot (\Lambda y)] \\ &= \frac{1}{2} [(\Lambda(x + y)) \cdot (\Lambda(x + y)) - (\Lambda x) \cdot (\Lambda x) - (\Lambda y) \cdot (\Lambda y)] \\ &= \frac{1}{2} [(x + y) \cdot (x + y) - x \cdot x - y \cdot y] = x \cdot y \ , \end{aligned}$$

for all $x, y \in \mathbb{R}^4$. Therefore, a matrix $\Lambda \in \mathrm{M}(4, \mathbb{R})$ is a Lorentz transformation if and only if

$$(\Lambda x) \cdot (\Lambda y) = x \cdot y \ ,$$

for all $x, y \in \mathbb{R}^4$.

Lemma 3.2 (The Lorentz group) $\mathcal{L}$, *equipped with operation of matrix multiplication, is a subgroup, the so called Lorentz group, of the general linear group* $GL(4, \mathbb{R})$.

Proof In the following, $\langle \, | \, \rangle$ denotes the Euclidean scalar product on $\mathbb{R}^4$. Then

$$x \cdot y = \langle x | (y_0, -y_1, -y_2, -y_3) \rangle = \langle x | \eta y \rangle$$

for all $x, y \in \mathbb{R}^4$, where

$$\eta := \begin{pmatrix} 1 & 0 & 0 & 0 \\ 0 & -1 & 0 & 0 \\ 0 & 0 & -1 & 0 \\ 0 & 0 & 0 & -1 \end{pmatrix} .$$

Hence $\Lambda \in M(4, \mathbb{R})$ is a Lorentz transformation if and only if

$$\left\langle x | \Lambda^t \eta \Lambda x \right\rangle = \langle \Lambda x | \eta \Lambda x \rangle = \Lambda x \cdot \Lambda x = x \cdot x = \langle x | \eta x \rangle \ ,$$

for all $x \in \mathbb{R}^4$ and hence if and only if

$$\Lambda^t \, \eta \, \Lambda = \eta \ , \tag{3.2}$$

and we note that $\Lambda^t \eta \Lambda$ and η are symmetric 4×4 matrices. Also, we used that if M is some symmetric real 4×4 matrix such that

$$\langle x | M x \rangle = 0 \ ,$$

for every $x \in \mathbb{R}^4$, then it follows that

$$\begin{aligned} 0 &= \langle x + y | M(x + y) \rangle = \langle x | M x \rangle + \langle x | M y \rangle + \langle y | M x \rangle + \langle y | M y \rangle \\ &= \langle x | M y \rangle + \langle y | M x \rangle = \langle x | M y \rangle + \langle M y | x \rangle = 2 \, \langle x | M y \rangle \end{aligned}$$

and hence that $\langle x | M y \rangle = 0$, for all $x, y \in \mathbb{R}^4$. As a consequence, $M = 0$. From (3.2), it follows that

$$\det(\Lambda) \in \{-1, 1\} \tag{3.3}$$

and that

$$\Lambda^{-1} = \eta \, \Lambda^t \eta = \begin{pmatrix} \Lambda_{00} & -\Lambda_{10} & -\Lambda_{20} & -\Lambda_{30} \\ -\Lambda_{01} & \Lambda_{11} & \Lambda_{21} & \Lambda_{31} \\ -\Lambda_{02} & \Lambda_{12} & \Lambda_{22} & \Lambda_{32} \\ -\Lambda_{03} & \Lambda_{13} & \Lambda_{23} & \Lambda_{33} \end{pmatrix} . \tag{3.4}$$

Also, by (3.1) it follows that the unit matrix is in $\mathcal{L}$, $\Lambda^{-1} \in \mathcal{L}$ and $\Lambda_1 \Lambda_2 \in \mathcal{L}$, for all $\Lambda, \Lambda_1, \Lambda_2 \in \mathcal{L}$. As a consequence, $\mathcal{L}$ is a subgroup of the general linear group $GL(4, \mathbb{R})$. □

Exercise 3.1 Show that $\Lambda^t \in \mathcal{L}$, for every $\Lambda \in \mathcal{L}$.

In the following, we assume the Lorentz group equipped with the topology induced by that of $GL(4, \mathbb{R})$, i.e., the topology generated by the topology of $GL(4, \mathbb{R})$ through intersection. The group operations of $\mathcal{L}$ are restrictions of the group operations of $GL(4, \mathbb{R})$ and hence continuous, with respect to the induced topology. In this way, $\mathcal{L}$ becomes a topological group. The same applies to subgroups of $\mathcal{L}$ that are defined in the following.

Exercise 3.2 Show that $\mathcal{L} \subset M(4, \mathbb{R})$ is closed and unbounded.

In the next step, we define the restricted Lorentz group $\mathcal{L}_+^\uparrow$, a subgroup of the Lorentz group that contains all orientation preserving transformations, i.e, $\Lambda \in \mathcal{L}$ such that $\det(\Lambda) = 1$, that are at the same time orthochronous, i.e., map future oriented time-like vectors into future oriented time-like vectors.

Lemma 3.3 (The restricted Lorentz group) *We define*

$$\mathcal{L}_+^\uparrow := \{\Lambda \in \mathcal{L} : \det(\Lambda) = 1, \Lambda_{00} \geqslant 0\} \ .$$

Then $\mathcal{L}_+^\uparrow$ is a subgroup, the so called restricted Lorentz group, of $\mathcal{L}$.

Proof Obviously, the unit matrix is contained in $\mathcal{L}_+^\uparrow$. Further, we note that it follows from (3.2) for $\Lambda \in \mathcal{L}$ that

$$1 = (\Lambda^t)_{00}\, \Lambda_{00} - \sum_{\alpha=1}^{3} (\Lambda^t)_{0\alpha}\, \Lambda_{\alpha 0} = (\Lambda_{00})^2 - \sum_{\alpha=1}^{3} (\Lambda_{\alpha 0})^2 \tag{3.5}$$

and hence for $\Lambda \in \mathcal{L}_+^\uparrow$ that

$$\Lambda_{00} = \left[1 + \sum_{\alpha=1}^{3} (\Lambda_{\alpha 0})^2 \right]^{1/2} \geqslant 1 \ . \tag{3.6}$$

For $\Lambda_1, \Lambda_2 \in \mathcal{L}_+^\uparrow$, it follows that

$$\det(\Lambda_1 \Lambda_2) = \det(\Lambda_1) \det(\Lambda_2) = 1 \ .$$

Further, by help of (3.6), we infer that

$$
\begin{aligned}
(\Lambda_1\Lambda_2)_{00} &= \sum_{a=0}^{3} (\Lambda_1)_{0a}(\Lambda_2)_{a0} \\
&= \left\{1+\sum_{\alpha=1}^{3}[(\Lambda_1)_{\alpha 0}]^2\right\}^{1/2} \left\{1+\sum_{\alpha=1}^{3}[(\Lambda_2)_{\alpha 0}]^2\right\}^{1/2} + \sum_{\alpha=1}^{3}(\Lambda_1)_{0\alpha}(\Lambda_2)_{\alpha 0} \\
&\geqslant \left\{1+\sum_{\alpha=1}^{3}[(\Lambda_1)_{\alpha 0}]^2\right\}^{1/2} \left\{1+\sum_{\alpha=1}^{3}[(\Lambda_2)_{\alpha 0}]^2\right\}^{1/2} \\
&\quad - \left\{\sum_{\alpha=1}^{3}[(\Lambda_1)_{\alpha 0}]^2\right\}^{1/2} \left\{\sum_{\alpha=1}^{3}[(\Lambda_2)_{\alpha 0}]^2\right\}^{1/2} \geqslant 0 \, .
\end{aligned}
$$

As a consequence, $\Lambda_1\Lambda_2 \in \mathcal{L}_+^{\uparrow}$. Further, it follows from (3.4) for $\Lambda \in \mathcal{L}_+^{\uparrow}$ that

$$
(\Lambda^{-1})_{00} = \Lambda_{00} \geqslant 0
$$

and hence that $\Lambda^{-1} \in \mathcal{L}_+^{\uparrow}$. Finally, it follows that $\mathcal{L}_+^{\uparrow}$ is a subgroup of $\mathcal{L}$. □

For later use, we note that it follows from (3.4) and (3.6) that

$$
\Lambda_{00} = (\Lambda^{-1})_{00} = \left\{1+\sum_{\alpha=1}^{3}[(\Lambda^{-1})_{\alpha 0}]^2\right\}^{1/2} = \left[1+\sum_{\alpha=1}^{3}(\Lambda_{0\alpha})^2\right]^{1/2} , \tag{3.7}
$$

for every $\Lambda \in \mathcal{L}_+^{\uparrow}$.

Exercise 3.3 Show that $\mathcal{L}_+^{\uparrow} \subset \mathrm{M}(4, \mathbb{R})$ is closed and unbounded.

In Part 2, we are going to construct a double cover $\Phi_3 : SL(2, \mathbb{C}) \to \mathcal{L}_+^{\uparrow}$ of $\mathcal{L}_+^{\uparrow}$ that is in particular continuous. Since, according to Proposition 1.24, $SL(2, \mathbb{C})$ is path-connected, it follows that $\mathcal{L}_+^{\uparrow}$, as an image of a path-connected set under a continuous map, is path-connected and hence also connected. Summarizing the topological information on $\mathcal{L}_+^{\uparrow}$ obtained so far, we have the following.

> Equipped with the induced topology and the operation of matrix multiplication, $\mathcal{L}_+^{\uparrow} \subset \mathrm{M}(4, \mathbb{R})$ is a topological group. In particular, $\mathcal{L}_+^{\uparrow} \subset \mathrm{M}(4, \mathbb{R})$ is unbounded, closed and path connected.

The following theorem gives a characterization of the elements of the restricted Lorentz group.

Theorem 3.4 (A decomposition of the members of $\mathcal{L}_+^\uparrow$) *For every $\Lambda \in \mathcal{L}_+^\uparrow$, there are $U, V \in SO(3)$ and $s \in \mathbb{R}$ such that*

$$\Lambda = \begin{pmatrix} 1 & 0 & 0 & 0 \\ 0 & U_{11} & U_{12} & U_{13} \\ 0 & U_{21} & U_{22} & U_{23} \\ 0 & U_{31} & U_{32} & U_{33} \end{pmatrix} \cdot \begin{pmatrix} \cosh(s) & 0 & 0 & \sinh(s) \\ 0 & 1 & 0 & 0 \\ 0 & 0 & 1 & 0 \\ \sinh(s) & 0 & 0 & \cosh(s) \end{pmatrix}$$
$$\cdot \begin{pmatrix} 1 & 0 & 0 & 0 \\ 0 & V_{11} & V_{12} & V_{13} \\ 0 & V_{21} & V_{22} & V_{23} \\ 0 & V_{31} & V_{32} & V_{33} \end{pmatrix} .$$

Proof For the proof, let $\Lambda \in \mathcal{L}_+^\uparrow$ and

$$x = \begin{pmatrix} x_0 \\ x_1 \\ x_2 \\ x_3 \end{pmatrix} := \Lambda \begin{pmatrix} 1 \\ 0 \\ 0 \\ 0 \end{pmatrix} , \; x_s := \begin{pmatrix} x_1 \\ x_2 \\ x_3 \end{pmatrix} .$$

We note that these definitions imply that

$$\begin{aligned} x \cdot x &= (x_0)^2 - |x_s|^2 \\ &= \Lambda\, {}^t(1,0,0,0) \cdot \Lambda\, {}^t(1,0,0,0) = {}^t(1,0,0,0) \cdot {}^t(1,0,0,0) = 1 , \\ x_0 &= \Lambda_{00} \geqslant 0 , \end{aligned}$$

where $|x_s|$ denotes the Euclidean length of x_s. According to Lemma 2.8, there is $V \in SO(3)$ such that

$$V x_s = \begin{pmatrix} 0 \\ 0 \\ |x_s| \end{pmatrix} .$$

As a consequence,

$$\hat{V} x = \begin{pmatrix} x_0 \\ 0 \\ 0 \\ |x_s| \end{pmatrix} ,$$

where

$$\hat{V} := \begin{pmatrix} 1 & 0 & 0 & 0 \\ 0 & V_{11} & V_{12} & V_{13} \\ 0 & V_{21} & V_{22} & V_{23} \\ 0 & V_{31} & V_{32} & V_{33} \end{pmatrix} \in \mathcal{L}_+^{\uparrow} \ .$$

Further, we define

$$s := -\ln(x_0 + |x_s|) \ .$$

Then,

$$\begin{aligned}
\cosh(s) &= \frac{1}{2}\left(\frac{1}{x_0 + |x_s|} + x_0 + |x_s|\right) \\
&= \frac{1}{2}(x_0 - |x_s| + x_0 + |x_s|) = x_0 \ , \\
\sinh(s) &= \frac{1}{2}\left(\frac{1}{x_0 + |x_s|} - x_0 - |x_s|\right) \\
&= \frac{1}{2}(x_0 - |x_s| - x_0 - |x_s|) = -|x_s|
\end{aligned}$$

and hence

$$\Lambda_s \hat{V} \Lambda \begin{pmatrix} 1 \\ 0 \\ 0 \\ 0 \end{pmatrix} = \begin{pmatrix} \cosh(s) & 0 & 0 & \sinh(s) \\ 0 & 1 & 0 & 0 \\ 0 & 0 & 1 & 0 \\ \sinh(s) & 0 & 0 & \cosh(s) \end{pmatrix} \begin{pmatrix} x_0 \\ 0 \\ 0 \\ |x_s| \end{pmatrix} = \begin{pmatrix} 1 \\ 0 \\ 0 \\ 0 \end{pmatrix} ,$$

where

$$\Lambda_s := \begin{pmatrix} \cosh(s) & 0 & 0 & \sinh(s) \\ 0 & 1 & 0 & 0 \\ 0 & 0 & 1 & 0 \\ \sinh(s) & 0 & 0 & \cosh(s) \end{pmatrix} \in \mathcal{L}_+^{\uparrow} ,$$

since $(\Lambda_s)_{00} \geqslant 0$, $\det(\Lambda_s) = 1$ and

$$\begin{aligned}
\Lambda_s y \cdot \Lambda_s y &= [\cosh(s) y_0 + \sinh(s) y_3]^2 - (y_1)^2 - (y_2)^2 \\
&\quad - [\sinh(s) y_0 + \cosh(s) y_3]^2 \\
&= \cosh^2(s)(y_0)^2 + 2\sinh(s)\cosh(s) y_0 y_3 + \sinh^2(s)(y_3)^2 \\
&\quad - [\sinh^2(s)(y_0)^2 + 2\sinh(s)\cosh(s) y_0 y_3 + \cosh^2(s)(y_3)^2] \\
&\quad - (y_1)^2 - (y_2)^2 \\
&= (y_0)^2 - (y_1)^2 - (y_2)^2 - (y_3)^2 = y \cdot y \ ,
\end{aligned}$$

for every $y \in \mathbb{R}^4$. Hence, it follows that

$$(\Lambda_s \hat{V} \Lambda)_{00} = 1\ ,\ (\Lambda_s \hat{V} \Lambda)_{10} = (\Lambda_s \hat{V} \Lambda)_{20} = (\Lambda_s \hat{V} \Lambda)_{30} = 0\ .$$

Since also

$$\eta\, (\Lambda_s \hat{V} \Lambda)^t\, \eta \begin{pmatrix} 1 \\ 0 \\ 0 \\ 0 \end{pmatrix} = (\Lambda_s \hat{V} \Lambda)^{-1} \begin{pmatrix} 1 \\ 0 \\ 0 \\ 0 \end{pmatrix} = \begin{pmatrix} 1 \\ 0 \\ 0 \\ 0 \end{pmatrix}\ ,$$

it follows that

$$(\Lambda_s \hat{V} \Lambda)^t \begin{pmatrix} 1 \\ 0 \\ 0 \\ 0 \end{pmatrix} = \begin{pmatrix} 1 \\ 0 \\ 0 \\ 0 \end{pmatrix}$$

and hence that

$$(\Lambda_s \hat{V} \Lambda)_{01} = 0\ ,\ (\Lambda_s \hat{V} \Lambda)_{02} = 0\ ,\ (\Lambda_s \hat{V} \Lambda)_{03} = 0\ .$$

Therefore,

$$\Lambda_s \hat{V} \Lambda = \begin{pmatrix} 1 & 0 & 0 & 0 \\ 0 & (\Lambda_s \hat{V} \Lambda)_{11} & (\Lambda_s \hat{V} \Lambda)_{12} & (\Lambda_s \hat{V} \Lambda)_{13} \\ 0 & (\Lambda_s \hat{V} \Lambda)_{21} & (\Lambda_s \hat{V} \Lambda)_{22} & (\Lambda_s \hat{V} \Lambda)_{23} \\ 0 & (\Lambda_s \hat{V} \Lambda)_{31} & (\Lambda_s \hat{V} \Lambda)_{32} & (\Lambda_s \hat{V} \Lambda)_{33} \end{pmatrix} \in \mathcal{L}_+^{\uparrow}\ .$$

As a consequence,

$$\begin{pmatrix} (\Lambda_s \hat{V} \Lambda)_{11} & (\Lambda_s \hat{V} \Lambda)_{12} & (\Lambda_s \hat{V} \Lambda)_{13} \\ (\Lambda_s \hat{V} \Lambda)_{21} & (\Lambda_s \hat{V} \Lambda)_{22} & (\Lambda_s \hat{V} \Lambda)_{23} \\ (\Lambda_s \hat{V} \Lambda)_{31} & (\Lambda_s \hat{V} \Lambda)_{32} & (\Lambda_s \hat{V} \Lambda)_{33} \end{pmatrix} \in SO(3)$$

and

$$\Lambda = \hat{V}^{-1} (\Lambda_s)^{-1} \begin{pmatrix} 1 & 0 & 0 & 0 \\ 0 & (\Lambda_s \hat{V} \Lambda)_{11} & (\Lambda_s \hat{V} \Lambda)_{12} & (\Lambda_s \hat{V} \Lambda)_{13} \\ 0 & (\Lambda_s \hat{V} \Lambda)_{21} & (\Lambda_s \hat{V} \Lambda)_{22} & (\Lambda_s \hat{V} \Lambda)_{23} \\ 0 & (\Lambda_s \hat{V} \Lambda)_{31} & (\Lambda_s \hat{V} \Lambda)_{32} & (\Lambda_s \hat{V} \Lambda)_{33} \end{pmatrix}\ .$$

□

The following corollary provides the transitivity of the action of $\mathcal{L}_+^{\uparrow}$ on the elements of

$$\{x \in \mathbb{R}^4 : x \cdot x = a^2, x_0 > 0\}\ ,$$

where $a > 0$.

Corollary 3.5 *Let $a > 0$. For every $x \in \mathbb{R}^4$ such that $x \cdot x = a^2$ as well as such that $x_0 > 0$, there is $\Lambda \in \mathcal{L}_+^{\uparrow}$, such that $\Lambda x = {}^t(a, 0, 0, 0)$.*

Proof According to the proof of Theorem 3.4, for every $x \in \mathbb{R}^4$ satisfying $x \cdot x = 1$ and $x_0 \geqslant 0$, there is $\Lambda \in \mathcal{L}_+^\uparrow$, such that $\Lambda x = {}^t(1, 0, 0, 0)$. If $a > 0$ and $y \in \mathbb{R}^4$ is such that $y \cdot y = a^2$, as well as such that $y_0 > 0$, then $x := a^{-1} y \in \mathbb{R}^4$, $x \cdot x = a^{-2}(y \cdot y) = 1$ and $x_0 = a^{-1} y_0 > 0$. Hence, there is $\Lambda \in \mathcal{L}_+^\uparrow$, such that $a^{-1}\Lambda y = \Lambda x = {}^t(1, 0, 0, 0)$, and it follows that $\Lambda y = {}^t(a, 0, 0, 0)$. □

The following exercises provide continuous one-parameter subgroups of the restricted Lorentz group, corresponding to rotations and Lorentz boosts.

Exercise 3.4 Show that by

$$M_1(s) := \begin{pmatrix} 1 & 0 & 0 & 0 \\ 0 & 1 & 0 & 0 \\ 0 & 0 & \cos(s) & \sin(s) \\ 0 & 0 & -\sin(s) & \cos(s) \end{pmatrix}, \quad M_2(s) := \begin{pmatrix} 1 & 0 & 0 & 0 \\ 0 & \cos(s) & 0 & -\sin(s) \\ 0 & 0 & 1 & 0 \\ 0 & \sin(s) & 0 & \cos(s) \end{pmatrix},$$
$$M_3(s) := \begin{pmatrix} 1 & 0 & 0 & 0 \\ 0 & \cos(s) & \sin(s) & 0 \\ 0 & -\sin(s) & \cos(s) & 0 \\ 0 & 0 & 0 & 1 \end{pmatrix}, \tag{3.8}$$

for every $s \in \mathbb{R}$, there are defined one-parameter groups $M_j : \mathbb{R} \to \mathcal{L}_+^\uparrow$, $j \in \{1, 2, 3\}$.

Exercise 3.5 Show that by

$$M_{01}(s) := \begin{pmatrix} \cosh(s) & \sinh(s) & 0 & 0 \\ \sinh(s) & \cosh(s) & 0 & 0 \\ 0 & 0 & 1 & 0 \\ 0 & 0 & 0 & 1 \end{pmatrix},$$
$$M_{02}(s) := \begin{pmatrix} \cosh(s) & 0 & \sinh(s) & 0 \\ 0 & 1 & 0 & 0 \\ \sinh(s) & 0 & \cosh(s) & 0 \\ 0 & 0 & 0 & 1 \end{pmatrix},$$
$$M_{03}(s) := \begin{pmatrix} \cosh(s) & 0 & 0 & \sinh(s) \\ 0 & 1 & 0 & 0 \\ 0 & 0 & 1 & 0 \\ \sinh(s) & 0 & 0 & \cosh(s) \end{pmatrix}, \tag{3.9}$$

for every $s \in \mathbb{R}$, there are defined one-parameter groups $M_{0j} : \mathbb{R} \to \mathcal{L}_+^\uparrow$, $j \in \{1, 2, 3\}$.

The following exercise provides a characterization of the elements of $T_E\mathcal{L}$.

Exercise 3.6 Show that

$$T_E\mathcal{L} = \left\{ \begin{pmatrix} 0 & a_1 & a_2 & a_3 \\ a_1 & 0 & -b_3 & b_2 \\ a_2 & b_3 & 0 & -b_1 \\ a_3 & -b_2 & b_1 & 0 \end{pmatrix} : a, b \in \mathbb{R}^3 \right\}$$

and hence that

$$\dim(T_E\mathcal{L}) = 6 \ .$$

The following exercise uses another characterization of $T_E\mathcal{L}$ to show that the matrix exponential function maps $T_E\mathcal{L}$ into $\mathcal{L}_+^\uparrow$.

Exercise 3.7 Show that

$$T_E\mathcal{L} = \{M \in \mathrm{M}(4, \mathbb{R}) : M^t = -\eta\, M \eta\} \ .$$

Use this fact to show that

$$\boxed{\exp(M) \in \mathcal{L}_+^\uparrow \ ,}$$

for every $M \in T_E\mathcal{L}$.

Exercise 3.8 Show that $T_E\mathcal{L}$ and

$$\{M \in \mathrm{M}(4, \mathbb{R}) : M^t = \eta \cdot M \cdot \eta\}$$

are invariant subspaces of the adjoint representation $\mathrm{Ad} : \mathcal{L} \to \mathrm{L}(\mathrm{M}(4, \mathbb{R}), \mathrm{M}(4, \mathbb{R}))$ of $\mathcal{L}$, where the latter is defined by

$$[\mathrm{Ad}(\Lambda)](M) := \Lambda \cdot M \cdot \Lambda^{-1} \ ,$$

for every $M \in \mathrm{M}(4, \mathbb{R})$ and $\Lambda \in \mathcal{L}$.

The following exercise gives an application of the Cayley-Hamilton theorem to the elements of $T_E\mathcal{L}$.

Exercise 3.9 Using the Cayley-Hamilton theorem, show that for every $a, b \in \mathbb{R}^3$,

$$A := \begin{pmatrix} 0 & a_1 & a_2 & a_3 \\ a_1 & 0 & -b_3 & b_2 \\ a_2 & b_3 & 0 & -b_1 \\ a_3 & -b_2 & b_1 & 0 \end{pmatrix} ,$$

Table 3.1 Properties of Λ_0, Λ_P, Λ_T and Λ_{PT}

$\det(\Lambda_0) = 1$	$(\Lambda_0)_{00} = 1 \geqslant 0$
$\det(\Lambda_{PT}) = 1$	$(\Lambda_{PT})_{00} = -1 \leqslant 0$
$\det(\Lambda_P) = -1$	$(\Lambda_P)_{00} = 1 \geqslant 0$
$\det(\Lambda_T) = -1$	$(\Lambda_T)_{00} = -1 \leqslant 0$

Table 3.2 Multiplication properties of Λ_0, Λ_P, Λ_T and Λ_{PT}, where the first factor is taken from the first column on the left and the second factor from the first row. Since this table is symmetric, the matrices Λ_0, Λ_P, Λ_T and Λ_{PT} commute pairwise

$\cdot$	Λ_0	Λ_P	Λ_T	Λ_{PT}
Λ_0	Λ_0	Λ_P	Λ_T	Λ_{PT}
Λ_P	Λ_P	Λ_0	Λ_{PT}	Λ_T
Λ_T	Λ_T	Λ_{PT}	Λ_0	Λ_P
Λ_{PT}	Λ_{PT}	Λ_T	Λ_P	Λ_0

we have that

$$A^4 + (\,|b|^2 - |a|^2)\, A^2 - (a \cdot b)^2\, E = 0\ .$$

Hence, it follows that $\exp(A)$ can be expressed as a polynomial of degree $\leqslant 3$ in A. The same is true for any function $f(A)$, where $f : U_R(0) \to \mathbb{C}$ is a holomorphic function and $R > \|A\|_{\text{op}}$.

In the following, we are going to derive a decomposition of $\mathcal{L}$ into disjoint sets. For this purpose, we define Λ_T an Λ_P, where T stands for "time reversal" and P for "parity," by

$$\Lambda_P := \begin{pmatrix} 1 & 0 & 0 & 0 \\ 0 & -1 & 0 & 0 \\ 0 & 0 & -1 & 0 \\ 0 & 0 & 0 & -1 \end{pmatrix}, \quad \Lambda_T := \begin{pmatrix} -1 & 0 & 0 & 0 \\ 0 & 1 & 0 & 0 \\ 0 & 0 & 1 & 0 \\ 0 & 0 & 0 & 1 \end{pmatrix}$$

as well as $\Lambda_0 := E$ and

$$\Lambda_{PT} := \Lambda_P \cdot \Lambda_T = \Lambda_T \cdot \Lambda_P = -E\ .$$

We note that $\Lambda_T = -\Lambda_P$,

$$(\Lambda_P x) \cdot (\Lambda_P x) = {}^t(x_0, -x_1, -x_2, -x_3) \cdot {}^t(x_0, -x_1, -x_2, -x_3) = x \cdot x\ ,$$
$$(\Lambda_T x) \cdot (\Lambda_T x) = {}^t(-\,x_0, x_1, x_2, x_3) \cdot {}^t(-\,x_0, x_1, x_2, x_3) = x \cdot x\ ,$$

for all $x \in \mathbb{R}^4$ and hence that $\Lambda_0, \Lambda_P, \Lambda_T, \Lambda_{PT} \in \mathcal{L}$, (Tables 3.1 and 3.2).

We note that $\mathcal{L}$ is the disjoint union of $\mathcal{L}_+^\uparrow$, $\mathcal{L}_+^\downarrow$, $\mathcal{L}_-^\uparrow$ and $\mathcal{L}_-^\downarrow$,

$$\mathcal{L} = \mathcal{L}_+^\uparrow \cup \mathcal{L}_+^\downarrow \cup \mathcal{L}_-^\uparrow \cup \mathcal{L}_-^\downarrow ,$$

where

$$\mathcal{L}_+^\downarrow := \{\Lambda \in \mathcal{L} : \det(\Lambda) = 1, \Lambda_{00} \leqslant 0\} ,$$
$$\mathcal{L}_-^\uparrow := \{\Lambda \in \mathcal{L} : \det(\Lambda) = -1, \Lambda_{00} \geqslant 0\} ,$$
$$\mathcal{L}_-^\downarrow := \{\Lambda \in \mathcal{L} : \det(\Lambda) = -1, \Lambda_{00} \leqslant 0\} .$$

This union is disjoint, as a consequence of (3.3) and since it follows from (3.5), that there is no Lorentz transformation Λ such that $\Lambda_{00} = 0$. In particular,

$$\Lambda_0 = E \in \mathcal{L}_+^\uparrow , \quad \Lambda_{PT} \in \mathcal{L}_+^\downarrow , \quad \Lambda_P \in \mathcal{L}_-^\uparrow , \quad \Lambda_T \in \mathcal{L}_-^\downarrow .$$

Further, we arrive at the following unique representation of the elements of $\mathcal{L}_+^\uparrow$, $\mathcal{L}_+^\downarrow$, $\mathcal{L}_-^\uparrow$ and $\mathcal{L}_-^\downarrow$, respectively.

Exercise 3.10 Show that

For every $\Lambda \in \mathcal{L}_+^\uparrow$, there is a uniquely determined $\bar{\Lambda} \in \mathcal{L}_+^\uparrow$, such that $\Lambda = \Lambda_0 \bar{\Lambda}$;
For every $\Lambda \in \mathcal{L}_+^\downarrow$, there is a uniquely determined $\bar{\Lambda} \in \mathcal{L}_+^\uparrow$, such that $\Lambda = \Lambda_{PT} \bar{\Lambda}$;
For every $\Lambda \in \mathcal{L}_-^\uparrow$, there is a uniquely determined $\bar{\Lambda} \in \mathcal{L}_+^\uparrow$, such that $\Lambda = \Lambda_P \bar{\Lambda}$;
For every $\Lambda \in \mathcal{L}_-^\downarrow$, there is a uniquely determined $\bar{\Lambda} \in \mathcal{L}_+^\uparrow$, such that $\Lambda = \Lambda_T \bar{\Lambda}$

As inverse images of the closed subset $\mathcal{L}_+^\uparrow$ of $M(4, \mathbb{R})$ under continuous maps and as images of the path connected set $\mathcal{L}_+^\uparrow$ under continuous maps, we arrive at the following.

Equipped with the induced topology, the subsets $\mathcal{L}_+^\downarrow$, $\mathcal{L}_-^\uparrow$ and $\mathcal{L}_-^\downarrow$ of $M(4, \mathbb{R})$ are unbounded, closed and path connected,

where the unboundedness of these subsets of $M(4, \mathbb{R})$ follows from the facts that

$$(\Lambda_P M_{01}(s))_{00} = \cosh(s) \ , \ \ (\Lambda_T M_{01}(s))_{00} = (\Lambda_{PT} M_{01}(s))_{00} = -\cosh(s) \ ,$$

for every $s \in \mathbb{R}$.

3.2 A Family of Unitary Representations of the Restricted Lorentz Group

The construction of representations of $\mathcal{L}_+^\uparrow$ follows that of the spinor representation $U_{1/2}$ of $SO(3)$. The latter is based on a natural representation U of O(3) from quantum mechanics. The additional component in the construction of $U_{1/2}$ is a finite dimensional unitary representation of $SO(3)$, governing the transformation of the spinor components. The representation $U_{1/2}$ is "induced" by U [40, 41]. The construction of U uses that the elements of O(3) induce orthogonal transformations of 3-dimensional Euclidean space, the latter a suitable data surface for the description of the evolution of non-relativistic quantum mechanical systems, and utilizes the invariance of the Lebesgue measure under such transformations.

For the case of $\mathcal{L}_+^\uparrow$, the construction of an equivalent U_a of U, provided in this section, where the dimensionless parameter $a \geqslant 0$ may be interpreted as a mass parameter, uses 3-dimensional hyperboloids H_{a+}, where $a \geqslant 0$, defined later on. In applications to relativistic quantum theory, a is a dimensionless multiple of the mass of the field. These hyperboloids are invariant under restricted Lorentz transformations. Further, for the case $a > 0$, H_{a+} is space-like, hence suitable for prescribing data for fields propagating in Minkowski space, and the Lorentz metric of Minkowski space induces a Riemannian metric of constant negative curvature $-1/a^2$ on H_{a+} (Fig. 3.1). In addition, restricted Lorentz transformations induce isometries of the thus obtained Riemannian space. The latter is relevant since the natural volume form on the Riemannian space is invariant under isometries, thus providing a suitable replacement for the Lebesgue measure that is used for the construction of the representation U of $SO(3)$. The chosen replacement approaches a limit for $a \to 0$. These geometric details are provided in the Appendix, see Sect. A.2. The subsequent construction covers at the same time the case $a = 0$, a case where the described geometric process leads to a degenerate "metric" on the forward light "cone" $H_{0+} \setminus \{^t(0, 0, 0, 0)\}$ with apex $^t(0, 0, 0, 0)$, with an identically vanishing "volume form" (Fig. 3.2). Indeed, it turns out that all cases can be treated on the same footing, without use of geometrical arguments. Still, to increase transparency, in the following we relate to the geometric approach, where appropriate.

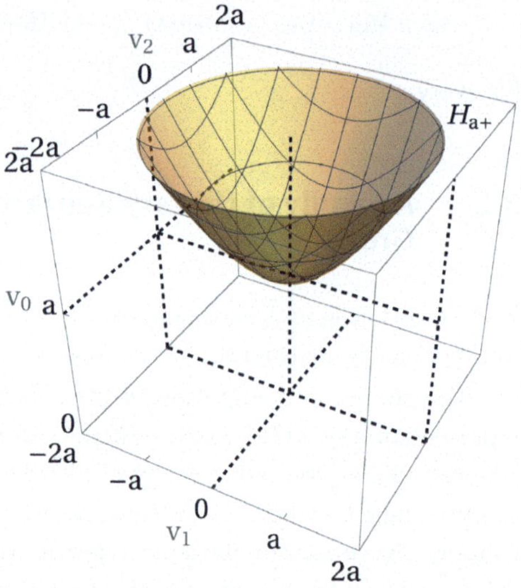

Fig. 3.1 Depiction of the analogue H_{a+} for $a > 0$ and dashed auxiliary lines in 3-dimensional Minkowski space

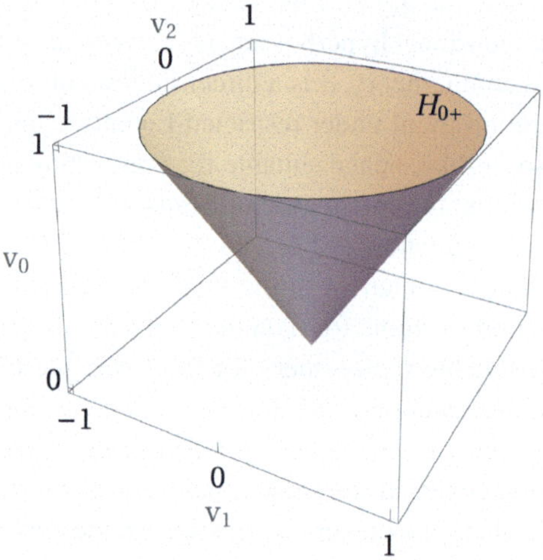

Fig. 3.2 Depiction of the analog H_{0+} in 3-dimensional Minkowski space

For $a \geqslant 0$,[1] we define the closed subset $H_{a+} \subset \mathbb{R}^4$, where we consider $\mathbb{R}^4$ equipped with the usual Euclidean topology, by

[1] In the relativistic quantum mechanical description of the free motion of a scalar particle of mass $m \in [0, \infty)$ in Minkowski space, see (3.24), the dimensionless constant a ($\in [0, \infty)$) is given by $a = mc/(\hbar\kappa) = 1/(\kappa\lambda_C)$, where λ_C denotes the reduced Compton wavelength of the particle and where $\kappa > 0$ is a scale factor with dimension 1/length, determining the scale in a position representation.

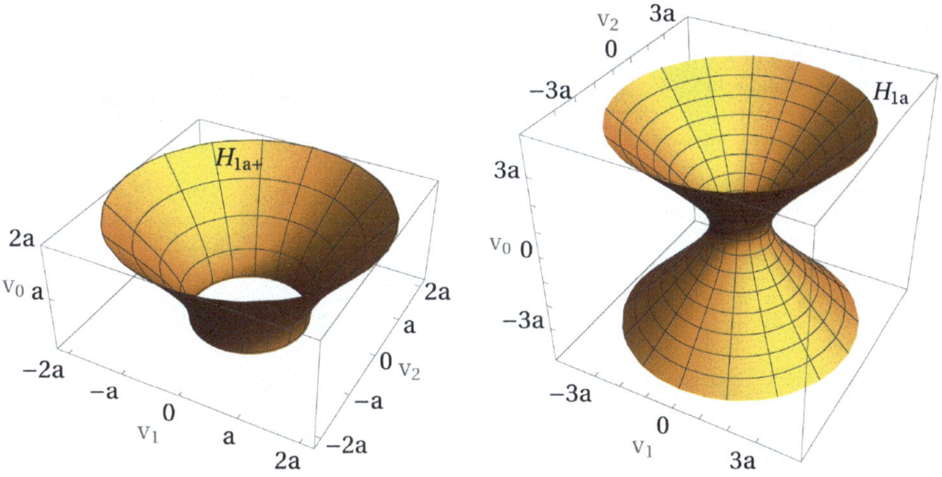

Fig. 3.3 Depiction of the analogues H_{1a+} and H_{1a} for $a > 0$ in 3-dimensional Minkowski space

$$\begin{aligned} H_{a+} &:= \{v \in \mathbb{R}^4 : v \cdot v = a^2, v_0 \geqslant 0\} \\ &= \{^t((a^2 + |v|^2)^{1/2}, v_1, v_2, v_3) \in \mathbb{R}^4 : v \in \mathbb{R}^3\} . \end{aligned}$$

We note that H_{0+} is given by a forward light "cone" with apex $^t(0, 0, 0, 0)$.

For use in the following exercises, we use a definition of tangent spaces at H_{a+} that employs the embedding of H_{a+} into $\mathbb{R}^4$, namely, we define the tangent space at a point $v \in H_{a+}$ to consist of all derivatives $\gamma'(0)$, of differentiable paths $\gamma : I \to \mathbb{R}^4$, where I is an open interval of $\mathbb{R}$ around 0, such that $Ran(\gamma) \subset H_{a+}$. For the definition of a differentiable (C^∞) manifold structure for H_{a+}, leading to an intrinsic definition of tangent spaces at H_{a+} in a differential geometric sense, see the Sect. A.2 in the Appendix.

Exercise 3.11 Show that the surface H_{a+} is space-like for every $a > 0$, i.e., every non-zero tangent vector $w \in \mathbb{R}^4$ of H_{a+} satisfies $w \cdot w < 0$.

Exercise 3.12 For $a > 0$, we define the closed subset $H_{1a+} \subset \mathbb{R}^4$, see (Fig. 3.3), where we consider $\mathbb{R}^4$ equipped with the usual Euclidean topology, by

$$\begin{aligned} H_{1a+} &:= \{v \in \mathbb{R}^4 : v \cdot v = -a^2\} \\ &= \{^t((|v|^2 - a^2)^{1/2}, v_1, v_2, v_3) \in \mathbb{R}^4 : v \in \mathbb{R}^3 \setminus U_a(0)\} . \end{aligned}$$

(a) Show that

$$H_{1a} := \{v \in \mathbb{R}^4 : v \cdot v = -a^2\} ,$$

see Fig. 3.3, is invariant under every Lorentz transformation, i.e.,

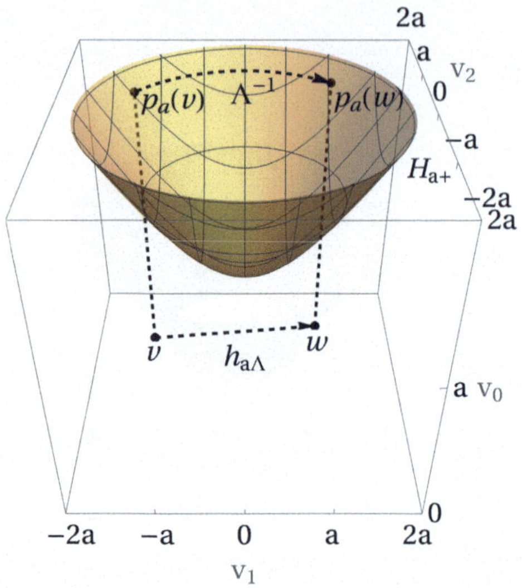

Fig. 3.4 Depiction of the action of the analog of h_Λ in 3-dimensional Minkowski space, where $w = h_\Lambda(\vec{v}\,)$

$$\Lambda H_{1a} = H_{1a} \; ,$$

for every $\Lambda \in \mathcal{L}$, but that there is $\Lambda \in \mathcal{L}_+^\uparrow$ such that

$$\Lambda H_{1a+} \not\subset H_{1a+} \; .$$

(b) Show that there are non-zero space-like and time-like tangent vectors of H_{1a+}.

For every $v \in H_{a+}$, it follows that

$$v_0^2 = v \cdot v + v_1^2 + v_2^2 + v_3^2 = a^2 + v_1^2 + v_2^2 + v_3^2$$

and as well as that

$$v_0 = (a^2 + v_1^2 + v_2^2 + v_3^2)^{1/2} \; .$$

Further, for every $\Lambda \in \mathcal{L}_+^\uparrow$ and $v \in H_{a+}$, we have that

$$a^2 = v \cdot v = (\Lambda v) \cdot (\Lambda v)$$

and that

$$
\begin{aligned}
(\Lambda v)_{00} &= \Lambda_{00} v_0 + \Lambda_{01} v_1 + \Lambda_{02} v_2 + \Lambda_{03} v_3 \\
&\geqslant \left[1 + \sum_{\alpha=1}^{3} (\Lambda_{0\alpha})^2\right]^{1/2} (a^2 + |\vec{v}|^2)^{1/2} - |\Lambda_{01} v_1 + \Lambda_{02} v_2 + \Lambda_{03} v_3| \\
&\geqslant \left[1 + \sum_{\alpha=1}^{3} (\Lambda_{0\alpha})^2\right]^{1/2} (a^2 + |\vec{v}|^2)^{1/2} - \left[\sum_{\alpha=1}^{3} (\Lambda_{0\alpha})^2\right]^{1/2} |\vec{v}| \\
&= \left[1 + \sum_{\alpha=1}^{3} (\Lambda_{0\alpha})^2\right]^{1/2} [\, (a^2 + |\vec{v}|^2)^{1/2} - |v|\,] \\
&\quad + \left\{\left[1 + \sum_{\alpha=1}^{3} (\Lambda_{0\alpha})^2\right]^{1/2} - \left[\sum_{\alpha=1}^{3} (\Lambda_{0\alpha})^2\right]^{1/2}\right\} |v| \\
&= \left[1 + \sum_{\alpha=1}^{3} (\Lambda_{0\alpha})^2\right]^{1/2} \frac{a^2}{(a^2 + |\vec{v}|^2)^{1/2} + |v|} \\
&\quad + \frac{1}{\left[1 + \sum_{\alpha=1}^{3} (\Lambda_{0\alpha})^2\right]^{1/2} + \left[\sum_{\alpha=1}^{3} (\Lambda_{0\alpha})^2\right]^{1/2}} |v| \\
&\geqslant \frac{1}{2\left[1 + \sum_{\alpha=1}^{3} (\Lambda_{0\alpha})^2\right]^{1/2}} |v| \geqslant 0 \ , \qquad (3.10)
\end{aligned}
$$

where $\vec{v} := {}^t(v_1, v_2, v_3) \in \mathbb{R}^3$ and (3.7) was used. Hence, it follows that $\Lambda H_{a+} \subset H_{a+}$, for every $\Lambda \in \mathcal{L}_+^{\uparrow}$, and therefore also that

$$\Lambda H_{a+} = H_{a+} \ ,$$

for every $\mathcal{L}_+^{\uparrow}$. Further, we define the parametrization $p_a : \mathbb{R}^3 \to H_{a+}$, of H_{a+} by

$$p_a(v) := {}^t((a^2 + |v|^2)^{1/2}, v_1, v_2, v_3) \ , \qquad (3.11)$$

for every $v \in \mathbb{R}^3$. We note that p_a is bijective with inverse $p_a^{-1} : H_{a+} \to \mathbb{R}^3$ given by[2]

$$p_a^{-1}(v) := {}^t(v_1, v_2, v_3) \ ,$$

for every $v \in H_{a+}$. Further, for every $\Lambda \in \mathcal{L}_+^{\uparrow}$, we define[3] $h_{a\Lambda} : \mathbb{R}^3 \to \mathbb{R}^3$ by

$$h_{a\Lambda}(v) := p_a^{-1}(\Lambda^{-1} \cdot p_a(v)) \ , \qquad (3.12)$$

---

[2] For $a > 0$, in the definition of a C^∞-manifold structure for H_{a+}, p_a^{-1} provides a global chart. Such a structure can be given for $V_+ \setminus \{{}^t(0, 0, 0, 0)\}$, by restricting p_a^{-1} in domain to $V_+ \setminus \{{}^t(0, 0, 0, 0)\}$ and in range to $\mathbb{R}^3 \setminus \{0\}$, see Sect. A.2 in the Appendix.

[3] $h_{a\Lambda}$ is the representation of the C^∞ map f_Λ in coordinate space, see Sect. A.2 in the Appendix.

for every $v \in \mathbb{R}^3$, (Fig. 3.4). We note that $h_{a\Lambda}$ is bijective with inverse

$$h_{a\Lambda}^{-1} = h_{a\Lambda^{-1}}$$

and that $h_{a\Lambda}|_{\mathbb{R}^3\backslash\{0\}} \in C^\infty(\mathbb{R}^3\backslash\{0\}, \mathbb{R}^3)$.

The following lemma is crucial for the proof that the representation U_a given below, see Theorem 3.7, is unitary.

Lemma 3.6 *For every $a \geqslant 0$, we have that*

$$\det(h'_{a\Lambda}(v)) = \frac{[\,|h_{a\Lambda}(v)|^2 + a^2\,]^{1/2}}{(\,|v|^2 + a^2)^{1/2}} = ,$$

for every $v \in \mathbb{R}^3 \setminus \{0\}$.

Proof In the following, in order to facilitate the application of the multiplicative property of determinants, we extend the map p_a. For this purpose, we define the open subset $U_+ \subset \mathbb{R}^4$ by

$$U_+ := \{v \in \mathbb{R}^4 : v \cdot v > 0, v_0 > 0\} .$$

We note that $U_+ \supset H_{a+}$, for every $a > 0$. Further, we define the auxiliary function $\mathfrak{p} : (0, \infty) \times \mathbb{R}^3 \to U_+$ by

$$\mathfrak{p}(w) := {}^t((w_0^2 + w_1^2 + w_2^2 + w_3^2)^{1/2}, w_1, w_2, w_3) ,$$

for every $w \in (0, \infty) \times \mathbb{R}^3$. Then, for every $a > 0$,

$$\mathfrak{p} \circ \iota_a = p_a ,$$

where the inclusion $\iota_a : \mathbb{R}^3 \hookrightarrow (0, \infty) \times \mathbb{R}^3$ is defined by $\iota_a(\vec{w}\,) := (a, w_1, w_2, w_3)$, for every $\vec{w} \in \mathbb{R}^3$. Also, for every $\Lambda \in \mathcal{L}_+^\uparrow$,

$$(\Lambda \cdot \mathfrak{p}(w)) \cdot (\Lambda \cdot \mathfrak{p}(w)) = \mathfrak{p}(w) \cdot \mathfrak{p}(w) > 0$$

and

$$\begin{aligned}
(\Lambda \cdot \mathfrak{p}(w))_{00} &= \Lambda_{00}\,[\mathfrak{p}(w)]_0 + \Lambda_{01}\,[\mathfrak{p}(w)]_1 + \Lambda_{02}\,[\mathfrak{p}(w)]_2 + \Lambda_{03}\,[\mathfrak{p}(w)]_3 \\
&\geqslant \left[1 + \sum_{\alpha=1}^{3}(\Lambda_{0\alpha})^2\right]^{1/2} (w_0^2 + |\vec{w}|^2)^{1/2} - |\Lambda_{01}w_1 + \Lambda_{02}w_2 + \Lambda_{03}w_3| \\
&\geqslant \left[1 + \sum_{\alpha=1}^{3}(\Lambda_{0\alpha})^2\right]^{1/2} (w_0^2 + |\vec{w}|^2)^{1/2} - \left[\sum_{\alpha=1}^{3}(\Lambda_{0\alpha})^2\right]^{1/2} |\vec{w}| > 0 ,
\end{aligned}$$

where $\vec{w} := {}^t(w_1, w_2, w_3) \in \mathbb{R}^3$ and (3.7) was used. Hence

$$\Lambda \cdot \mathfrak{p}((0,\infty) \times \mathbb{R}^3) \subset U_+ \ .$$

Furthermore, since

$$\begin{aligned}\mathfrak{p}((v \cdot v)^{1/2}, v_1, v_2, v_3) &= {}^t((v \cdot v + v_1^2 + v_2^2 + v_3^2)^{1/2}, v_1, v_2, v_3) \\ &= {}^t(v_0, v_1, v_2, v_3) = v \ ,\end{aligned}$$

for every $v \in U_+$, and

$${}^t((\mathfrak{p}(w) \cdot \mathfrak{p}(w))^{1/2}, [\mathfrak{p}(w)]_1, [\mathfrak{p}(w)]_2, [\mathfrak{p}(w)]_3) = {}^t((w_0^2)^{1/2}, w_1, w_2, w_3) = w \ ,$$

for every $w \in (0,\infty) \times \mathbb{R}^3$, it follows that $\mathfrak{p}$ is a C^∞-diffeomorphism with inverse $\mathfrak{p}^{-1} : U_+ \to (0,\infty) \times \mathbb{R}^3$, given by

$$\mathfrak{p}^{-1}(v) = {}^t((v \cdot v)^{1/2}, v_1, v_2, v_3) \ ,$$

for every $v \in U_+$. We note that

$$\mathfrak{p}'(w) = \begin{pmatrix} \frac{w_0}{[\mathfrak{p}(w)]_0} & \frac{w_1}{[\mathfrak{p}(w)]_0} & \frac{w_2}{[\mathfrak{p}(w)]_0} & \frac{w_3}{[\mathfrak{p}(w)]_0} \\ 0 & 1 & 0 & 0 \\ 0 & 0 & 1 & 0 \\ 0 & 0 & 0 & 1 \end{pmatrix} , \ \det(\mathfrak{p}'(w)) = \frac{w_0}{[\mathfrak{p}(w)]_0} \ ,$$

$$(\mathfrak{p}^{-1})'(v) = \begin{pmatrix} \frac{v_0}{(v\cdot v)^{1/2}} & -\frac{v_1}{(v\cdot v)^{1/2}} & -\frac{v_2}{(v\cdot v)^{1/2}} & -\frac{v_3}{(v\cdot v)^{1/2}} \\ 0 & 1 & 0 & 0 \\ 0 & 0 & 1 & 0 \\ 0 & 0 & 0 & 1 \end{pmatrix} ,$$

$$\det((\mathfrak{p}^{-1})'(v)) = \frac{v_0}{(v \cdot v)^{1/2}} \ ,$$

for every $w \in (0,\infty) \times \mathbb{R}^3$ and $v \in U_+$. In addition, for $\Lambda \in \mathcal{L}_+^\uparrow$, we define the C^∞-map $\mathfrak{h}_\Lambda : (0,\infty) \times \mathbb{R}^3 \to (0,\infty) \times \mathbb{R}^3$ by

$$\mathfrak{h}_\Lambda(w) := \mathfrak{p}^{-1}(\Lambda^{-1} \cdot \mathfrak{p}(w)) \ ,$$

for every $w \in (0,\infty) \times \mathbb{R}^3$. We note that

$$\begin{aligned}\mathfrak{h}_\Lambda(w) &= (\mathfrak{h}_\Lambda \circ \iota_{w_0})(\vec{w}) = \mathfrak{p}^{-1}(\Lambda^{-1} \cdot p_{w_0}(\vec{w})) \\ &= {}^t([(\Lambda^{-1} \cdot \mathfrak{p}(w)) \cdot (\Lambda^{-1} \cdot \mathfrak{p}(w))]^{1/2}, [h_{w_0\Lambda}(\vec{w})]_1, [h_{w_0\Lambda}(\vec{w})]_2, [h_{w_0\Lambda}(\vec{w})]_3) \\ &= {}^t([\mathfrak{p}(w) \cdot \mathfrak{p}(w)]^{1/2}, [h_{w_0\Lambda}(\vec{w})]_1, [h_{w_0\Lambda}(\vec{w})]_2, [h_{w_0\Lambda}(\vec{w})]_3) \\ &= {}^t(w_0, [h_{w_0\Lambda}(\vec{w})]_1, [h_{w_0\Lambda}(\vec{w})]_2, [h_{w_0\Lambda}(\vec{w})]_3) \ ,\end{aligned}$$

for every $w \in (0, \infty) \times \mathbb{R}^3$, where $\vec{w} := {}^t(w_1, w_2, w_3)$. Hence,

$$\mathfrak{h}'_\Lambda(w) = \begin{pmatrix} 1 & 0 & 0 & 0 \\ \frac{\partial \mathfrak{h}_{\Lambda 1}}{\partial w_0}(w) & \frac{\partial h_{w_0 \Lambda 1}}{\partial v_1}(\vec{w}) & \frac{\partial h_{w_0 \Lambda 1}}{\partial v_2}(\vec{w}) & \frac{\partial h_{w_0 \Lambda 1}}{\partial v_3}(\vec{w}) \\ \frac{\partial \mathfrak{h}_{\Lambda 2}}{\partial w_0}(\vec{w}) & \frac{\partial h_{w_0 \Lambda 2}}{\partial v_1}(\vec{w}) & \frac{\partial h_{w_0 \Lambda 2}}{\partial v_2}(\vec{w}) & \frac{\partial h_{w_0 \Lambda 2}}{\partial v_3}(\vec{w}) \\ \frac{\partial \mathfrak{h}_{\Lambda 3}}{\partial w_0}(\vec{w}) & \frac{\partial h_{w_0 \Lambda 3}}{\partial v_1}(\vec{w}) & \frac{\partial h_{w_0 \Lambda 3}}{\partial v_2}(\vec{w}) & \frac{\partial h_{w_0 \Lambda 3}}{\partial v_3}(\vec{w}) \end{pmatrix}, \tag{3.13}$$

and

$$\det(\mathfrak{h}'_\Lambda(w)) = \det(h'_{w_0 \Lambda}(\vec{w})) .$$

Since,

$$\begin{aligned} \det(\mathfrak{h}'_\Lambda(w)) &= \det((\mathfrak{p}^{-1})'(\Lambda^{-1} \cdot \mathfrak{p}(w))) \cdot \det(\Lambda^{-1}) \cdot \det(\mathfrak{p}'(w)) \\ &= \frac{(\Lambda^{-1} \cdot \mathfrak{p}(w))_0}{((\Lambda^{-1} \cdot \mathfrak{p}(w)) \cdot (\Lambda^{-1} \cdot \mathfrak{p}(w)))^{1/2}} \cdot \frac{w_0}{[\mathfrak{p}(w)]_0} \\ &= \frac{(\Lambda^{-1} \cdot \mathfrak{p}(w))_0}{(\mathfrak{p}(w) \cdot \mathfrak{p}(w))^{1/2}} \cdot \frac{w_0}{[\mathfrak{p}(w)]_0} = \frac{(\Lambda^{-1} \cdot \mathfrak{p}(w))_0}{[\mathfrak{p}(w)]_0} , \end{aligned}$$

for every $w \in (0, \infty) \times \mathbb{R}^3$, finally, we have for every $a > 0$ that

$$\begin{aligned} &\det(h'_{a\Lambda}(v)) \\ &= \frac{(\Lambda^{-1})_{00}\,(a^2 + |v|^2)^{1/2} + (\Lambda^{-1})_{01}\, v_1 + (\Lambda^{-1})_{02}\, v_2 + (\Lambda^{-1})_{03}\, v_3}{(a^2 + |v|^2)^{1/2}} , \end{aligned}$$

for every $v \in \mathbb{R}^3$. In the following, we consider the case that $a = 0$. For every $w \in (0, \infty) \times \mathbb{R}^3$, it follows that

$$\begin{aligned} \mathfrak{h}'_\Lambda(w) = &\begin{pmatrix} \frac{(\Lambda^{-1} \cdot \mathfrak{p}(w))_0}{w_0} & -\frac{(\Lambda^{-1} \cdot \mathfrak{p}(w))_1}{w_0} & -\frac{(\Lambda^{-1} \cdot \mathfrak{p}(w))_2}{w_0} & -\frac{(\Lambda^{-1} \cdot \mathfrak{p}(w))_3}{w_0} \\ 0 & 1 & 0 & 0 \\ 0 & 0 & 1 & 0 \\ 0 & 0 & 0 & 1 \end{pmatrix} \\ &\cdot \Lambda^{-1} \cdot \begin{pmatrix} \frac{w_0}{[\mathfrak{p}(w)]_0} & \frac{w_1}{[\mathfrak{p}(w)]_0} & \frac{w_2}{[\mathfrak{p}(w)]_0} & \frac{w_3}{[\mathfrak{p}(w)]_0} \\ 0 & 1 & 0 & 0 \\ 0 & 0 & 1 & 0 \\ 0 & 0 & 0 & 1 \end{pmatrix} \end{aligned}$$

$$
= \begin{pmatrix} \frac{(\Lambda^{-1}\cdot\mathfrak{p}(w))_0}{w_0} & -\frac{(\Lambda^{-1}\cdot\mathfrak{p}(w))_1}{w_0} & -\frac{(\Lambda^{-1}\cdot\mathfrak{p}(w))_2}{w_0} & -\frac{(\Lambda^{-1}\cdot\mathfrak{p}(w))_3}{w_0} \\ 0 & 1 & 0 & 0 \\ 0 & 0 & 1 & 0 \\ 0 & 0 & 0 & 1 \end{pmatrix}
$$

$$
\cdot \begin{pmatrix} (\Lambda^{-1})_{00} & (\Lambda^{-1})_{01} & (\Lambda^{-1})_{02} & (\Lambda^{-1})_{03} \\ (\Lambda^{-1})^{1}_{10} & (\Lambda^{-1})_{11} & (\Lambda^{-1})_{12} & (\Lambda^{-1})_{13} \\ (\Lambda^{-1})_{20} & (\Lambda^{-1})_{21} & (\Lambda^{-1})_{22} & (\Lambda^{-1})_{23} \\ (\Lambda^{-1})_{30} & (\Lambda^{-1})_{31} & (\Lambda^{-1})_{32} & (\Lambda^{-1})_{33} \end{pmatrix} \cdot \begin{pmatrix} \frac{w_0}{[\mathfrak{p}(w)]_0} & \frac{w_1}{[\mathfrak{p}(w)]_0} & \frac{w_2}{[\mathfrak{p}(w)]_0} & \frac{w_3}{[\mathfrak{p}(w)]_0} \\ 0 & 1 & 0 & 0 \\ 0 & 0 & 1 & 0 \\ 0 & 0 & 0 & 1 \end{pmatrix}
$$

$$
= \begin{pmatrix} \frac{(\Lambda^{-1}\cdot\mathfrak{p}(w))_0}{w_0} & -\frac{(\Lambda^{-1}\cdot\mathfrak{p}(w))_1}{w_0} & -\frac{(\Lambda^{-1}\cdot\mathfrak{p}(w))_2}{w_0} & -\frac{(\Lambda^{-1}\cdot\mathfrak{p}(w))_3}{w_0} \\ 0 & 1 & 0 & 0 \\ 0 & 0 & 1 & 0 \\ 0 & 0 & 0 & 1 \end{pmatrix}
$$

$$
\cdot \begin{pmatrix} \frac{(\Lambda^{-1})_{00} w_0}{[\mathfrak{p}(w)]_0} & \frac{(\Lambda^{-1})_{00} w_1}{[\mathfrak{p}(w)]_0} + (\Lambda^{-1})_{01} & \frac{(\Lambda^{-1})_{00} w_2}{[\mathfrak{p}(w)]_0} + (\Lambda^{-1})_{02} & \frac{(\Lambda^{-1})_{00} w_3}{[\mathfrak{p}(w)]_0} + (\Lambda^{-1})_{03} \\ \frac{(\Lambda^{-1})_{10} w_0}{[\mathfrak{p}(w)]_0} & \frac{(\Lambda^{-1})_{10} w_1}{[\mathfrak{p}(w)]_0} + (\Lambda^{-1})_{11} & \frac{(\Lambda^{-1})_{10} w_2}{[\mathfrak{p}(w)]_0} + (\Lambda^{-1})_{12} & \frac{(\Lambda^{-1})_{10} w_3}{[\mathfrak{p}(w)]_0} + (\Lambda^{-1})_{13} \\ \frac{(\Lambda^{-1})_{20} w_0}{[\mathfrak{p}(w)]_0} & \frac{(\Lambda^{-1})_{20} w_1}{[\mathfrak{p}(w)]_0} + (\Lambda^{-1})_{21} & \frac{(\Lambda^{-1})_{20} w_2}{[\mathfrak{p}(w)]_0} + (\Lambda^{-1})_{22} & \frac{(\Lambda^{-1})_{20} w_3}{[\mathfrak{p}(w)]_0} + (\Lambda^{-1})_{23} \\ \frac{(\Lambda^{-1})_{30} w_0}{[\mathfrak{p}(w)]_0} & \frac{(\Lambda^{-1})_{30} w_1}{[\mathfrak{p}(w)]_0} + (\Lambda^{-1})_{31} & \frac{(\Lambda^{-1})_{30} w_2}{[\mathfrak{p}(w)]_0} + (\Lambda^{-1})_{32} & \frac{(\Lambda^{-1})_{30} w_3}{[\mathfrak{p}(w)]_0} + (\Lambda^{-1})_{33} \end{pmatrix}
$$

$$
= \begin{pmatrix} 1 & 0 & 0 & 0 \\ \frac{(\Lambda^{-1})_{10} w_0}{[\mathfrak{p}(w)]_0} & \frac{(\Lambda^{-1})_{10} w_1}{[\mathfrak{p}(w)]_0} + (\Lambda^{-1})_{11} & \frac{(\Lambda^{-1})_{10} w_2}{[\mathfrak{p}(w)]_0} + (\Lambda^{-1})_{12} & \frac{(\Lambda^{-1})_{10} w_3}{[\mathfrak{p}(w)]_0} + (\Lambda^{-1})_{13} \\ \frac{(\Lambda^{-1})_{20} w_0}{[\mathfrak{p}(w)]_0} & \frac{(\Lambda^{-1})_{20} w_1}{[\mathfrak{p}(w)]_0} + (\Lambda^{-1})_{21} & \frac{(\Lambda^{-1})_{20} w_2}{[\mathfrak{p}(w)]_0} + (\Lambda^{-1})_{22} & \frac{(\Lambda^{-1})_{20} w_3}{[\mathfrak{p}(w)]_0} + (\Lambda^{-1})_{23} \\ \frac{(\Lambda^{-1})_{30} w_0}{[\mathfrak{p}(w)]_0} & \frac{(\Lambda^{-1})_{30} w_1}{[\mathfrak{p}(w)]_0} + (\Lambda^{-1})_{31} & \frac{(\Lambda^{-1})_{30} w_2}{[\mathfrak{p}(w)]_0} + (\Lambda^{-1})_{32} & \frac{(\Lambda^{-1})_{30} w_3}{[\mathfrak{p}(w)]_0} + (\Lambda^{-1})_{33} \end{pmatrix}.
$$

Hence, $\mathfrak{h}'_\Lambda$ has an extension to a continuous map $\overline{\mathfrak{h}'_\Lambda} : ([0, \infty) \times \mathbb{R}^3) \setminus \{{}^t(0, 0, 0, 0)\} \to \mathbb{R}^4$, given by

$$
\overline{\mathfrak{h}'_\Lambda}(w) =
$$

$$
\begin{pmatrix} 1 & 0 & 0 & 0 \\ \frac{(\Lambda^{-1})_{10} w_0}{(w_0^2+|\vec{w}|^2)^{1/2}} & \frac{(\Lambda^{-1})_{10} w_1}{(w_0^2+|\vec{w}|^2)^{1/2}} + (\Lambda^{-1})_{11} & \frac{(\Lambda^{-1})_{10} w_2}{(w_0^2+|\vec{w}|^2)^{1/2}} + (\Lambda^{-1})_{12} & \frac{(\Lambda^{-1})_{10} w_3}{(w_0^2+|\vec{w}|^2)^{1/2}} + (\Lambda^{-1})_{13} \\ \frac{(\Lambda^{-1})_{20} w_0}{(w_0^2+|\vec{w}|^2)^{1/2}} & \frac{(\Lambda^{-1})_{20} w_1}{(w_0^2+|\vec{w}|^2)^{1/2}} + (\Lambda^{-1})_{21} & \frac{(\Lambda^{-1})_{20} w_2}{(w_0^2+|\vec{w}|^2)^{1/2}} + (\Lambda^{-1})_{22} & \frac{(\Lambda^{-1})_{20} w_3}{(w_0^2+|\vec{w}|^2)^{1/2}} + (\Lambda^{-1})_{23} \\ \frac{(\Lambda^{-1})_{30} w_0}{(w_0^2+|\vec{w}|^2)^{1/2}} & \frac{(\Lambda^{-1})_{30} w_1}{(w_0^2+|\vec{w}|^2)^{1/2}} + (\Lambda^{-1})_{31} & \frac{(\Lambda^{-1})_{30} w_2}{(w_0^2+|\vec{w}|^2)^{1/2}} + (\Lambda^{-1})_{32} & \frac{(\Lambda^{-1})_{30} w_3}{(w_0^2+|\vec{w}|^2)^{1/2}} + (\Lambda^{-1})_{33} \end{pmatrix},
$$

for every $w \in ([0, \infty) \times \mathbb{R}^3) \setminus \{{}^t(0, 0, 0, 0)\}$, where $\vec{w} := {}^t(w_1, w_2, w_3)$ and we assume component-wise convergence for matrices. In particular, it follows from (3.13) that component-wise

$$\lim_{a \to 0} h'_{a\Lambda}(v) = \begin{pmatrix} \frac{(\Lambda^{-1})_{10} v_1}{|v|} + (\Lambda^{-1})_{11} & \frac{(\Lambda^{-1})_{10} v_2}{|v|} + (\Lambda^{-1})_{12} & \frac{(\Lambda^{-1})_{10} v_3}{|v|} + (\Lambda^{-1})_{13} \\ \frac{(\Lambda^{-1})_{20} v_1}{|v|} + (\Lambda^{-1})_{21} & \frac{(\Lambda^{-1})_{20} v_2}{|v|} + (\Lambda^{-1})_{22} & \frac{(\Lambda^{-1})_{20} v_3}{|v|} + (\Lambda^{-1})_{23} \\ \frac{(\Lambda^{-1})_{30} v_1}{|v|} + (\Lambda^{-1})_{31} & \frac{(\Lambda^{-1})_{30} v_2}{|v|} + (\Lambda^{-1})_{32} & \frac{(\Lambda^{-1})_{30} v_3}{|v|} + (\Lambda^{-1})_{33} \end{pmatrix} ,$$

for every $v \in \mathbb{R}^3 \backslash \{0\}$. On the other hand, we have that

$$p_0(v) = {}^t(|v|, v_1, v_2, v_3) ,$$

for every $v \in \mathbb{R}^3$ and that

$$p_0^{-1}(v) = {}^t(v_1, v_2, v_3) ,$$

for every $v \in H_{0+} = V_+$. Hence, for every $\Lambda \in \mathcal{L}_+^{\uparrow}$, $h_{0\Lambda} : \mathbb{R}^3 \to \mathbb{R}^3$, given by

$$h_{0\Lambda}(v) = p_0^{-1}(\Lambda^{-1} \cdot p_0(v)) ,$$

for every $v \in \mathbb{R}^3$, is differentiable on $\mathbb{R}^3 \setminus \{0\}$, with derivative

$$h'_{0\Lambda}(v) = \begin{pmatrix} 0 & 1 & 0 & 0 \\ 0 & 0 & 1 & 0 \\ 0 & 0 & 0 & 1 \end{pmatrix} \cdot \Lambda^{-1} \cdot \begin{pmatrix} \frac{v_1}{|v|} & \frac{v_2}{|v|} & \frac{v_3}{|v|} \\ 1 & 0 & 0 \\ 0 & 1 & 0 \\ 0 & 0 & 1 \end{pmatrix}$$

$$= \begin{pmatrix} 0 & 1 & 0 & 0 \\ 0 & 0 & 1 & 0 \\ 0 & 0 & 0 & 1 \end{pmatrix} \cdot \begin{pmatrix} \frac{(\Lambda^{-1})_{00} v_1}{|v|} + (\Lambda^{-1})_{01} & \frac{(\Lambda^{-1})_{00} v_2}{|v|} + (\Lambda^{-1})_{02} & \frac{(\Lambda^{-1})_{00} v_3}{|v|} + (\Lambda^{-1})_{03} \\ \frac{(\Lambda^{-1})_{10} v_1}{|v|} + (\Lambda^{-1})_{11} & \frac{(\Lambda^{-1})_{10} v_2}{|v|} + (\Lambda^{-1})_{12} & \frac{(\Lambda^{-1})_{10} v_3}{|v|} + (\Lambda^{-1})_{13} \\ \frac{(\Lambda^{-1})_{20} v_1}{|v|} + (\Lambda^{-1})_{21} & \frac{(\Lambda^{-1})_{20} v_2}{|v|} + (\Lambda^{-1})_{22} & \frac{(\Lambda^{-1})_{20} v_3}{|v|} + (\Lambda^{-1})_{23} \\ \frac{(\Lambda^{-1})_{30} v_1}{|v|} + (\Lambda^{-1})_{31} & \frac{(\Lambda^{-1})_{30} v_2}{|v|} + (\Lambda^{-1})_{32} & \frac{(\Lambda^{-1})_{30} v_3}{|v|} + (\Lambda^{-1})_{33} \end{pmatrix}$$

$$= \begin{pmatrix} \frac{(\Lambda^{-1})_{10} v_1}{|v|} + (\Lambda^{-1})_{11} & \frac{(\Lambda^{-1})_{10} v_2}{|v|} + (\Lambda^{-1})_{12} & \frac{(\Lambda^{-1})_{10} v_3}{|v|} + (\Lambda^{-1})_{13} \\ \frac{(\Lambda^{-1})_{20} v_1}{|v|} + (\Lambda^{-1})_{21} & \frac{(\Lambda^{-1})_{20} v_2}{|v|} + (\Lambda^{-1})_{22} & \frac{(\Lambda^{-1})_{20} v_3}{|v|} + (\Lambda^{-1})_{23} \\ \frac{(\Lambda^{-1})_{30} v_1}{|v|} + (\Lambda^{-1})_{31} & \frac{(\Lambda^{-1})_{30} v_2}{|v|} + (\Lambda^{-1})_{32} & \frac{(\Lambda^{-1})_{30} v_3}{|v|} + (\Lambda^{-1})_{33} \end{pmatrix} ,$$

for every $v \in \mathbb{R}^3 \setminus \{0\}$. As a consequence, component-wise, we have that

$$\lim_{a\to 0} h'_{a\Lambda}(v) = h'_{0\Lambda}(v) ,$$

for every $v \in \mathbb{R}^3 \setminus \{0\}$ and it follows with the help of the continuity of the determinant function that

$$\begin{aligned}
\det(h'_{0\Lambda}(v)) &= \lim_{a\to 0} \det(h'_{a\Lambda}(v)) \\
&= \lim_{a\to 0} \frac{(\Lambda^{-1})_{00}\,(a^2+|v|^2)^{1/2} + (\Lambda^{-1})_{01}\,v_1 + (\Lambda^{-1})_{02}\,v_2 + (\Lambda^{-1})_{03}\,v_3}{(a^2+|v|^2)^{1/2}} \\
&= \frac{(\Lambda^{-1})_{00}\,|v| + (\Lambda^{-1})_{01}\,v_1 + (\Lambda^{-1})_{02}\,v_2 + (\Lambda^{-1})_{03}\,v_3}{|v|} ,
\end{aligned}$$

for every $v \in \mathbb{R}^3 \setminus \{0\}$. Further, for $a \geqslant 0$, since

$$[\Lambda^{-1}\cdot p_a(v)]\cdot[\Lambda^{-1}\cdot p_a(v)] = p_a(v)\cdot p_a(v) = a^2 ,$$

we have that

$$\begin{aligned}
&\left[(\Lambda^{-1})_{00}\,(a^2+|v|^2)^{1/2} + (\Lambda^{-1})_{01}\,v_1 + (\Lambda^{-1})_{02}\,v_2 + (\Lambda^{-1})_{03}\,v_3\right]^2 \\
&= a^2 + \sum_{k=1}^{3}\left[(\Lambda^{-1})_{k0}\,(a^2+|v|^2)^{1/2} + (\Lambda^{-1})_{k1}\,v_1 + (\Lambda^{-1})_{k2}\,v_2 + (\Lambda^{-1})_{k3}\,v_3\right]^2 \\
&= |h_{a\Lambda}(v)|^2 + a^2
\end{aligned}$$

and hence that

$$\begin{aligned}
&(\Lambda^{-1})_{00}\,(a^2+|v|^2)^{1/2} + (\Lambda^{-1})_{01}\,v_1 + (\Lambda^{-1})_{02}\,v_2 + (\Lambda^{-1})_{03}\,v_3 \\
&= [\,|h_{a\Lambda}(v)|^2 + a^2]^{1/2} ,
\end{aligned}$$

for every $v \in \mathbb{R}^3$, where we used (3.10). For later use, we note that from the latter equation and (3.10), it follows that

$$|h_{a\Lambda}(v)| + a \geqslant [\,|h_{a\Lambda}(v)|^2 + a^2]^{1/2} \geqslant \frac{1}{2\left\{1+\sum_{\alpha=1}^{3}[(\Lambda^{-1})_{0\alpha}]^2\right\}^{1/2}}\,|v| , \tag{3.14}$$

for every $v \in \mathbb{R}^3$. □

For $a \geqslant 0$, we define in the following the function $v_a^0 : \mathbb{R}^3 \to \mathbb{R}$ by

$$v_a^0(v) := (|v|^2 + a^2)^{1/2} ,$$

for every $v \in \mathbb{R}^3$. Then, $1/v_a^0$ is a.e.-defined on $\mathbb{R}^3$, strictly positive, continuous and hence also Lebesgue measurable. In addition, $1/(1/v_a^0)$ is a.e.-defined on $\mathbb{R}^3$, continuous and hence also Lebesgue measurable. Further, $1/v_a^0$ and $1/(1/v_a^0)$ are locally Lebesgue integrable. According to integration theory,

by

$$\varphi_a(I) := \int_{\mathbb{R}^3} \chi_I \cdot (v_a^0)^{-1} \, dv^3 \, ,$$

for every bounded interval I of $\mathbb{R}^3$, there is defined an additive, monotone and regular interval function φ_a, where v^3 denotes the Lebesgue measure on $\mathbb{R}^3$, such that

- a subset $N \subset \mathbb{R}^3$ is a φ_a-zero set if and only if it is a v^3-zero set
- and that g is φ_a-integrable if and only if $(v_a^0)^{-1} g$ is v^3-integrable and that in this case we have that

$$\int_{\mathbb{R}^3} g \, d\varphi_a = \int_{\mathbb{R}^3} g \cdot (v_a^0)^{-1} \, dv^3 \, .$$

Exercise 3.13 Show that $C_0^\infty(\mathbb{R}^3, \mathbb{C})$ is dense in $L^2_{\mathbb{C}}(\mathbb{R}^3, \varphi_a)$, for every $a \geqslant 0$.

The previously gathered information is sufficient for the construction of an unitary representation U_a of $\mathcal{L}_+^\uparrow$ that is also fundamental for the definition of spinor representations of $\mathcal{L}_+^\uparrow$ for all spins. In the latter sense, the representation U_a is associated with ("mass" a) and spin 0.

Theorem 3.7 (An unitary representation of the restricted Lorentz group) *Let $a \geqslant 0$.*

(i) For every $\Lambda \in \mathcal{L}_+^\uparrow$, by

$$U_a(\Lambda) f := f \circ h_{a\Lambda} \, ,$$

for every $f \in L^2_{\mathbb{C}}(\mathbb{R}^3, \varphi_a)$, there is defined a unitary linear operator $U_a(\Lambda)$ on $L^2_{\mathbb{C}}(\mathbb{R}^3, \varphi_a)$.

(ii) The map $U_a : \mathcal{L}_+^\uparrow \to L(L^2_{\mathbb{C}}(\mathbb{R}^3, \varphi_a), L^2_{\mathbb{C}}(\mathbb{R}^3, \varphi_a))$, that associates with every $\Lambda \in \mathcal{L}_+^\uparrow$ the corresponding $U_a(\Lambda)$, is a strongly continuous unitary representation of $\mathcal{L}_+^\uparrow$, where we assume component-wise convergence for matrices.

Proof Let $\Lambda \in \mathcal{L}_+^\uparrow$. If $f \in L^2_{\mathbb{C}}(\mathbb{R}^3, \varphi_a)$, then f is φ_a-a.e. defined on $\mathbb{R}^3$ and φ_a-measurable and hence also v^3-a.e. defined on $\mathbb{R}^3$ and v^3-measurable. As a consequence of the proof of the change of variable theorem for the Lebesgue integral, $f \circ h_{a\Lambda}$ is a.e.-defined on $\mathbb{R}^3$ and

v^3-measurable and hence also φ_a-a.e. defined on $\mathbb{R}^3$ and φ_a-measurable. Further, by the change of variable theorem for the Lebesgue integral, it follows that

$$\begin{aligned}\int_{\mathbb{R}^3} |f|^2 \, d\varphi_a &= \int_{\mathbb{R}^3} |f|^2 \cdot (v_a^0)^{-1} \, dv^3 \\ &= \int_{\mathbb{R}^3} \left([\,|f|^2 \cdot (v_a^0)^{-1}] \circ h_{a\Lambda} \right) \cdot |\det(h'_{a\Lambda})| \, dv^3 \\ &= \int_{\mathbb{R}^3} \left([\,|f|^2 \cdot (v_a^0)^{-1}] \circ h_{a\Lambda} \right) \cdot \frac{v_a^0 \circ h_{a\Lambda}}{v_a^0} \, dv^3 \\ &= \int_{\mathbb{R}^3} |f \circ h_{a\Lambda}|^2 \cdot (v_a^0)^{-1} \, dv^3 \ .\end{aligned}$$

Hence, $|f \circ h_{a\Lambda}|^2$ is φ_a-integrable, $f \circ h_{a\Lambda} \in L^2_{\mathbb{C}}(\mathbb{R}^3, \varphi_a)$ and

$$\int_{\mathbb{R}^3} |f|^2 \, d\varphi_a = \int_{\mathbb{R}^3} |f \circ h_{a\Lambda}|^2 \, d\varphi_a \ .$$

Therefore, it follows that by

$$U_a(\Lambda) f := f \circ h_{a\Lambda} \ ,$$

for every $f \in L^2_{\mathbb{C}}(\mathbb{R}^3, \varphi_a)$, there is defined an isometry $U_a(\Lambda) : L^2_{\mathbb{C}}(\mathbb{R}^3, \varphi_a) \to L^2_{\mathbb{C}}(\mathbb{R}^3, \varphi_a)$. Further, $U_a(\Lambda)$ is obviously linear, and it follows from the polarization identities for scalar products on complex vector spaces that $U_a(\Lambda)$ preserves the scalar product. Also,

$$\begin{aligned}U_a(\Lambda)\, U_a(\Lambda^{-1}) f &= [\, U_a(\Lambda^{-1}) f \,] \circ h_{a\Lambda} = f \circ h_{a\Lambda^{-1}} \circ h_{a\Lambda} = f \circ h_{a\Lambda}^{-1} \circ h_{a\Lambda} \\ &= f = U_a(E) f \ ,\end{aligned}$$

for every $f \in L^2_{\mathbb{C}}(\mathbb{R}^3, \varphi_a)$, where E denotes the 4×4 unit matrix. Hence, it follows that $U_a(\Lambda)$ is surjective and altogether a unitary linear operator on $L^2_{\mathbb{C}}(\mathbb{R}^3, \varphi_a)$. Also, the previous shows that

$$U_a(E) = \mathrm{id}_{L^2_{\mathbb{C}}(\mathbb{R}^3, \varphi_a)} \ , \quad U_a(\Lambda^{-1}) = [U_a(\Lambda)]^{-1} \ .$$

Further, since

$$\begin{aligned}h_{a\Lambda_1\Lambda_2}(v) &= p_a^{-1}((\Lambda_1\Lambda_2)^{-1} \cdot p_a(v)) = p_a^{-1}(\Lambda_2^{-1}\Lambda_1^{-1} \cdot p_a(v)) \\ &= p_a^{-1}(\Lambda_2^{-1} p_a(p_a^{-1}(\Lambda_1^{-1} \cdot p_a(v)))) = p_a^{-1}(\Lambda_2^{-1} p_a(h_{a\Lambda_1}(v))) \\ &= h_{a\Lambda_2}(h_{a\Lambda_1}(v)) = (h_{a\Lambda_2} \circ h_{a\Lambda_1})(v) \ ,\end{aligned}$$

for every $v \in \mathbb{R}^3$ and hence

$$h_{a\Lambda_1\Lambda_2} = h_{a\Lambda_2} \circ h_{a\Lambda_1} \ ,$$

it follows for $\Lambda_1, \Lambda_2 \in \mathcal{L}_+^\uparrow$ that

$$\begin{aligned} U_a(\Lambda_1\Lambda_2)f &= f \circ h_{a\Lambda_1\Lambda_2} = f \circ h_{a\Lambda_2} \circ h_{a\Lambda_1} = [U_a(\Lambda_2)f] \circ h_{a\Lambda_1} \\ &= U_a(\Lambda_1)U_a(\Lambda_2)f \ , \end{aligned}$$

for every $f \in L^2_{\mathbb{C}}(\mathbb{R}^3, \varphi_a)$ and therefore that

$$U_a(\Lambda_1\Lambda_2) = U_a(\Lambda_1) \circ U_a(\Lambda_2) \ ,$$

for all $\Lambda_1, \Lambda_2 \in \mathcal{L}_+^\uparrow$. As a consequence, U_a is an unitary representation of $\mathcal{L}_+^\uparrow$. In the next step, we are going to show that U_a is strongly continuous, i.e., if $\Lambda_1, \Lambda_2, \dots$ is a sequence in $\mathcal{L}_+^\uparrow$ that is component-wise convergent to $\Lambda \in \mathcal{L}_+^\uparrow$, then

$$\lim_{\nu\to\infty} U_a(\Lambda_\nu)f = U_a(\Lambda)f \ ,$$

for every $f \in L^2_{\mathbb{C}}(\mathbb{R}^3, \varphi_a)$. That the latter is true for $f \in C_0^\infty(\mathbb{R}^3, \mathbb{C})$ can be seen as follows. Since $f \in C_0^\infty(\mathbb{R}^3, \mathbb{C})$, there is $R > 0$, such that $\text{supp}(f) \subset U_R(0)$. Further, since

$$\lim_{\nu\to\infty} \Lambda_{\nu lm} = \Lambda_{lm} \ ,$$

for all $l, m \in \{0, \dots, 3\}$, there is $C \geqslant 0$ such that

$$|\Lambda_{\nu lm}| \leqslant C \ , \ |\Lambda_{lm}| \leqslant C \ ,$$

for all $l, m \in \{0, \dots, 3\}$ and $\nu \in \mathbb{N}^*$. Using (3.4), we infer that

$$\lim_{\nu\to\infty} (\Lambda^{-1})_{\nu lm} = (\Lambda^{-1})_{lm} \ ,$$

for all $l, m \in \{0, \dots, 3\}$ as well as that

$$|(\Lambda^{-1})_{\nu lm}| \leqslant C \ , \ |(\Lambda^{-1})_{lm}| \leqslant C \ ,$$

for all $l, m \in \{0, \dots, 3\}$ and $\nu \in \mathbb{N}^*$. Hence, it follows for $\nu \in \mathbb{N}^*$ and $v \in \mathbb{R}^3$ that

$$|h_{a\Lambda_\nu}(v)| + a \geqslant \frac{1}{2\left\{1 + \sum_{\alpha=1}^3 [(\Lambda^{-1})_{\nu 0\alpha}]^2\right\}^{1/2}} |v| \geqslant \frac{1}{2(1+3C^2)^{1/2}} |v| \ ,$$

and analogously that

$$|h_{a\Lambda}(v)| + a \geqslant \frac{1}{2(1+3C^2)^{1/2}} |v| \ ,$$

where we used (3.14). Therefore, we have that

$$|h_{a\Lambda_\nu}(v)| \geqslant R \ , \ |h_{a\Lambda}(v)| \geqslant R \ ,$$

and hence that

$$[U_a(\Lambda_\nu)f](v) = f \circ h_{a\Lambda_\nu}(v) = 0 \ , \ [U_a(\Lambda)f](v) = f \circ h_{a\Lambda}(v) = 0 \ ,$$

for every $v \in \mathbb{R}^3$ such that

$$|v| \geqslant \rho := 2\,(1 + 3C^2)^{1/2}(R + a) \ .$$

As a consequence,

$$|f \circ h_{a\Lambda_\nu}| \leqslant \|f\|_\infty \, \chi_{U_\rho(0)} \ , \ |f \circ h_{a\Lambda}| \leqslant \|f\|_\infty \, \chi_{U_\rho(0)} \ .$$

For the following, we note that for $v \in \mathbb{R}^4$

$$\begin{aligned}
&|\Lambda_\nu^{-1} v - \Lambda^{-1} v|^2 \\
&= \sum_{l=0}^{3} |\sum_{m=0}^{3} [(\Lambda^{-1})_{\nu l m} - (\Lambda^{-1})_{l m}] \cdot v_m|^2 \\
&= \sum_{l=0}^{3} \left(|\sum_{m=0}^{3} [(\Lambda^{-1})_{\nu l m} - (\Lambda^{-1})_{l m}] \cdot v_m| \right)^2 \\
&\leqslant \sum_{l=0}^{3} \left(\sum_{m=0}^{3} |(\Lambda^{-1})_{\nu l m} - (\Lambda^{-1})_{l m}| \cdot |v_m| \right)^2 \\
&\leqslant \sum_{l=0}^{3} \sum_{m=0}^{3} |(\Lambda^{-1})_{\nu l m} - (\Lambda^{-1})_{l m}|^2 \cdot |v|^2 \\
&= \left[\sum_{l=0}^{3} \sum_{m=0}^{3} |(\Lambda^{-1})_{\nu l m} - (\Lambda^{-1})_{l m}|^2 \right] \cdot |v|^2 \ ,
\end{aligned}$$

where $|\Lambda_\nu^{-1} v - \Lambda^{-1} v|$ and $|v|$ denote the Euclidean norm of $\Lambda_\nu^{-1} v - \Lambda^{-1} v$ and v, respectively, implying that

$$\lim_{\nu \to \infty} \Lambda_\nu^{-1} v = \Lambda^{-1} v \ ,$$

for every $v \in \mathbb{R}^4$ and $\mathbb{R}^4$ is equipped with Euclidean norm. Hence, it follows that

$$\left(|f \circ h_{a\Lambda_\nu} - f \circ h_{a\Lambda}|^2\right)_{\nu \in \mathbb{N}^*}$$

is an everywhere pointwise to 0 convergent sequence of φ_a-integrable functions whose members are dominated by the φ_a-integrable function

$$4\,\|f\|_\infty^2 \, \chi_{U_\rho(0)} \ .$$

Hence it follows from Lebesgue's dominated convergence theorem that

$$\lim_{\nu \to \infty} \|f \circ h_{a\Lambda_\nu} - f \circ h_{a\Lambda}\|_2^2 = 0 \ ,$$

where $\| \ \|_2$ denotes the norm on $L^2_{\mathbb{C}}(\mathbb{R}^3, \varphi_a)$, and therefore also that

$$\lim_{\nu \to \infty} U_a(\Lambda_\nu) f = U_a(\Lambda) f \ .$$

If $f \in L^2_{\mathbb{C}}(\mathbb{R}^3, \varphi_a)$ and $\varepsilon > 0$, then, since $C_0^\infty(\mathbb{R}^3, \mathbb{C})$ is dense in $L^2_{\mathbb{C}}(\mathbb{R}^3, \varphi_a)$, there is $g \in C_0^\infty(\mathbb{R}^3, \mathbb{C})$ such that

$$\|g - f\|_2 < \frac{\varepsilon}{3} \ ,$$

and there is $\nu_0 \in \mathbb{N}^*$ such that for $\nu \in \mathbb{N}^*$ satisfying $\nu \geqslant \nu_0$

$$\|U_a(\Lambda_\nu) g - U_a(\Lambda) g\|_2 < \frac{\varepsilon}{3} \ .$$

Then, it follows for $\nu \in \mathbb{N}^*$ satisfying $\nu \geqslant \nu_0$ that

$$\begin{aligned}
&\|U_a(\Lambda_\nu) f - U_a(\Lambda) f\|_2 \\
&\leqslant \|U_a(\Lambda_\nu) f - U_a(\Lambda_\nu) g\|_2 + \|U_a(\Lambda_\nu) g - U_a(\Lambda) g\|_2 + \|U_a(\Lambda) g - U_a(\Lambda) f\|_2 \\
&\leqslant \|f - g\|_2 + \|U_a(\Lambda_\nu) g - U_a(\Lambda) g\|_2 + \|g - f\|_2 < \varepsilon
\end{aligned}$$

and hence that

$$\lim_{\nu \to \infty} U_a(\Lambda_\nu) f = U_a(\Lambda) f \ .$$

□

3.2.1 Generators Associated with Rotations about the Spatial Coordinate Axes

In the following, we analyze the generators of U_a for one-parameter subgroups associated with rotations. For this purpose, let $a \geqslant 0$. If

$$M : (\mathbb{R}, +) \to \mathcal{L}_+^\uparrow$$

is a continuous group homomorphism, i.e., such that

$$M(s_1 + s_2) = M(s_1) \cdot M(s_2) \ ,$$

for all $s_1, s_2 \in \mathbb{R}$ and such that, for every sequence $s_1, s_2, \ldots$ in $\mathbb{R}$ that is convergent to $s \in \mathbb{R}$, the corresponding sequence $M(s_1), M(s_2), \ldots$ converges component-wise to $M(s)$, then $U_a \circ M$ is a strongly continuous one-parameter unitary group. According to Stone's theorem, there is a unique densely-defined, linear and self-adjoint operator A_M in $X := L^2_{\mathbb{C}}(\mathbb{R}^3, \varphi_a)$ such that

$$\exp(isA_M) = (U_a \circ M)(s) \ , \tag{3.15}$$

for every $s \in \mathbb{R}$ and, in particular, that $A_M : D(A_M) \to X$ is given by

$$D(A_M) = \{f \in X : \lim_{s \to 0, s \neq 0} \frac{1}{s} [(U_a \circ M)(s) - \mathrm{id}_X] f \text{ exists}\} \tag{3.16}$$

and for every $f \in D(A_M)$

$$A_M f = \frac{1}{i} \lim_{s \to 0, s \neq 0} \frac{1}{s} [(U_a \circ M)(s) - \mathrm{id}_X] f \ . \tag{3.17}$$

In the following, our main cases of interest are the one-parameter groups of rotations about the coordinate axes,

$$M_j : \mathbb{R} \to \mathcal{L}_+^{\uparrow} \ ,$$

$j \in \{1, 2, 3\}$, where

$$M_1(s) := \begin{pmatrix} 1 & 0 & 0 & 0 \\ 0 & 1 & 0 & 0 \\ 0 & 0 & \cos(s) & \sin(s) \\ 0 & 0 & -\sin(s) & \cos(s) \end{pmatrix} , \ M_2(s) := \begin{pmatrix} 1 & 0 & 0 & 0 \\ 0 & \cos(s) & 0 & -\sin(s) \\ 0 & 0 & 1 & 0 \\ 0 & \sin(s) & 0 & \cos(s) \end{pmatrix} ,$$

$$M_3(s) := \begin{pmatrix} 1 & 0 & 0 & 0 \\ 0 & \cos(s) & \sin(s) & 0 \\ 0 & -\sin(s) & \cos(s) & 0 \\ 0 & 0 & 0 & 1 \end{pmatrix} , \tag{3.18}$$

for every $s \in \mathbb{R}$, compare Exercise 3.4. Subsequently, we analyze $U_a \circ M_3$. If $f \in C_0^1(\mathbb{R}^3, \mathbb{C})$, then it follows from the mean value theorem in several variables the existence of $C \geqslant 0$ such that

$$|f(w) - f(v)| \leqslant C\, |w - v| \ ,$$

for all $v, w \in \mathbb{R}^3$. Further, if $R > 0$ is such that $\mathrm{supp}(f) \subset U_R(0)$, we conclude for $s \in \mathbb{R}$, $v = (v_1, v_2, v_3) \in \mathbb{R}^3$ that

$$\begin{aligned} &\left|[(U_a \circ M_3)(s) f](v) - f(v)\right|^2 = \left|[f \circ h_{aM_3(s)}](v) - f(v)\right|^2 \\ &= \left|f(p_a^{-1}(M_3(-s) \cdot p_a(v))) - f(v)\right|^2 . \end{aligned}$$

Since, $p_a : \mathbb{R}^3 \to H_{a+}$ is given by

$$p_a(v) := {}^t((a^2 + |v|^2)^{1/2}, v_1, v_2, v_3) \ ,$$

for every $v \in \mathbb{R}^3$, and we have that

$$p_a^{-1}(v) = {}^t(v_1, v_2, v_3) \ ,$$

for every $v \in H_{a+}$, it follows that

$$M_3(-s)p_a(v)$$
$$= {}^t((a^2+|v|^2)^{1/2}, \cos(s)v_1 - \sin(s)v_2, \sin(s)v_1 + \cos(s)v_2, v_3)$$

and hence that

$$\begin{aligned} h_{aM_3(-s)}(v) &= p_a^{-1} M_3(-s)p_a(v) \\ &= p_a^{-1}\, {}^t((a^2+|v|^2)^{1/2}, \cos(s)v_1 - \sin(s)v_2, \sin(s)v_1 + \cos(s)v_2, v_3) \\ &= {}^t(\cos(s)v_1 - \sin(s)v_2, \sin(s)v_1 + \cos(s)v_2, v_3)\ . \end{aligned}$$

As a consequence, the support of $(U_a \circ M_3)(s)f$ is contained in $U_R(0)$, and we have for $v \in U_R(0)$ that

$$\begin{aligned} &\left|[(U_a \circ M_3)(s)f](v) - f(v)\right|^2 \\ &= \left|f(\cos(s)v_1 - \sin(s)v_2, \sin(s)v_1 + \cos(s)v_2, v_3) - f(v_1, v_2, v_3)\right|^2 \\ &\leqslant C^2 \cdot \left|((\cos(s)-1)v_1 - \sin(s)v_2, \sin(s)v_1 + (\cos(s)-1)v_2, 0)\right|^2 \\ &= C^2\,[(\cos(s)-1)^2 + \sin^2(s)](v_1^2+v_2^2) = 2C^2\,[1-\cos(s)](v_1^2+v_2^2) \\ &= 4C^2 \sin^2\left(\frac{s}{2}\right)(v_1^2+v_2^2) \leqslant C^2 R^2 s^2 \end{aligned}$$

and hence that

$$\left|\frac{1}{s}\,[(U_a \circ M_3)(s)f - f]\right|^2 \leqslant C^2\,R^2\,\chi_{U_R(0)}\ ,$$

for every $s \in \mathbb{R}^*$. Further,

$$\lim_{s\to 0, s\neq 0} \frac{1}{s}\,\{[(U_a \circ M_3)(s) - \mathrm{id}_X]\,f\}\,(v) = \left(v_1\,\frac{\partial f}{\partial v_2} - v_2\,\frac{\partial f}{\partial v_1}\right)(v)$$

for every $v \in \mathbb{R}^3$, where v_1 and v_2 denote the coordinate projection of $\mathbb{R}^3$ onto the first and second coordinate, respectively. As a consequence, if $s_1, s_2, \dots$ is a sequence in $\mathbb{R}^*$ that is convergent to 0, then

$$\left((v_a^0)^{-1}\left|\frac{1}{s_\nu}\,[(U_a \circ M_3)(s_\nu)f - f] - \left(v_1\,\frac{\partial f}{\partial v_2} - v_2\,\frac{\partial f}{\partial v_1}\right)\right|^2\right)_{\nu\in\mathbb{N}^*}$$

is a sequence of integrable functions that is everywhere on $\mathbb{R}^3$ convergent to the 0-function on $\mathbb{R}^3$ and whose members are dominated by the integrable function

$$(v_a^0)^{-1}\left[2C^2\,R^2\,\chi_{U_R(0)} + 2\left|v_1\,\frac{\partial f}{\partial v_2} - v_2\,\frac{\partial f}{\partial v_1}\right|^2\right]\ .$$

Hence, it follows from Lebesgue's dominated convergence theorem that

$$\lim_{\nu\to\infty}\int_{\mathbb{R}^3}\left|\frac{1}{s_\nu}[(U_a\circ M_3)(s_\nu)f-f]-\left(v_1\frac{\partial f}{\partial v_2}-v_2\frac{\partial f}{\partial v_1}\right)\right|^2(v_a^0)^{-1}\,dv^3=0\ ,$$

i.e., that

$$\lim_{\nu\to\infty}\left\|\frac{1}{s_\nu}[(U_a\circ M_3)(s_\nu)f-f]-\left(v_1\frac{\partial f}{\partial v_2}-v_2\frac{\partial f}{\partial v_1}\right)\right\|_2=0\ .$$

We conclude that $C_0^1(\mathbb{R}^3,\mathbb{C})\subset D(A_{M_3})$ a well as that

$$A_{M_3}f=\frac{1}{i}\left(v_1\frac{\partial f}{\partial v_2}-v_2\frac{\partial f}{\partial v_1}\right)\ ,$$

for every $f\in C_0^1(\mathbb{R}^3,\mathbb{C})$. In addition, $C_0^1(\mathbb{R}^3,\mathbb{C})$ is dense in X and left invariant by $U_a\circ M_3$, i.e., $(U_a\circ M_3)(s)(C_0^1(\mathbb{R}^3,\mathbb{C}))\subset C_0^1(\mathbb{R}^3,\mathbb{C})$, for every $s\in\mathbb{R}$. Hence, $C_0^1(\mathbb{R}^3,\mathbb{C})$ is a core for A_{M_3}. We define the operator $\hat{L}_3$ that is associated to the 3rd component of angular momentum by

$$\hat{L}_3:=\hbar\,A_{M_3}\ .$$

Analogously, we define the operators $\hat{L}_1$ and $\hat{L}_2$ that are associated to the 1st and 2nd component of angular momentum, respectively, by

$$\hat{L}_1:=\hbar\,A_{M_1}\ ,\ \ \hat{L}_2:=\hbar\,A_{M_2}\ .$$

We leave the details of the derivation of the corresponding properties of these operators to the reader.

Exercise 3.14 Show that

$$M_3(s)=\begin{pmatrix}1&0&0&0\\0&\cos(s)&\sin(s)&0\\0&-\sin(s)&\cos(s)&0\\0&0&0&1\end{pmatrix}=\exp(sA)\ ,$$

for every $s\in\mathbb{R}$, where

$$A:=\begin{pmatrix}0&0&0&0\\0&0&1&0\\0&-1&0&0\\0&0&0&0\end{pmatrix}\ .$$

Exercise 3.15 With the help of the result of Exercise 3.14, show that

$$\begin{pmatrix} 1 & 0 & 0 & 0 \\ 0 & 1 & 0 & 0 \\ 0 & 0 & \cos(s) & \sin(s) \\ 0 & 0 & -\sin(s) & \cos(s) \end{pmatrix} = M_1(s) = U_1^{-1} M_3(s) U_1 = \exp(sB) \ ,$$

$$\begin{pmatrix} 1 & 0 & 0 & 0 \\ 0 & \cos(s) & 0 & -\sin(s) \\ 0 & 0 & 1 & 0 \\ 0 & \sin(s) & 0 & \cos(s) \end{pmatrix} = M_2(s) = U_2^{-1} M_3(s) U_2 = \exp(sC) \ ,$$

for every $s \in \mathbb{R}$, where

$$B := U_1^{-1} A U_1 = \begin{pmatrix} 0 & 0 & 0 & 0 \\ 0 & 0 & 0 & 0 \\ 0 & 0 & 0 & 1 \\ 0 & 0 & -1 & 0 \end{pmatrix} \ , \ C := U_2^{-1} A U_2 = \begin{pmatrix} 0 & 0 & 0 & 0 \\ 0 & 0 & 0 & -1 \\ 0 & 0 & 0 & 0 \\ 0 & 1 & 0 & 0 \end{pmatrix}$$

and

$$U_1 := \begin{pmatrix} 1 & 0 & 0 & 0 \\ 0 & 0 & 0 & 1 \\ 0 & 0 & -1 & 0 \\ 0 & 1 & 0 & 0 \end{pmatrix} \in \mathcal{L}_+^{\uparrow} \ , \ U_2 := \begin{pmatrix} 1 & 0 & 0 & 0 \\ 0 & -1 & 0 & 0 \\ 0 & 0 & 0 & 1 \\ 0 & 0 & 1 & 0 \end{pmatrix} \in \mathcal{L}_+^{\uparrow} \ .$$

Verify that $U_1, U_2 \in \mathcal{L}_+^{\uparrow}$ and that

$$U_1^{-1} = U_1 \ , \ U_2^{-1} = U_2 \ .$$

Exercise 3.16 Show that, for every $j \in \{1, 2\}$ and $s \in \mathbb{R}$, we have that

$$U_a(M_j(s)) = \exp\Big(i \, \frac{s}{\hbar} \, U_a(U_j) \circ \hat{L}_3 \circ U_a(U_j)\Big) \ ,$$

where $U_1, U_2 \in \mathcal{L}_+^{\uparrow}$ are defined in Exercise 3.15. Further, show that

$$[U_a(U_1) f](v) = f(v_3, -v_2, v_1) \ , \ [U_a(U_2) f](v) = f(-v_1, v_3, v_2) \ ,$$

for every $f \in X$ and almost all $v = {}^t(v_1, v_2, v_3) \in \mathbb{R}^3$. Finally, for $j \in \{1, 2\}$, show that $U_a(U_j) C_0^1(\mathbb{R}^3, \mathbb{C}) = C_0^1(\mathbb{R}^3, \mathbb{C})$ and for $f \in C_0^1(\mathbb{R}^3, \mathbb{C})$ that

$$U_a(U_1) \circ \hat{L}_3 \circ U_a(U_1) f = \frac{\hbar}{i} \left(v_2 \, \frac{\partial f}{\partial v_3} - v_3 \, \frac{\partial f}{\partial v_2} \right) \ ,$$

$$U_a(U_2) \circ \hat{L}_3 \circ U_a(U_2) f = \frac{\hbar}{i} \left(v_3 \, \frac{\partial f}{\partial v_1} - v_1 \, \frac{\partial f}{\partial v_3} \right) \ .$$

Hence, we arrive at the following result.

Connection Between Rotations About The Spatial Coordinate Axes and the Components of Angular Momentum

For every $k \in \{1, 2, 3\}$, the following representation is true

$$\exp\left(i \frac{s}{\hbar} \hat{L}_k\right) f = (U_a \circ M_k)(s) \,, \tag{3.19}$$

for every $f \in X := L^2_{\mathbb{C}}(\mathbb{R}^3, \varphi_a)$ and $s \in \mathbb{R}$, where for $k \in \{1, 2, 3\}$, $\hat{L}_k : D(\hat{L}_k) \to X$ is the densely-defined, linear and self-adjoint operator in X, given by

$$D(\hat{L}_k) = \{f \in X : \lim_{s \to 0, s \neq 0} \frac{1}{s} [(U_a \circ M_k)(s) - \mathrm{id}_X] f \text{ exists}\}$$

and

$$\hat{L}_k f := \frac{\hbar}{i} \lim_{s \to 0, s \neq 0} \frac{1}{s} [(U_a \circ M_k)(s) - \mathrm{id}_X] f \,,$$

for every $f \in D(\hat{L}_k)$ and where $M_1 : (\mathbb{R}, +) \to \mathcal{L}_+^\uparrow$, $M_2 : (\mathbb{R}, +) \to \mathcal{L}_+^\uparrow$ and $M_3 : (\mathbb{R}, +) \to \mathcal{L}_+^\uparrow$ are given by

$$M_1(s) = \begin{pmatrix} 1 & 0 & 0 & 0 \\ 0 & 1 & 0 & 0 \\ 0 & 0 & \cos(s) & \sin(s) \\ 0 & 0 & -\sin(s) & \cos(s) \end{pmatrix} ,$$

$$M_2(s) = \begin{pmatrix} 1 & 0 & 0 & 0 \\ 0 & \cos(s) & 0 & -\sin(s) \\ 0 & 0 & 1 & 0 \\ 0 & \sin(s) & 0 & \cos(s) \end{pmatrix} ,$$

$$M_3(s) = \begin{pmatrix} 1 & 0 & 0 & 0 \\ 0 & \cos(s) & \sin(s) & 0 \\ 0 & -\sin(s) & \cos(s) & 0 \\ 0 & 0 & 0 & 1 \end{pmatrix} ,$$

for every $s \in \mathbb{R}$. In particular, $C_0^1(\mathbb{R}^3, \mathbb{C})$ is a core for $\hat{L}_1$, $\hat{L}_2$ and $\hat{L}_3$, and we have that

$$\hat{L}_1 f = \frac{\hbar}{i} \left(v_2 \frac{\partial f}{\partial v_3} - v_3 \frac{\partial f}{\partial v_2} \right) , \quad \hat{L}_2 f = \frac{\hbar}{i} \left(v_3 \frac{\partial f}{\partial v_1} - v_1 \frac{\partial f}{\partial v_3} \right) ,$$

$$\hat{L}_3 f = \frac{\hbar}{i} \left(v_1 \frac{\partial f}{\partial v_2} - v_2 \frac{\partial f}{\partial v_1} \right) ,$$

for every $f \in C_0^1(\mathbb{R}^3, \mathbb{C})$, where v_1, v_2 and v_3 denote the coordinate projection of $\mathbb{R}^3$ onto the first, second and third coordinate, respectively.

For the determination of the spectra of $\hat{L}_1$, $\hat{L}_2$ and $\hat{L}_3$, we use the result of the following exercise.

Exercise 3.17 For $a \geqslant 0$, show that $\mathcal{V}_a : L^2_{\mathbb{C}}(\mathbb{R}^3, \varphi_a) \to L^2_{\mathbb{C}}(\mathbb{R}^3)$, defined by

$$\mathcal{V}_a f := (v_a^0)^{-1/2} f \ ,$$

for every $f \in L^2_{\mathbb{C}}(\mathbb{R}^3, \varphi_a)$ is a Hilbert space isomorphism, with inverse $\mathcal{V}_a^{-1} : L^2_{\mathbb{C}}(\mathbb{R}^3) \to L^2_{\mathbb{C}}(\mathbb{R}^3, \varphi_a)$, given by

$$\mathcal{V}_a^{-1} f := (v_a^0)^{1/2} f \ ,$$

for every $f \in L^2_{\mathbb{C}}(\mathbb{R}^3)$.

Using the Hilbert space isomorphism $\mathcal{V}_a : L^2_{\mathbb{C}}(\mathbb{R}^3, \varphi_a) \to L^2_{\mathbb{C}}(\mathbb{R}^3)$ from Exercise 3.17, it follows for the case that $a > 0$ that $(v_a^0)^{1/2} \in C^1(\mathbb{R}^3, \mathbb{C})$ and hence for $f \in C_0^1(\mathbb{R}^3, \mathbb{C})$ that $\mathcal{V}_a^{-1} f = (v_a^0)^{1/2} f \in C_0^1(\mathbb{R}^3, \mathbb{C})$ and

$$\begin{aligned} \mathcal{V}_a \hat{L}_3 \mathcal{V}_a^{-1} f &= \frac{\hbar}{i} (v_a^0)^{-1/2} \left(v_1 \frac{\partial}{\partial v_2} - v_2 \frac{\partial}{\partial v_1} \right) (v_a^0)^{1/2} f \\ &= \frac{\hbar}{i} \left(v_1 \frac{\partial}{\partial v_2} - v_2 \frac{\partial}{\partial v_1} \right) f + \frac{\hbar}{i} f \cdot \left(v_1 \frac{\partial}{\partial v_2} - v_2 \frac{\partial}{\partial v_1} \right) (v_a^0)^{1/2} \\ &= \frac{\hbar}{i} \left(v_1 \frac{\partial}{\partial v_2} - v_2 \frac{\partial}{\partial v_1} \right) f \ , \end{aligned}$$

for every $f \in C_0^1(\mathbb{R}^3, \mathbb{C})$. Hence, it follows with the help of results from quantum mechanics about angular momentum operators, see Sect. 2.4.3 in [7], that $\hat{L}_3$ has a pure point spectrum given by $\hbar.\mathbb{Z}$, consisting of eigenvalues of infinite multiplicity. For the case $a = 0$, we have that $(v_a^0)^{1/2} = |\ |^{1/2}$ and follow the reasoning that lead to the representation of $\hat{L}_3$. If $f \in C_0^1(\mathbb{R}^3, \mathbb{C})$, then it follows from the mean value theorem in several variables the existence of $C \geqslant 0$ such that

$$|f(w) - f(v)| \leqslant C\,|w - v| \ ,$$

for all $v, w \in \mathbb{R}^3$, Hence, we have that

$$|(|\ |^{1/2} f)(w) - (|\ |^{1/2} f)(v)| \leqslant C\,|w - v|^{3/2} \ ,$$

for all $v, w \in \mathbb{R}^3$. Further, if $R > 0$ is such that $\text{supp}(f) \subset U_R(0)$ and $s \in \mathbb{R}^*$ such that $|s| \leqslant 1$, the support of $(U_a \circ M_3)(s) |\;|^{1/2} f$ is contained in $U_R(0)$, and we have for $v \in U_R(0)$ that

$$\begin{aligned}
&\left|[(U_a \circ M_3)(s)|\;|^{1/2} f](v) - (|\;|^{1/2} f)(v)\right|^2 \\
&= \big|(|\;|^{1/2} f)(\cos(s)v_1 - \sin(s)v_2, \sin(s)v_1 + \cos(s)v_2, v_3) \\
&\quad - (|\;|^{1/2} f)(v_1, v_2, v_3)\big|^2 \\
&\leqslant C^2 \cdot \left|((\cos(s) - 1)v_1 - \sin(s)v_2, \sin(s)v_1 + (\cos(s) - 1)v_2, 0)\right|^3 \\
&= C^2 \,[(\cos(s) - 1)^2 + \sin^2(s)]^{3/2} \,(v_1^2 + v_2^2)^{3/2} \\
&= 2^{3/2} C^2 \,[1 - \cos(s)]^{3/2} (v_1^2 + v_2^2)^{3/2} \\
&= 2^3 C^2 \left|\sin\left(\frac{s}{2}\right)\right|^3 (v_1^2 + v_2^2)^{3/2} \leqslant C^2 R^3 |s|^3
\end{aligned}$$

and hence that

$$\left|\frac{1}{s}\,[(U_a \circ M_3)(s) |\;|^{1/2} f] - |\;|^{1/2} f]\right|^2 \leqslant C^2\, R^3\, \chi_{U_R(0)} \;.$$

Further,

$$\lim_{s \to 0, s \neq 0} \frac{1}{s} \left\{[(U_a \circ M_3)(s) - \text{id}_X]|\;|^{1/2} f\right\}(v) = |v|^{1/2} \left(v_1 \frac{\partial f}{\partial v_2} - v_2 \frac{\partial f}{\partial v_1}\right)(v) \;,$$

for every $v \in \mathbb{R}^3 \setminus \{0\}$, where v_1 and v_2 denote the coordinate projection of $\mathbb{R}^3$ onto the first and second coordinate, respectively. As a consequence, if $s_1, s_2, \ldots$ is a sequence in $[-1, 1] \setminus \{0\}$ that is convergent to 0, then

$$\begin{aligned}
&\Bigg((v_a^0)^{-1} \bigg|\frac{1}{s_\nu}\,[(U_a \circ M_3)(s_\nu) |\;|^{1/2} f - |\;|^{1/2} f] \\
&\qquad - |\;|^{1/2} \left(v_1 \frac{\partial f}{\partial v_2} - v_2 \frac{\partial f}{\partial v_1}\right)\bigg|^2\Bigg)_{\nu \in \mathbb{N}^*}
\end{aligned}$$

is a sequence of integrable functions that is a.e. on $\mathbb{R}^3$ convergent to the 0-function on $\mathbb{R}^3$ and whose members are dominated by the integrable function

$$(v_a^0)^{-1} \left[2C^2\, R^3\, \chi_{U_R(0)} + 2R \left|v_1 \frac{\partial f}{\partial v_2} - v_2 \frac{\partial f}{\partial v_1}\right|^2\right] \;.$$

Hence, it follows from Lebesgue's dominated convergence theorem that

$$\lim_{\nu\to\infty}\int_{\mathbb{R}^3}\left|\frac{1}{s_\nu}\left[(U_a\circ M_3)(s_\nu)|\ |^{1/2}f-|\ |^{1/2}f\right]\right.$$
$$\left.-|\ |^{1/2}\left(v_1\frac{\partial f}{\partial v_2}-v_2\frac{\partial f}{\partial v_1}\right)\right|^2(v_a^0)^{-1}\,dv^3=0\ ,$$

i.e., that

$$\lim_{\nu\to\infty}\left\|\frac{1}{s_\nu}\left[(U_a\circ M_3)(s_\nu)|\ |^{1/2}f-|\ |^{1/2}f\right]-|\ |^{1/2}\left(v_1\frac{\partial f}{\partial v_2}-v_2\frac{\partial f}{\partial v_1}\right)\right\|_2=0\ .$$

We conclude that $|\ |^{1/2}\,C_0^1(\mathbb{R}^3,\mathbb{C})\subset D(\hat{L}_3)$ a well as that

$$\hat{L}_3|\ |^{1/2}f=\frac{\hbar}{i}|\ |^{1/2}\left(v_1\frac{\partial f}{\partial v_2}-v_2\frac{\partial f}{\partial v_1}\right)\ ,$$

for every $f\in C_0^1(\mathbb{R}^3,\mathbb{C})$. Hence, it follows also in the case $a=0$ that

$$\begin{aligned}\mathcal{V}_a\hat{L}_3\mathcal{V}_a^{-1}f&=\frac{\hbar}{i}(v_a^0)^{-1/2}\left(v_1\frac{\partial}{\partial v_2}-v_2\frac{\partial}{\partial v_1}\right)(v_a^0)^{1/2}f\\&=\frac{\hbar}{i}\left(v_1\frac{\partial}{\partial v_2}-v_2\frac{\partial}{\partial v_1}\right)f\ ,\end{aligned}$$

for every $f\in C_0^1(\mathbb{R}^3,\mathbb{C})$ and, with the help of results from quantum mechanics about angular momentum operators, see Sect. 2.4.3 in [7], that $\hat{L}_3$ has a pure point spectrum given by $\hbar.\mathbb{Z}$, consisting of eigenvalues of infinite multiplicity.

Since, according to Exercise 3.16, $\hat{L}_1$ and $\hat{L}_3$ as well as $\hat{L}_2$ and $\hat{L}_3$ are unitarily equivalent, also $\hat{L}_1$ and $\hat{L}_2$ have a pure point spectrum given by $\hbar.\mathbb{Z}$, consisting of eigenvalues of infinite multiplicity. Hence, we arrive at the following result.

> Each of the operators $\hat{L}_1$, $\hat{L}_2$ and $\hat{L}_3$ has a pure point spectrum given by $\hbar.\mathbb{Z}$, consisting of eigenvalues of infinite multiplicity.

3.2.2 Generators Associated with Lorentz Boosts

In the following, we analyze the generators of U_a for one-parameter subgroups associated with Lorentz boosts. For this purpose, let $a\geqslant 0$. Subsequently, our main cases of interest are the one-parameter groups of Lorentz boosts,

$$M_{0j}:\mathbb{R}\to\mathcal{L}_+^\uparrow\ ,$$

$j \in \{1, 2, 3\}$, where

$$M_{01}(s) := \begin{pmatrix} \cosh(s) & \sinh(s) & 0 & 0 \\ \sinh(s) & \cosh(s) & 0 & 0 \\ 0 & 0 & 1 & 0 \\ 0 & 0 & 0 & 1 \end{pmatrix},$$

$$M_{02}(s) := \begin{pmatrix} \cosh(s) & 0 & \sinh(s) & 0 \\ 0 & 1 & 0 & 0 \\ \sinh(s) & 0 & \cosh(s) & 0 \\ 0 & 0 & 0 & 1 \end{pmatrix},$$

$$M_{03}(s) := \begin{pmatrix} \cosh(s) & 0 & 0 & \sinh(s) \\ 0 & 1 & 0 & 0 \\ 0 & 0 & 1 & 0 \\ \sinh(s) & 0 & 0 & \cosh(s) \end{pmatrix}, \tag{3.20}$$

for every $s \in \mathbb{R}$, compare Exercise 3.5.

Subsequently, we analyze $U_a \circ M_{03}$. Again, if $f \in C_0^1(\mathbb{R}^3, \mathbb{C})$, then it follows from the mean value theorem in several variables the existence of $C \geqslant 0$ such that

$$|f(w) - f(v)| \leqslant C\,|w - v| \,,$$

for all $v, w \in \mathbb{R}^3$. Hence,

$$\begin{aligned} &\left|[(U_a \circ M_{03})(s) f](v) - f(v)\right|^2 = \left|[f \circ h_{aM_{03}(s)}](v) - f(v)\right|^2 \\ &\leqslant C^2 \left|h_{aM_{03}(s)}(v) - v\right|^2 = C^2 \left|p_a^{-1}(M_{03}(-s) \cdot p_a(v)) - v\right|^2 , \end{aligned}$$

for $v \in \mathbb{R}^3$. Further, we conclude for $s \in \mathbb{R}$, $v = (v_1, v_2, v_3) \in \mathbb{R}^3$ that

$$\begin{aligned} &\left|[(U_a \circ M_{03})(s) f](v) - f(v)\right|^2 = \left|[f \circ h_{aM_{03}(s)}](v) - f(v)\right|^2 \\ &= \left|f(p_a^{-1}(M_{03}(-s) \cdot p_a(v))) - f(v)\right|^2 . \end{aligned}$$

Since, $p_a : \mathbb{R}^3 \to H_{a+}$ is given by

$$p_a(v) := {}^t((a^2 + |v|^2)^{1/2}, v_1, v_2, v_3) \,,$$

for every $v \in \mathbb{R}^3$, and we have that

$$p_a^{-1}(v) = {}^t(v_1, v_2, v_3) \,,$$

for every $v \in H_{a+}$, it follows that

$$\begin{aligned}&M_{03}(-s)p_a(v)\\&={}^t(\cosh(s)(a^2+|v|^2)^{1/2}-\sinh(s)v_3,\, v_1,\, v_2,\\&\qquad -\sinh(s)(a^2+|v|^2)^{1/2}+\cosh(s)v_3)\end{aligned}$$

and hence that

$$\begin{aligned}h_{aM_{03}(-s)}(v)&=p_a^{-1}M_{03}(-s)p_a(v)\\&={}^t(v_1,\, v_2,\, -\sinh(s)(a^2+|v|^2)^{1/2}+\cosh(s)v_3)\ .\end{aligned}$$

We note that if $v_3 \neq 0$, we have that

$$\begin{aligned}&-\sinh(s)(a^2+|v|^2)^{1/2}+\cosh(s)v_3\\&=\sinh(s)|v_3|-\sinh(s)(a^2+|v|^2)^{1/2}-\sinh(s)|v_3|+\cosh(s)v_3\\&=\sinh(s)\left[|v_3|-(a^2+|v|^2)^{1/2}\right]-\sinh(s)|v_3|+\cosh(s)v_3\\&=\sinh(s)\,\frac{\left[|v_3|-(a^2+|v|^2)^{1/2}\right]\cdot\left[|v_3|+(a^2+|v|^2)^{1/2}\right]}{|v_3|+(a^2+|v|^2)^{1/2}}\\&\quad-\sinh(s)|v_3|+\cosh(s)v_3\\&=\sinh(s)\,\frac{v_3^2-(a^2+|v|^2)}{|v_3|+(a^2+|v|^2)^{1/2}}-\sinh(s)|v_3|+\cosh(s)v_3\\&=-\sinh(s)\,\frac{a^2+v_1^2+v_2^2}{|v_3|+(a^2+|v|^2)^{1/2}}-\sinh(s)|v_3|+\cosh(s)v_3\\&=-\sinh(s)\,\frac{a^2+v_1^2+v_2^2}{|v_3|+(a^2+|v|^2)^{1/2}}+\begin{cases}-\sinh(s)v_3+\cosh(s)v_3 & \text{if } v_3>0\\ \sinh(s)v_3+\cosh(s)v_3 & \text{if } v_3<0\end{cases}\\&=-\sinh(s)\,\frac{a^2+v_1^2+v_2^2}{|v_3|+(a^2+v_1^2+v_2^2+v_3^2)^{1/2}}+\begin{cases}e^{-s}v_3 & \text{if } v_3>0\\ e^{s}v_3 & \text{if } v_3<0\end{cases}.\end{aligned}$$

Hence, it follows that, if

$$v_3 > e^s[R+\sinh(|s|)(a^2+v_1^2+v_2^2)^{1/2}]\ ,$$

then

$$-\sinh(s)(a^2+|v|^2)^{1/2}+\cosh(s)v_3 > R$$

and if

$$v_3 < -e^{-s}[R+\sinh(|s|)\,(a^2+v_1^2+v_2^2)^{1/2}]\ ,$$

then

$$-\sinh(s)(a^2+|v|^2)^{1/2}+\cosh(s)v_3 < -R\ .$$

As a consequence, if $v_1^2+v_2^2 \geqslant R^2$, then

$$f \circ h_{aM_{03}(s)}(v_1, v_2, \cdot)$$

vanishes and hence

$$\text{supp}(f \circ h_{aM_{03}(s)}) \subset B_R(0) \times \mathbb{R} .$$

Further, if $v_1^2 + v_2^2 \leqslant R^2$, then

$$f \circ h_{aM_{03}(s)}(v_1, v_2, \cdot)$$

vanishes on the set

$$(-\infty, -e^{-s}[R + \sinh(|s|)\,(a^2 + R^2)^{1/2}]) \cup (e^s[R + \sinh(|s|)(a^2 + R^2)^{1/2}], \infty)$$

and hence,

$$\begin{aligned}&\text{supp}(f \circ h_{aM_{03}(s)}) \subset \\ &B_R(0) \times [-e^{-s}[R + \sinh(|s|)\,(a^2 + R^2)^{1/2}, e^s[R + \sinh(|s|)(a^2 + R^2)^{1/2}]] .\end{aligned}$$

As a consequence, $(U_a \circ M_{03})(s)f \in C_0^1(\mathbb{R}^3, \mathbb{C})$, and

$$\begin{aligned}
&\left|[(U_a \circ M_{03})(s)f](v) - f(v)\right|^2 \\
&= \left|f(v_1, v_2, -\sinh(s)(a^2 + |v|^2)^{1/2} + \cosh(s)v_3) - f(v_1, v_2, v_3)\right|^2 \\
&\leqslant C^2 \cdot \left|\sinh(s)(a^2 + |v|^2)^{1/2} + [1 - \cosh(s)]v_3\right|^2 \\
&= C^2 \cdot \left|2\sinh\left(\frac{s}{2}\right)\cosh\left(\frac{s}{2}\right)(a^2 + |v|^2)^{1/2} - 2\sinh^2\left(\frac{s}{2}\right) v_3\right|^2 \\
&= 4C^2 \sinh^2\left(\frac{s}{2}\right) \cdot \left|\cosh\left(\frac{s}{2}\right)(a^2 + |v|^2)^{1/2} - \sinh\left(\frac{s}{2}\right) v_3\right|^2 \\
&\leqslant 4C^2 \sinh^2\left(\frac{s}{2}\right) \cdot \left|\cosh\left(\frac{s}{2}\right)(a^2 + |v|^2)^{1/2} + \sinh\left(\frac{|s|}{2}\right)|v_3|\right|^2 \\
&\leqslant 4C^2 \sinh^2\left(\frac{s}{2}\right) \cdot \left|\cosh\left(\frac{s}{2}\right) + \sinh\left(\frac{|s|}{2}\right)\right|^2 (a^2 + |v|^2) \\
&= 4C^2 e^{|s|} \sinh^2\left(\frac{s}{2}\right)(a^2 + |v|^2) .
\end{aligned}$$

Taking into account, see above, that $(U_a \circ M_{03})(s)f - f$ has a compact support for every $s \in \mathbb{R}$, it follows the existence for $C' \geqslant 0$ and $R' \geqslant 0$ such that

$$\left|\frac{1}{s}[(U_a \circ M_{03})(s)f - f]\right|^2 \leqslant C'^2 \chi_{U_{R'}(0)} ,$$

for every $s \in (0, \tau]$, where $\tau > 0$. Further,

$$\lim_{s \to 0, s \neq 0} \frac{1}{s}\{[(U_a \circ M_{03})(s) - \text{id}_X] f\}(v) = -\left(v_a^0 \frac{\partial f}{\partial v_3}\right)(v) ,$$

for every $v \in \mathbb{R}^3$. As a consequence, if $s_1, s_2, \ldots$ is a sequence in $\mathbb{R}^*$ that is convergent to 0 and by using that for every $\alpha, \beta \in \mathbb{C}$, we have that

$$|\alpha+\beta|^2 \leqslant (|\alpha|+|\beta|)^2 = |\alpha|^2+|\beta|^2+2\,|\alpha|\cdot|\beta| \leqslant 2\,(|\alpha|^2+|\beta|^2) \ ,$$

it follows that

$$\left((v_a^0)^{-1}\left|\frac{1}{s_\nu}\,[(U_a\circ M_{03})(s_\nu)f - f] + v_a^0\,\frac{\partial f}{\partial v_3}\right|^2\right)_{\nu\in\mathbb{N}^*}$$

is a sequence of integrable functions that is everywhere on $\mathbb{R}^3$ convergent to the 0-function on $\mathbb{R}^3$ and whose members are dominated by an integrable function. Hence, it follows from Lebesgue's dominated convergence theorem that

$$\lim_{\nu\to\infty}\int_{\mathbb{R}^3}\left|\frac{1}{s_\nu}\,[(U_a\circ M_{03})(s_\nu)f - f] + v_a^0\,\frac{\partial f}{\partial v_3}\right|^2 (v_a^0)^{-1}\,dv^3 = 0 \ ,$$

i.e., that

$$\lim_{\nu\to\infty}\left\|\frac{1}{s_\nu}\,[(U_a\circ M_{03})(s_\nu)f - f] + v_a^0\,\frac{\partial f}{\partial v_3}\right\|_2 = 0 \ .$$

We conclude, see the beginning of Sect. 3.2.1, (3.15), (3.16) and (3.17), that $C_0^1(\mathbb{R}^3,\mathbb{C}) \subset D(A_{M_{03}})$ a well as that

$$A_{M_{03}}f = i\,v_a^0\,\frac{\partial f}{\partial v_3} \ ,$$

for every $f \in C_0^1(\mathbb{R}^3,\mathbb{C})$. In addition, $C_0^1(\mathbb{R}^3,\mathbb{C})$ is dense in X and left invariant by $U_a \circ M_{03}$, i.e., $(U_a\circ M_{03})(s)(C_0^1(\mathbb{R}^3,\mathbb{C})) \subset C_0^1(\mathbb{R}^3,\mathbb{C})$, for every $s \in \mathbb{R}$. Hence, $C_0^1(\mathbb{R}^3,\mathbb{C})$ is a core for $A_{M_{03}}$.[4] We define the corresponding operator $\hat{L}_{03}$ by

$$\hat{L}_{03} := \hbar\,A_{M_{03}} \ .$$

Analogously, we define the operators $\hat{L}_{01}$ and $\hat{L}_{02}$. We leave the details of the derivation of the corresponding properties of these operators to the reader.

Exercise 3.18 Show that

$$M_{03}(s) = \begin{pmatrix} \cosh(s) & 0 & 0 & \sinh(s) \\ 0 & 1 & 0 & 0 \\ 0 & 0 & 1 & 0 \\ \sinh(s) & 0 & 0 & \cosh(s) \end{pmatrix} = \exp(sA) \ ,$$

[4] We note that the previous arguments apply to $C_0^2(\mathbb{R}^3,\mathbb{C})$, i.e., $C_0^2(\mathbb{R}^3,\mathbb{C})$ is a core for $A_{M_{03}}$. This fact is going to be used, later on.

for every $s \in \mathbb{R}$, where

$$A := \begin{pmatrix} 0 & 0 & 0 & 1 \\ 0 & 0 & 0 & 0 \\ 0 & 0 & 0 & 0 \\ 1 & 0 & 0 & 0 \end{pmatrix} .$$

Exercise 3.19 With the help of the result of Exercise 3.18, show that

$$\begin{pmatrix} \cosh(s) & \sinh(s) & 0 & 0 \\ \sinh(s) & \cosh(s) & 0 & 0 \\ 0 & 0 & 1 & 0 \\ 0 & 0 & 0 & 1 \end{pmatrix} = M_{01}(s) = U_{13}^{-1} M_{03}(s) U_{13} = \exp(sB) ,$$

$$\begin{pmatrix} \cosh(s) & 0 & \sinh(s) & 0 \\ 0 & 1 & 0 & 0 \\ \sinh(s) & 0 & \cosh(s) & 0 \\ 0 & 0 & 0 & 1 \end{pmatrix} = M_{02}(s) = U_{23}^{-1} M_{03}(s) U_{23} = \exp(sC) ,$$

for every $s \in \mathbb{R}$, where

$$B := U_{13}^{-1} A U_{13} = \begin{pmatrix} 0 & 1 & 0 & 0 \\ 1 & 0 & 0 & 0 \\ 0 & 0 & 0 & 0 \\ 0 & 0 & 0 & 0 \end{pmatrix} , \quad C := U_{23}^{-1} A U_{23} = \begin{pmatrix} 0 & 0 & 1 & 0 \\ 0 & 0 & 0 & 0 \\ 1 & 0 & 0 & 0 \\ 0 & 0 & 0 & 0 \end{pmatrix}$$

and[5]

$$U_{13} := \begin{pmatrix} 1 & 0 & 0 & 0 \\ 0 & 0 & 0 & 1 \\ 0 & 0 & -1 & 0 \\ 0 & 1 & 0 & 0 \end{pmatrix} \in \mathcal{L}_+^{\uparrow} , \quad U_{23} := \begin{pmatrix} 1 & 0 & 0 & 0 \\ 0 & -1 & 0 & 0 \\ 0 & 0 & 0 & 1 \\ 0 & 0 & 1 & 0 \end{pmatrix} \in \mathcal{L}_+^{\uparrow} .$$

Verify that $U_{13}, U_{23} \in \mathcal{L}_+^{\uparrow}$ and that

$$U_{13}^{-1} = U_{13} , \quad U_{23}^{-1} = U_{23} .$$

Exercise 3.20 Show that, for every $j \in \{1, 2\}$ and $s \in \mathbb{R}$, we have that

$$U_a(M_{0j}(s)) = \exp\Big(i \, \frac{s}{\hbar} \, U_a(U_{j3}) \circ \hat{L}_{03} \circ U_a(U_{j3})\Big) ,$$

where $U_{13}, U_{23} \in \mathcal{L}_+^{\uparrow}$ are defined in Exercise 3.19. Further, show that

$$[U_a(U_{13})f](v) = f(v_3, -v_2, v_1) , \quad [U_a(U_{23})f](v) = f(-v_1, v_3, v_2) ,$$

[5] Note that U_{13} and U_{23} coincide with U_1 and U_2, respectively, from Exercise 3.15.

for every $f \in X$ and almost all $v = {}^t(v_1, v_2, v_3) \in \mathbb{R}^3$. Finally, for $j \in \{1, 2\}$, show that $U_a(U_{j3})C_0^1(\mathbb{R}^3, \mathbb{C}) = C_0^1(\mathbb{R}^3, \mathbb{C})$ and for $f \in C_0^1(\mathbb{R}^3, \mathbb{C})$ that

$$U_a(U_{13}) \circ \hat{L}_{03} \circ U_a(U_{13})f = i\hbar\, v_a^0 \frac{\partial f}{\partial v_1}\ ,\ \ U_a(U_{23}) \circ \hat{L}_{03} \circ U_a(U_{23})f = i\hbar\, v_a^0 \frac{\partial f}{\partial v_2}\ .$$

Hence, we arrive at the following result.

Angular Momenta Associated With Lorentz Boosts

For every $k \in \{1, 2, 3\}$, the following representation is true

$$\exp\Big(i\,\frac{s}{\hbar}\,\hat{L}_{0k}\Big)\, f = (U_a \circ M_{0k})(s)\ , \tag{3.21}$$

for every $f \in X := L^2_{\mathbb{C}}(\mathbb{R}^3, \varphi_a)$ and $s \in \mathbb{R}$, where for $k \in \{1, 2, 3\}$, $\hat{L}_{0k} : D(\hat{L}_{0k}) \to X$ is the densely-defined, linear and self-adjoint operator in X, given by

$$D(\hat{L}_{0k}) = \{f \in X : \lim_{s\to 0, s\neq 0} \frac{1}{s}\,[\,(U_a \circ M_{0k})(s) - \mathrm{id}_X]\, f \text{ exists}\}$$

and

$$\hat{L}_{0k} f := \frac{\hbar}{i} \lim_{s\to 0, s\neq 0} \frac{1}{s}\,[\,(U_a \circ M_{0k})(s) - \mathrm{id}_X]\, f\ ,$$

for every $f \in D(\hat{L}_k)$ and where $M_{01} : (\mathbb{R}, +) \to \mathcal{L}_+^{\uparrow}$, $M_{02} : (\mathbb{R}, +) \to \mathcal{L}_+^{\uparrow}$ and $M_{03} : (\mathbb{R}, +) \to \mathcal{L}_+^{\uparrow}$ are given by

$$M_{01}(s) = \begin{pmatrix} \cosh(s) & \sinh(s) & 0 & 0 \\ \sinh(s) & \cosh(s) & 0 & 0 \\ 0 & 0 & 1 & 0 \\ 0 & 0 & 0 & 1 \end{pmatrix},$$

$$M_{02}(s) = \begin{pmatrix} \cosh(s) & 0 & \sinh(s) & 0 \\ 0 & 1 & 0 & 0 \\ \sinh(s) & 0 & \cosh(s) & 0 \\ 0 & 0 & 0 & 1 \end{pmatrix},$$

$$M_{03}(s) = \begin{pmatrix} \cosh(s) & 0 & 0 & \sinh(s) \\ 0 & 1 & 0 & 0 \\ 0 & 0 & 1 & 0 \\ \sinh(s) & 0 & 0 & \cosh(s) \end{pmatrix},$$

for every $s \in \mathbb{R}$. In particular, $C_0^1(\mathbb{R}^3, \mathbb{C})$ is a core for $\hat{L}_{01}$, $\hat{L}_{02}$ and $\hat{L}_{03}$, and we have that

$$\hat{L}_{01} f = i\hbar\, v_a^0 \frac{\partial f}{\partial v_1}\ , \quad \hat{L}_{02} f = i\hbar\, v_a^0 \frac{\partial f}{\partial v_2}\ , \quad \hat{L}_{03} f = i\hbar\, v_a^0 \frac{\partial f}{\partial v_3}\ ,$$

for every $f \in C_0^1(\mathbb{R}^3, \mathbb{C})$, where v_1, v_2 and v_3 denote the coordinate projection of $\mathbb{R}^3$ onto the first, second and third coordinate, respectively.

For the determination of the spectra of $\hat{L}_{01}$, $\hat{L}_{02}$ and $\hat{L}_{03}$, we use the result of the following exercise.

Exercise 3.21 (i) Define

$$\mathcal{U}_a := \begin{cases} \mathbb{R}^3 & \text{if } a > 0 \\ \mathbb{R}^3 \setminus (\{0\} \times \{0\} \times \mathbb{R}) & \text{if } a = 0 \end{cases} .$$

Show that the map $\mathscr{h}_a : \mathcal{U}_a \to \mathcal{U}_a$, defined by

$$\mathscr{h}_a(w) := {}^t(w_1, w_2, (w_1^2 + w_2^2 + a^2)^{1/2} \sinh(w_3))\ ,$$

for every $w \in \mathcal{U}_a$, is a C^∞-diffeomorphism with inverse $\mathscr{h}_a^{-1} : \mathcal{U}_a \to \mathcal{U}_a$ given by

$$\mathscr{h}_a^{-1}(v) := {}^t\!\left(v_1, v_2, \operatorname{arsinh}\left(\frac{v_3}{(v_1^2 + v_2^2 + a^2)^{1/2}}\right)\right)\ ,$$

for every $v \in \mathcal{U}_a$ and such that

$$(v_a^0 \circ \mathscr{h}_a)^{-1}\, |\det(\mathscr{h}_a')\,| = 1\ .$$

(ii) Show that by $\mathcal{W}_a f := f \circ \mathscr{h}_a$, for every $f \in L^2_{\mathbb{C}}(\mathbb{R}^3, \varphi_a)$, there is defined a Hilbert space isomorphism $\mathcal{W}_a : L^2_{\mathbb{C}}(\mathbb{R}^3, \varphi_a) \to L^2_{\mathbb{C}}(\mathbb{R}^3)$, with corresponding inverse $\mathcal{W}_a^{-1} : L^2_{\mathbb{C}}(\mathbb{R}^3) \to L^2_{\mathbb{C}}(\mathbb{R}^3, \varphi_a)$ given by $\mathcal{W}_a^{-1} f = f \circ \mathscr{h}_a^{-1}$, for every $f \in L^2_{\mathbb{C}}(\mathbb{R}^3)$.

Using the Hilbert space isomorphism $\mathcal{W}_a : L^2_{\mathbb{C}}(\mathbb{R}^3, \varphi_a) \to L^2_{\mathbb{C}}(\mathbb{R}^3)$ from Exercise 3.21, it follows for the case that $a > 0$ and $f \in C_0^1(\mathbb{R}^3, \mathbb{C})$ that $\mathcal{W}_a^{-1} f \in C_0^1(\mathbb{R}^3, \mathbb{C})$ and

$$\mathcal{W}_a \hat{L}_{03} \mathcal{W}_a^{-1} f = i\hbar\, \mathcal{W}_a v_a^0 \frac{\partial (f \circ \mathscr{h}_a^{-1})}{\partial v_3} = i\hbar\, \mathcal{W}_a v_a^0 \sum_{j=1}^{3} \left(\frac{\partial f}{\partial w_j} \circ \mathscr{h}_a^{-1}\right) \cdot \frac{\partial (\mathscr{h}_a^{-1})_j}{\partial v_3}$$

$$= i\hbar\, (v_a^0 \circ \mathscr{h}_a)\, \frac{\partial f}{\partial w_3} \cdot \left(\frac{\partial (\mathscr{h}_a^{-1})_3}{\partial v_3} \circ \mathscr{h}_a\right)\ .$$

Since

$$\frac{\partial(\hbar_a^{-1})_3}{\partial v_3}(\hbar_a(w)) = \frac{1}{\sqrt{w_1^2 + w_2^2 + a^2\cosh(w_3)}} = \frac{1}{v_a^0(\hbar_a(w))} ,$$

for every $w \in \mathbb{R}^3$, we have that

$$\mathcal{W}_a \hat{L}_{03} \mathcal{W}_a^{-1} f = i\hbar \frac{\partial f}{\partial w_3} .$$

Hence, with the help of results from quantum mechanics about momentum operators, see Sect. 1.4 in [7], it follows that $\hat{L}_{03}$ has an absolutely continuous spectrum given by all real numbers. For the case that $a = 0$ and $f \in C_0^2(\mathbb{R}^3, \mathbb{C})$, we have that

$$\mathcal{W}_{1/\nu}^{-1} f = f \circ \hbar_{1/\nu}^{-1} \in C_0^2(\mathbb{R}^3, \mathbb{C})$$

and that

$$\begin{aligned} i\hbar\, v_0^0 \frac{\partial \mathcal{W}_{1/\nu}^{-1} f}{\partial v_3} &= \frac{v_0^0}{v_{1/\nu}^0} i\hbar\, v_{1/\nu}^0 \frac{\partial (f \circ \hbar_{1/\nu}^{-1})}{\partial v_3} = i\hbar \frac{v_0^0}{v_{1/\nu}^0} \left(\frac{\partial f}{\partial w_3} \circ \hbar_{1/\nu}^{-1} \right) \\ &= i\hbar \frac{v_0^0}{v_{1/\nu}^0} \mathcal{W}_{1/\nu}^{-1} \frac{\partial f}{\partial w_3} , \end{aligned}$$

for every $\nu \in \mathbb{N}^*$. Further, for $g \in C_0^1(\mathbb{R}^3, \mathbb{C})$, it follows that $\mathcal{W}_{1/1}^{-1} g = g \circ \hbar_{1/1}^{-1}, \mathcal{W}_{1/2}^{-1} g = g \circ \hbar_{1/2}^{-1}, \dots$ is a sequence in $C_0^1(\mathbb{R}^3, \mathbb{C}) \subset L_{\mathbb{C}}^2(\mathbb{R}^3, \varphi_0)$ that is a.e. on $\mathbb{R}^3$ pointwise convergent to $g \circ \hbar_0^{-1} = \mathcal{W}_0^{-1} g \in L_{\mathbb{C}}^2(\mathbb{R}^3, \varphi_0)$. Since $g \in C_0^1(\mathbb{R}^3, \mathbb{C})$, it follows from the mean value theorem in several variables the existence of $C \geqslant 0$ such that

$$|g(w_1) - g(w_2)| \leqslant C\, |w_1 - w_2| ,$$

for all $w_1, w_2 \in \mathbb{R}^3$. In a first step, we are going to investigate the support properties of $g \circ \hbar_\nu^{-1}$, for $\nu \in \mathbb{N}^*$. If $R > 0$ is such that $\operatorname{supp}(g) \subset U_R(0)$ and $v \in \mathbb{R}^3$, then

$$\begin{aligned} |\hbar_{1/\nu}^{-1}(v)|^2 &= \left| {}^t\!\left(v_1, v_2, \operatorname{arsinh}\left(\frac{v_3}{(v_1^2 + v_2^2 + (1/\nu^2))^{1/2}} \right) \right) \right|^2 \\ &= v_1^2 + v_2^2 + \operatorname{arsinh}^2\left(\frac{|v_3|}{(v_1^2 + v_2^2 + (1/\nu^2))^{1/2}} \right) \\ &\geqslant v_1^2 + v_2^2 + \left[\ln\left(1 + \frac{|v_3|}{(v_1^2 + v_2^2 + (1/\nu^2))^{1/2}} \right) \right]^2 , \end{aligned}$$

where we used that

$$\operatorname{arsinh}(x) = \operatorname{arsinh}(x) - \operatorname{arsinh}(0) = \int_0^x \frac{1}{\sqrt{1+y^2}}\,dy \geqslant \int_0^x \frac{1}{1+y}\,dy$$
$$= \ln(1+x) - \ln(1) = \ln(1+x) \ ,$$

for every $x \geqslant 0$. Hence, if $v_1^2 + v_2^2 \geqslant R^2$, then $g(h_{1/\nu}^{-1}(v)) = 0$. If $v_1^2 + v_2^2 \leqslant R^2$ and $|v_3| \geqslant e^R\,(R^2+1)^{1/2}$, then

$$1 + \frac{|v_3|}{(v_1^2 + v_2^2 + (1/\nu^2))^{1/2}} \geqslant 1 + \frac{|v_3|}{(R^2+1)^{1/2}} \geqslant 1 + e^R \geqslant e^R$$

and hence

$$\ln\left(1 + \frac{|v_3|}{(v_1^2 + v_2^2 + (1/\nu^2))^{1/2}}\right) \geqslant R$$

as well as $g(h_{1/\nu}^{-1}(v)) = 0$. As a consequence, we have that

$$g(h_{1/\nu}^{-1}(v)) = 0 \ , \ \text{ if } |v_3| \geqslant e^R\,(R^2+1)^{1/2} \ .$$

Taking the limit $\nu \to \infty$, we also have that

$$g(h_0^{-1}(v)) = 0 \ , \ \text{ if } v \in \mathbb{R}^3 \setminus (\{0\} \times \{0\} \times \mathbb{R}) \text{ and } |v_3| \geqslant e^R\,(R^2+1)^{1/2} \ .$$

Further, for $v \in \mathbb{R}^3 \setminus (\{0\} \times \{0\} \times \mathbb{R})$ such that $|v_3| \leqslant e^R\,(R^2+1)^{1/2}$, it follows that

$$\begin{aligned}
&|(g \circ h_{1/\nu}^{-1})(v) - (g \circ h_0^{-1})(v)| \leqslant C\,|h_{1/\nu}^{-1}(v) - h_0^{-1}(v)| \\
&= C\left|\operatorname{arsinh}\left(\frac{v_3}{(v_1^2 + v_2^2 + (1/\nu)^2)^{1/2}}\right) - \operatorname{arsinh}\left(\frac{v_3}{(v_1^2 + v_2^2)^{1/2}}\right)\right| \\
&\leqslant C\left|\frac{v_3}{(v_1^2 + v_2^2 + (1/\nu)^2)^{1/2}} - \frac{v_3}{(v_1^2 + v_2^2)^{1/2}}\right| \\
&= C\,|v_3|\,\frac{|(v_1^2 + v_2^2)^{1/2} - (v_1^2 + v_2^2 + (1/\nu)^2)^{1/2}|}{(v_1^2 + v_2^2)^{1/2} \cdot (v_1^2 + v_2^2 + (1/\nu)^2)^{1/2}} \\
&= \frac{C\,|v_3|/\nu^2}{(v_1^2 + v_2^2)^{1/2}(v_1^2 + v_2^2 + (1/\nu)^2)^{1/2}[(v_1^2 + v_2^2)^{1/2} + (v_1^2 + v_2^2 + (1/\nu)^2)^{1/2}|]} \\
&\leqslant \frac{C}{2\nu^2}\,\frac{|v_3|}{(v_1^2 + v_2^2)^{3/2}} \leqslant \frac{1}{2\nu^2}\,C\,e^R\,(R^2+1)^{1/2}\,\frac{1}{(v_1^2 + v_2^2)^{3/2}} \\
&\leqslant \frac{1}{2}\,C\,e^R\,(R^2+1)^{1/2}\,\frac{1}{(v_1^2 + v_2^2)^{3/2}} \ ,
\end{aligned}$$

where we used the mean-value theorem in one variable. As a consequence,

$$
\begin{aligned}
&|(g \circ \hbar_{1/\nu}^{-1})(v) - (g \circ \hbar_0^{-1})| \\
&\leqslant \frac{1}{2}\, C\, e^R\, (R^2+1)^{1/2}\, \frac{1}{(v_1^2+v_2^2)^{3/2}}\, \chi_{\mathbb{R}^2 \times (-e^R\,(R^2+1)^{1/2}, e^R\,(R^2+1)^{1/2})} \ ,
\end{aligned}
$$

a.e. on $\mathbb{R}^3$, where v_1 and v_2 denote the coordinate projections of $\mathbb{R}^3$ onto the first and second component, respectively. Since we have also that

$$
|(g \circ \hbar_{1/\nu}^{-1})(v) - (g \circ \hbar_0^{-1})(v)| \leqslant 2\, \|g\|_\infty \ ,
$$

for every $v \in \mathbb{R}^3 \setminus (\{0\} \times \{0\} \times \mathbb{R})$, it follows that

$$
\begin{aligned}
&|(g \circ \hbar_{1/\nu}^{-1})(v) - (g \circ \hbar_0^{-1})| \\
&\leqslant \left[\frac{1}{2}\, C\, e^R\, (R^2+1)^{1/2} + 2\, \|g\|_\infty\right] \\
&\quad \cdot \min\left\{1, \frac{1}{(v_1^2+v_2^2)^{3/2}}\right\} \chi_{\mathbb{R}^2 \times (-e^R\,(R^2+1)^{1/2}, e^R\,(R^2+1)^{1/2})} \ ,
\end{aligned}
$$

a.e. on $\mathbb{R}^3$. Hence,

$$
|\ |^{-1} \cdot |(g \circ \hbar_{1/1}^{-1})(v) - (g \circ \hbar_0^{-1})|^2 \, , \ |\ |^{-1} \cdot |(g \circ \hbar_{1/2}^{-1})(v) - (g \circ \hbar_0^{-1})|^2 \, , \ \dots
$$

is a sequence of integrable functions that is a.e. on $\mathbb{R}^3$ pointwise convergent to the zero-function and that is dominated by an integrable function. Hence it follows from Lebesgue's dominated convergence theorem that

$$
\lim_{\nu \to \infty} \|\mathcal{W}_{1/\nu}^{-1} g - \mathcal{W}_0^{-1} g\|_2 = 0 \ ,
$$

where $\|\ \|_2$ denotes the norm on $L^2_{\mathbb{C}}(\mathbb{R}^3, \varphi_0)$. In addition, we note for $\nu \in \mathbb{N}^*$ that

$$
\frac{v_0^0}{v_{1/\nu}^0} = \frac{|\ |}{[|\ |^2 + (1/\nu)^2]^{1/2}} \in L^\infty_{\mathbb{C}}(\mathbb{R}^3, \varphi_0) \ ,
$$

$\|v_0^0 / v_{1/\nu}^0\|_\infty \leqslant 1$ and that

$$
\lim_{\nu \to \infty} \frac{v_0^0}{v_{1/\nu}^0} = 1 \ ,
$$

a.e. on $\mathbb{R}^3$. Hence, also

$$
\begin{aligned}
&|\ |^{-1} \cdot \left| \frac{|\ |}{[|\ |^2 + (1/\nu)^2]^{1/2}} \, (g \circ \hbar_{1/1}^{-1})(v) - (g \circ \hbar_0^{-1}) \right|^2 \ , \\
&|\ |^{-1} \cdot \left| \frac{|\ |}{[|\ |^2 + (1/\nu)^2]^{1/2}} \, (g \circ \hbar_{1/2}^{-1})(v) - (g \circ \hbar_0^{-1}) \right|^2 \ , \ \dots
\end{aligned}
$$

is a sequence of integrable functions that is a.e. on $\mathbb{R}^3$ pointwise convergent to the zero-function and that is dominated by an integrable function, and it follows from Lebesgue's dominated convergence theorem that

$$\lim_{\nu\to\infty} \left\| \frac{v_0^0}{v_{1/\nu}^0} \mathcal{W}_{1/\nu}^{-1} g - \mathcal{W}_0^{-1} g \right\|_2 = 0 .$$

As a consequence, we have that

$$\lim_{\nu\to\infty} \mathcal{W}_{1/\nu}^{-1} f = \mathcal{W}_0^{-1} f \ , \ \mathcal{W}_0^{-1} \frac{\partial f}{\partial w_3} = \lim_{\nu\to\infty} \frac{v_0^0}{v_{1/\nu}^0} \mathcal{W}_{1/\nu}^{-1} \frac{\partial f}{\partial w_3} = \lim_{\nu\to\infty} v_0^0 \frac{\partial \mathcal{W}_{1/\nu}^{-1} f}{\partial v_3} \ ,$$

in $L^2_{\mathbb{C}}(\mathbb{R}^3, \varphi_0)$. Hence,

$$\lim_{\nu\to\infty} \mathcal{W}_{1/\nu}^{-1} f = \mathcal{W}_0^{-1} f \ , \ \lim_{\nu\to\infty} \hat{L}_{03} \mathcal{W}_{1/\nu}^{-1} f = \mathcal{W}_0^{-1} \, i\hbar \frac{\partial f}{\partial w_3} \ ,$$

in $L^2_{\mathbb{C}}(\mathbb{R}^3, \varphi_0)$ and, since $\hat{L}_{03}$ is closed,

$$\mathcal{W}_0^{-1} f \in D(\hat{L}_{03}) \ , \ \hat{L}_{03} \mathcal{W}_0^{-1} f = \mathcal{W}_0^{-1} \, i\hbar \frac{\partial f}{\partial w_3} \ .$$

The latter implies that

$$\mathcal{W}_0 \hat{L}_{03} \mathcal{W}_0^{-1} f = i\hbar \frac{\partial f}{\partial w_3} \ ,$$

for every $f \in C_0^2(\mathbb{R}^3, \mathbb{C})$. Hence, with the help of results from quantum mechanics about momentum operators, see Sect. 1.4 in [7], and using that $C_0^2(\mathbb{R}^3, \mathbb{C})$ is a core for these operators, it follows that $\hat{L}_{03}$ has an absolutely continuous spectrum given by all real numbers. Since, according to Exercise 3.20, $\hat{L}_{01}$ and $\hat{L}_{03}$ as well as $\hat{L}_{02}$ and $\hat{L}_{03}$ are unitarily equivalent, also the spectra of $\hat{L}_{01}$ and $\hat{L}_{02}$ are given by all real numbers. Hence, we arrive at the following result.

Each of the operators $\hat{L}_{01}$, $\hat{L}_{02}$ and $\hat{L}_{03}$ has an absolutely continuous spectrum consisting of all real numbers. In addition, if $a > 0$, we have that

$$\mathcal{W}_a \hat{L}_{03} \mathcal{W}_a^{-1} f = i\hbar \frac{\partial f}{\partial w_3} \ ,$$

for every $f \in C_0^1(\mathbb{R}^3, \mathbb{C})$, and, if $a = 0$, we have that

$$\mathcal{W}_0 \hat{L}_{03} \mathcal{W}_0^{-1} f = i\hbar \frac{\partial f}{\partial w_3} ,$$

for every $f \in C_0^2(\mathbb{R}^3, \mathbb{C})$.

3.3 Basic Properties of the Poincaré Group

Our next goal is the extension of U_a to a representation $\hat{U}_a$ that includes the translation group, i.e., an extension of U_a to a representation of the restricted Poincaré group. As a result generators, corresponding to one-parameter unitary groups that are associated with translations, are going to be maximal multiplication operators. Since such generators are momentum operators, in this way, the representation space $L^2_{\mathbb{C}}(\mathbb{R}^3, \varphi_a)$ acquires the role of a momentum space.

In the following, we define the Poincaré group as a semi-direct group of the translation group and the Lorentz group,

$$\mathcal{P} := \mathbb{R}^4 \times \mathcal{L} ,$$

equipped $\mathcal{P}$ with the product $\cdot : \mathcal{P} \times \mathcal{P} \to \mathcal{P}$, defined by

$$(a_1, \Lambda_1) \cdot (a_2, \Lambda_2) := (a_1 + \Lambda_1 a_2, \Lambda_1 \Lambda_2) ,$$

for all $(a_1, \Lambda_1), (a_2, \Lambda_2) \in \mathcal{P}$.

Lemma 3.8 (The Poincaré group) $(\mathcal{P}, \cdot)$ *is a group, the so called Poincaré group.*

Proof For (a, Λ), (a_1, Λ_1), (a_2, Λ_2) and $(a_3, \Lambda_3) \in \mathcal{P}$, it follows that

$$\begin{aligned}
&(a_1, \Lambda_1) \cdot [(a_2, \Lambda_2) \cdot (a_3, \Lambda_3)] = (a_1, \Lambda_1) \cdot (a_2 + \Lambda_2 a_3, \Lambda_2 \Lambda_3) \\
&= (a_1 + \Lambda_1 (a_2 + \Lambda_2 a_3), \Lambda_1 \Lambda_2 \Lambda_3) = (a_1 + \Lambda_1 a_2 + \Lambda_1 \Lambda_2 a_3, \Lambda_1 \Lambda_2 \Lambda_3) , \\
&[(a_1, \Lambda_1) \cdot (a_2, \Lambda_2)] \cdot (a_3, \Lambda_3) = (a_1 + \Lambda_1 a_2, \Lambda_1 \Lambda_2) \cdot (a_3, \Lambda_3) \\
&= (a_1 + \Lambda_1 a_2 + \Lambda_1 \Lambda_2 a_3, \Lambda_1 \Lambda_2 \Lambda_3) , \\
&(0, E) \cdot (a, \Lambda) = (0 + Ea, E\Lambda) = (a, \Lambda) = (a + \Lambda \cdot 0, \Lambda E) = (a, \Lambda) \cdot (0, E) , \\
&(-\Lambda^{-1} a, \Lambda^{-1}) \cdot (a, \Lambda) = (-\Lambda^{-1} a + \Lambda^{-1} a, \Lambda^{-1} \cdot \Lambda) = (0, E) , \\
&(a, \Lambda) \times (-\Lambda^{-1} a, \Lambda^{-1}) = (a + \Lambda (-\Lambda^{-1} a), \Lambda \Lambda^{-1}) = (0, E) .
\end{aligned}$$

Hence, $(\mathcal{P}, \cdot)$ is a group with unit element $(0, E)$, and for every $(a, \Lambda) \in \mathcal{P}$, the corresponding inverse element is given by $(-\Lambda^{-1} a, \Lambda^{-1})$. □

In the following, we assume the real vector space $\mathbb{R}^4 \times \mathrm{M}(4, \mathbb{R})$ equipped with the norm

$$\|(m, M)\| := (\,|m|^2 + \|M\|_{op}^2)^{1/2}\ ,$$

for every $(m, M) \in \mathbb{R}^4 \times \mathrm{M}(4, \mathbb{R})$, where $|\ |$ denotes the Euclidean norm on $\mathbb{R}^4$. Since $(\mathbb{R}^4, |\ |)$ and $(\mathrm{M}(4, \mathbb{R}), \|\ \|_{op})$ are complete, $(\mathbb{R}^4 \times \mathrm{M}(4, \mathbb{R}), \|\ \|)$ is a Banach space, see e.g. Lemma 12.2.7 in the Appendix of [7]. In particular, the coordinate projections onto $\mathbb{R}^4$ and $\mathrm{M}(4, \mathbb{R})$ are continuous. In addition to $\|\ \|$, we are also going to use the equivalent maximum norm $\|\ \|_\infty$ on $\mathbb{R}^4 \times \mathrm{M}(4, \mathbb{R})$, where $\|(m, M)\|_\infty$ is defined as the maximum of the absolute values of the components of m and M, corresponding to $(m, M) \in \mathbb{R}^4 \times \mathrm{M}(4, \mathbb{R})$. That the norms $\|\ \|$ and $\|\ \|_\infty$ are equivalent can be seen as follows.[6] For $(m, M) \in \mathbb{R}^4 \times \mathrm{M}(4, \mathbb{R})$, we have that

$$|m_k| \leqslant |m|\ ,\ \ |M_{jk}| \leqslant \|M\|_\infty \leqslant 2\,\|M\|_{op}\ ,$$

for $k \in \{1, \dots, 4\}$, where we used (1.1), and hence that

$$\|(m, M)\|_\infty \leqslant 2\,(|m| + \|M\|_{op}) \leqslant 2\sqrt{2}\,(|m|^2 + \|M\|_{op}^2)^{1/2} = 2\sqrt{2}\,\|(m, M)\|\ .$$

Further,

$$|m| \leqslant 2\,\|m\|_\infty\ ,\ \ \|M\|_{op} \leqslant 8\,\|M\|_\infty\ ,$$

where we used (1.1), and hence

$$|m| + \|M\|_{op} \leqslant 2\,\|m\|_\infty + 8\,\|M\|_\infty \leqslant 10\,\|(m, M)\|_\infty\ .$$

The latter implies that

$$\|(m, M)\| = (\,|m|^2 + \|M\|_{op}^2)^{1/2} \leqslant |m| + \|M\|_{op} \leqslant 10\,\|(m, M)\|_\infty\ .$$

As consequence, we have that

$$\|(m, M)\| \leqslant 10\,\|(m, M)\|_\infty \leqslant 20\sqrt{2}\,\|(m, M)\|\ ,$$

for every $(m, M) \in \mathbb{R}^4 \times \mathrm{M}(4, \mathbb{R})$, and hence that $\|\ \|$ and $\|\ \|_\infty$ are equivalent.

> We equip the Poincaré group with the topology induced by that of $(\mathbb{R}^4 \times \mathrm{M}(4, \mathbb{R}), \|\ \|)$. Then $\mathscr{P}$ is a Hausdorff topological space, with a countable basis.

[6] This can also be concluded from a general result, see [39], that all norms defined on a finite dimensional vector space are equivalent.

We note that the multiplication is continuous, since if $(a_1, \Lambda_1), (a_2, \Lambda_2) \in \mathcal{P}$, $((a_{1\nu}, \Lambda_{1\nu}))_{\nu\in\mathbb{N}}$ and $((a_{2\nu}, \Lambda_{2\nu}))_{\nu\in\mathbb{N}}$ are sequences in $\mathbb{R}^4 \times \mathrm{M}(4, \mathbb{R})$ that are convergent to (a_1, Λ_1) and $(a_2, \Lambda_2) \in \mathcal{P}$, respectively, then

$$\begin{aligned}
&\|(a_{1\nu}, \Lambda_{1\nu}) \cdot (a_{2\nu}, \Lambda_{2\nu}) - (a_1, \Lambda_1) \cdot (a_2, \Lambda_2)\|^2 \\
&= \|(a_{1\nu} + \Lambda_{1\nu} a_{2\nu}, \Lambda_{1\nu}\Lambda_{2\nu}) - (a_1 + \Lambda_1 a_2, \Lambda_1\Lambda_2)\|^2 \\
&= \|(a_{1\nu} - a_1 + \Lambda_{1\nu} a_{2\nu} - \Lambda_1 a_2, \Lambda_{1\nu}\Lambda_{2\nu} - \Lambda_1\Lambda_2)\|^2 \\
&= |a_{1\nu} - a_1 + \Lambda_{1\nu} a_{2\nu} - \Lambda_1 a_2|^2 + \|\Lambda_{1\nu}\Lambda_{2\nu} - \Lambda_1\Lambda_2\|_{op}^2 \\
&= |a_{1\nu} - a_1 + (\Lambda_{1\nu} - \Lambda_1)(a_{2\nu} - a_2) + (\Lambda_{1\nu} - \Lambda_1)a_2 + \Lambda_1(a_{2\nu} - a_2)|^2 \\
&\quad + \|(\Lambda_{1\nu} - \Lambda_1)(\Lambda_{2\nu} - \Lambda_2) + \Lambda_1(\Lambda_{2\nu} - \Lambda_2) + (\Lambda_{1\nu} - \Lambda_1)\Lambda_2\|_{op}^2 \\
&\leqslant 4\,[\,|a_{1\nu} - a_1|^2 + |(\Lambda_{1\nu} - \Lambda_1)(a_{2\nu} - a_2)|^2 + |(\Lambda_{1\nu} - \Lambda_1)a_2|^2 \\
&\quad + |\Lambda_1(a_{2\nu} - a_2)|^2 + \|(\Lambda_{1\nu} - \Lambda_1)(\Lambda_{2\nu} - \Lambda_2)\|_{op}^2 + \|\Lambda_1(\Lambda_{2\nu} - \Lambda_2)\|_{op}^2 \\
&\quad + \|(\Lambda_{1\nu} - \Lambda_1)\Lambda_2\|_{op}^2] \\
&\leqslant 4\,[\,|a_{1\nu} - a_1|^2 + \|\Lambda_{1\nu} - \Lambda_1\|_{op}^2\, |a_{2\nu} - a_2|^2 + \|\Lambda_{1\nu} - \Lambda_1\|_{op}^2\, |a_2|^2 \\
&\quad + \|\Lambda_1\|_{op}^2\, |a_{2\nu} - a_2|^2 + \|\Lambda_{1\nu} - \Lambda_1\|_{op}^2\, \|\Lambda_{2\nu} - \Lambda_2\|_{op}^2 + \|\Lambda_1\|_{op}^2\, \|\Lambda_{2\nu} - \Lambda_2\|_{op}^2 \\
&\quad + \|\Lambda_{1\nu} - \Lambda_1\|_{op}^2\, \|\Lambda_2\|_{op}^2]\,,
\end{aligned}$$

for every $\nu \in \mathbb{N}$ and hence

$$\lim_{\nu\to\infty} (a_{1\nu}, \Lambda_{1\nu}) \cdot (a_{2\nu}, \Lambda_{2\nu}) = (a_1, \Lambda_1) \cdot (a_2, \Lambda_2)\,.$$

In addition,

$$\begin{aligned}
&\|(a_{1\nu}, \Lambda_{1\nu})^{-1} - (a_1, \Lambda_1)^{-1}\|_{op}^2 = \|(-\Lambda_{1\nu}^{-1} a_{1\nu}, \Lambda_{1\nu}^{-1}) - (-\Lambda_1^{-1} a_1, \Lambda_1^{-1})\|_{op}^2 \\
&= |\Lambda_{1\nu}^{-1} a_{1\nu} - \Lambda_1^{-1} a_1|^2 + \|\Lambda_{1\nu}^{-1} - \Lambda_1^{-1}\|_{op}^2 \\
&= |(\Lambda_{1\nu}^{-1} - \Lambda_{1\nu}^{-1})(a_{1\nu} - a_1) + \Lambda_1^{-1}(a_{1\nu} - a_1) + (\Lambda_{1\nu}^{-1} - \Lambda_1^{-1})\, a_1|^2 \\
&\quad + \|\Lambda_{1\nu}^{-1} - \Lambda_1^{-1}\|_{op}^2 \\
&\leqslant 4\,[\,\|\Lambda_{1\nu}^{-1} - \Lambda_{1\nu}^{-1}\|_{op}^2\, |a_{1\nu} - a_1|^2 + \|\Lambda_1^{-1}\|_{op}^2\, |a_{1\nu} - a_1|^2 \\
&\quad + \|\Lambda_{1\nu}^{-1} - \Lambda_1^{-1}\|_{op}^2\, |a_1|^2\,] + \|\Lambda_{1\nu}^{-1} - \Lambda_1^{-1}\|_{op}^2
\end{aligned}$$

for every $\nu \in \mathbb{N}$ and hence

$$\lim_{\nu\to\infty} (a_{1\nu}, \Lambda_{1\nu})^{-1} = (a_1, \Lambda_1)^{-1}\,.$$

Therefore, the group operations, i.e., multiplication and inversion, of $\mathcal{P}$ are continuous, with respect to the induced topology. In this way, $\mathcal{P}$ becomes a topological group. The same applies to subgroups of $\mathcal{P}$ that are defined in the following.

Exercise 3.22 Show that $\mathcal{P} \subset \mathbb{R}^4 \times \mathrm{M}(4, \mathbb{R})$ is closed and unbounded.

Lemma 3.9 (The restricted Poincaré group) *We define*

$$\mathcal{P}_+^\uparrow := \mathbb{R}^4 \times \mathcal{L}_+^\uparrow .$$

Then $\mathcal{P}_+^\uparrow$ is a subgroup, the so called restricted Poincaré group, of $\mathcal{P}$.

Proof Obviously, $(0, E) \in \mathcal{P}_+^\uparrow$ and if $(a_1, \Lambda_1), (a_2, \Lambda_2) \in \mathcal{P}_+^\uparrow$, then

$$(a_1, \Lambda_1) \cdot (a_2, \Lambda_2) = (a_1 + \Lambda_1 a_2, \Lambda_1 \Lambda_2) \in \mathcal{P}_+^\uparrow . \tag{3.22}$$

Finally, if $(a, \Lambda) \in \mathcal{P}_+^\uparrow$, then $(a, \Lambda)^{-1} = (-\Lambda^{-1} a, \Lambda^{-1}) \in \mathcal{P}_+^\uparrow$. □

Exercise 3.23 Show that $\mathcal{P}_+^\uparrow \subset \mathbb{R}^4 \times \mathrm{M}(4, \mathbb{R})$ is closed, path-connected and unbounded.

As a semi-direct product, $\mathcal{P}$ is not a matrix group and hence the results of Sect. 1.1 do not apply. Of course, there is a straightforward definition of tangent spaces to $\mathcal{P}$, as subspaces of $\mathbb{R}^4 \times \mathrm{M}(4, \mathbb{R})$. Since the elements of $\mathbb{R}^4 \times \mathrm{M}(4, \mathbb{R})$ are neither matrices nor linear maps, they are not part of the domain of an operator exponential map. On the other hand, $\mathcal{P}$ is isomorphic to a subgroup of $GL(\mathbb{R}^4 \times \mathrm{M}(4, \mathbb{R}))$, see Exercise 1.1, and in this way, the results of Sect. 1.2 are applicable, see the subsequent exercise. In the following, this fact will not be used.

Exercise 3.24 In the following, $(\mathbb{R}^4)^{\mathbb{R}^4}$ denotes the vector space of all maps from $\mathbb{R}^4$ to $\mathbb{R}^4$ and $\mathrm{A}(\mathbb{R}^4)$ the subspace of $(\mathbb{R}^4)^{\mathbb{R}^4}$ of affine transformations,

$$\mathrm{A}(\mathbb{R}^4) := \{m + M \cdot id_{\mathbb{R}^4} : (m, M) \in \mathbb{R}^4 \times \mathrm{M}(4, \mathbb{R})\} .$$

(a) Show that by

$$[R((a, \Lambda))f](x) := f(\Lambda^{-1} \cdot (x - a)) .$$

for every $f \in (\mathbb{R}^4)^{\mathbb{R}^4}$, $x \in \mathbb{R}^4$ and every $(a, \Lambda) \in \mathcal{P}$, there is defined a representation $R : \mathcal{P} \to L((\mathbb{R}^4)^{\mathbb{R}^4}, (\mathbb{R}^4)^{\mathbb{R}^4})$ of $\mathcal{P}$.

b) Show that $\mathrm{A}(\mathbb{R}^4)$ is an invariant subspace of R, such that through the restriction of $R((a, \Lambda))$ in domain and in range to $\mathrm{A}(\mathbb{R}^4)$, for every $(a, \Lambda) \in \mathcal{P}$, we arrive at a representation $R_0 : \mathcal{P} \to \mathrm{L}(\mathrm{A}(\mathbb{R}^4), \mathrm{A}(\mathbb{R}^4))$ of $\mathcal{P}$.

c) Show that by

$$[T((a, \Lambda))]((m, M)) := (m - M\Lambda^{-1}a, M\Lambda^{-1}) ,$$

for every $(m, M) \in \mathbb{R}^4 \times \mathrm{M}(4, \mathbb{R})$ and $(a, \Lambda) \in \mathcal{P}$, there is defined a faithful representation $T : \mathcal{P} \to L(\mathbb{R}^4 \times \mathrm{M}(4, \mathbb{R}), \mathbb{R}^4 \times \mathrm{M}(4, \mathbb{R}))$ of $\mathcal{P}$.

Associated with the decomposition of $\mathcal{L}$,

$$\mathcal{L} = \mathcal{L}_+^{\uparrow} \,\dot{\cup}\, \mathcal{L}_+^{\downarrow} \,\dot{\cup}\, \mathcal{L}_-^{\uparrow} \,\dot{\cup}\, \mathcal{L}_-^{\downarrow} ,$$

is the decomposition of $\mathcal{P}$ given by

$$\mathcal{P} = \mathcal{P}_+^{\uparrow} \,\dot{\cup}\, \mathcal{P}_+^{\downarrow} \,\dot{\cup}\, \mathcal{P}_-^{\uparrow} \,\dot{\cup}\, \mathcal{P}_-^{\downarrow} ,$$

where

$$\mathcal{P}_+^{\downarrow} := \mathbb{R}^4 \times \mathcal{L}_+^{\downarrow} , \ \mathcal{P}_-^{\uparrow} := \mathbb{R}^4 \times \mathcal{L}_+^{\uparrow} , \ \mathcal{P}_-^{\downarrow} := \mathbb{R}^4 \times \mathcal{L}_-^{\downarrow} .$$

Moreover, we note that

$$(0, \Lambda_0) \in \mathcal{P}_+^{\uparrow} , \ (0, \Lambda_P) \in \mathcal{P}_-^{\uparrow} , \ (0, \Lambda_T) \in \mathcal{P}_-^{\downarrow} , \ (0, \Lambda_{PT}) \in \mathcal{P}_+^{\downarrow} .$$

Exercise 3.25 Show that

For every $(a, \Lambda) \in \mathcal{P}_+^{\uparrow}$, there is a uniquely determined $(\bar{a}, \bar{\Lambda}) \in \mathcal{P}_+^{\uparrow}$, such that $(a, \Lambda) = (0, \Lambda_0) \cdot (\bar{a}, \bar{\Lambda})$;
For every $(a, \Lambda) \in \mathcal{P}_+^{\downarrow}$, there is a uniquely determined $(\bar{a}, \bar{\Lambda}) \in \mathcal{P}_+^{\uparrow}$, such that $(a, \Lambda) = (0, \Lambda_{PT}) \cdot (\bar{a}, \bar{\Lambda})$;
For every $(a, \Lambda) \in \mathcal{P}_-^{\uparrow}$, there is a uniquely determined $(\bar{a}, \bar{\Lambda}) \in \mathcal{P}_+^{\uparrow}$, such that $(a, \Lambda) = (0, \Lambda_P) \cdot (\bar{a}, \bar{\Lambda})$;
For every $(a, \Lambda) \in \mathcal{P}_-^{\downarrow}$, there is a uniquely determined $(\bar{a}, \bar{\Lambda}) \in \mathcal{P}_+^{\uparrow}$, such that $(a, \Lambda) = (0, \Lambda_T) \cdot (\bar{a}, \bar{\Lambda})$

As inverse images of the closed subset $\mathcal{P}_+^{\uparrow}$ of $\mathbb{R}^4 \times \mathrm{M}(4, \mathbb{R})$ under continuous maps and as images of the path connected set $\mathcal{P}_+^{\uparrow}$ under continuous maps, we arrive at the following.

Equipped with the induced topology, the subsets $\mathcal{P}_{+}^{\downarrow}$, $\mathcal{P}_{-}^{\uparrow}$ and $\mathcal{P}_{-}^{\downarrow}$ of $\mathbb{R}^4 \times \mathrm{M}(4, \mathbb{R})$ are unbounded, closed and path connected,

where the unboundedness of these subsets of $\mathbb{R}^4 \times \mathrm{M}(4, \mathbb{R})$ follows from the facts that for $\nu \in \mathbb{N}$, we have that

$$({}^t(\nu, 0, 0, 0), \Lambda_P) \in \mathcal{P}_{-}^{\uparrow}\ , \quad ({}^t(\nu, 0, 0, 0), \Lambda_T) \in \mathcal{P}_{-}^{\downarrow}\ , \quad ({}^t(\nu, 0, 0, 0), \Lambda_{PT}) \in \mathcal{P}_{+}^{\downarrow}\ .$$

3.4 An Extension to a Strongly Continuous Representation of the Restricted Poincaré Group

In the next step, we extend U_a to a representation $\hat{U}_a$ of the restricted Poincaré group. As a result generators, corresponding to one-parameter unitary groups that are associated with translations, appear as maximal multiplication operators. Since such generators are momentum operators, in this way, the representation space $L^2_{\mathbb{C}}(\mathbb{R}^3, \varphi_a)$ acquires the role of a momentum space.

Theorem 3.10 (An unitary representation of the restricted Poincaré group) *Let $a \geqslant 0$.*

(i) For every $(a, \Lambda) \in \mathcal{P}_{+}^{\uparrow}$, by

$$\begin{aligned}\hat{U}_a(a, \Lambda) f &:= \exp(i\, (a_0\, v_a^0 - \vec{a} \cdot \mathrm{id}_{\mathbb{R}^3}))\, U_a(\Lambda) f \\ &= \exp(i\, a \cdot p_a)\, \{f \circ [\, p_a^{-1} \circ (\Lambda^{-1} \cdot p_a)]\} \\ &= [\exp(i\ (\Lambda^{-1} \cdot a) \cdot \mathrm{id}_{H_a}) \cdot (f \circ p_a^{-1})] \circ (\Lambda^{-1} \cdot p_a)\ ,\end{aligned}$$

for every $f \in L^2_{\mathbb{C}}(\mathbb{R}^3, \varphi_a)$, where $\vec{a} := {}^t(a_1, a_2, a_3)$ and $\vec{a} \cdot \mathrm{id}_{\mathbb{R}^3} : \mathbb{R}^3 \to \mathbb{R}^3$ is defined by $(\vec{a} \cdot \mathrm{id}_{\mathbb{R}^3})(v) := \vec{a} \cdot v$, for every $v \in \mathbb{R}^3$, there is defined a unitary linear operator $\hat{U}_a(a, \Lambda)$ on $L^2_{\mathbb{C}}(\mathbb{R}^3, \varphi_a)$.

(ii) The map $\hat{U}_a : \mathcal{P}_{+}^{\uparrow} \to L(L^2_{\mathbb{C}}(\mathbb{R}^3, \varphi_a), L^2_{\mathbb{C}}(\mathbb{R}^3, \varphi_a))$, that associates with every $(a, \Lambda) \in \mathcal{P}_{+}^{\uparrow}$ the corresponding $\hat{U}_a(a, \Lambda)$, is an unitary representation of $\mathcal{P}_{+}^{\uparrow}$. In particular, $\hat{U}_a$ is strongly continuous i.e., if $(a_1, \Lambda_1), (a_2, \Lambda_2), \ldots$ is a sequence in $\mathcal{P}_{+}^{\uparrow}$ that is component-wise convergent to $(a, \Lambda) \in \mathcal{P}_{+}^{\uparrow}$, then

$$\lim_{\nu \to \infty} \hat{U}_a(a_\nu, \Lambda_\nu) f = \hat{U}_a(a, \Lambda) f\ ,$$

for every $f \in L^2_{\mathbb{C}}(\mathbb{R}^3, \varphi_a)$.

Proof In the following, for every complex-valued function h that is a.e. defined on $\mathbb{R}^3$ as well as φ_a-measurable, T_h denotes the maximal multiplication operator in $L^2_{\mathbb{C}}(\mathbb{R}^3, \varphi_a)$, defined by $T_h f := h \cdot f$, for every $f \in L^2_{\mathbb{C}}(\mathbb{R}^3, \varphi_a)$.
Part (i): For $(a, \Lambda) \in \mathcal{P}_+^{\uparrow}$, since $\exp(i\,(a_0\, v_a^0 - \vec{a} \cdot id_{\mathbb{R}^3}))$ is bounded and continuous and hence also φ_a-measurable, the corresponding maximal multiplication operator $T_{\exp(i\,(a_0\, v_a^0 - \vec{a}\cdot id_{\mathbb{R}^3}))}$ is continuous. Further, we have that

$$\left(T_{\exp(i\,(a_0\, v_a^0 - \vec{a}\cdot id_{\mathbb{R}^3}))}\right)^* = T_{\exp(-i\,(a_0\, v_a^0 - \vec{a}\cdot id_{\mathbb{R}^3}))} = \left(T_{\exp(i\,(a_0\, v_a^0 - \vec{a}\cdot id_{\mathbb{R}^3}))}\right)^{-1}$$

and hence that $T_{\exp(i\,(a_0\, v_a^0 - \vec{a}\cdot id_{\mathbb{R}^3}))}$ is unitary. Further,

$$\hat{U}_a(a, \Lambda) = T_{\exp(i\,(a_0\, v_a^0 - \vec{a}\cdot id_{\mathbb{R}^3}))} U_a(\Lambda)\ ,$$

as a composition of unitary linear operators, is an unitary linear operator on $L^2_{\mathbb{C}}(\mathbb{R}^3, \varphi_a)$.
Part (ii): In a first step, we note that for every $\Lambda \in \mathcal{L}_+^{\uparrow}$ and $v \in \mathbb{R}^3$, it follows that

$$\begin{aligned}
&(a^2 + |h_{a\Lambda}(v)|^2)^{1/2} = (a^2 + |p_a^{-1}(\Lambda^{-1}\cdot p_a(v))|^2)^{1/2} \\
&= \left\{a^2 + [(\Lambda^{-1}\cdot p_a(v))^0]^2 - [(\Lambda^{-1}\cdot p_a(v))^0]^2 + |p_a^{-1}(\Lambda^{-1}\cdot p_a(v))|^2\right\}^{1/2} \\
&= \left\{a^2 + [(\Lambda^{-1}\cdot p_a(v))^0]^2 - (\Lambda^{-1}\cdot p_a(v))\cdot(\Lambda^{-1}\cdot p_a(v))\right\}^{1/2} \\
&= \left\{a^2 + [(\Lambda^{-1}\cdot p_a(v))^0]^2 - p_a(v)\cdot p_a(v))\right\}^{1/2} \\
&= \left\{a^2 + [(\Lambda^{-1}\cdot p_a(v))^0]^2 - a^2\right\}^{1/2} = (\Lambda^{-1}\cdot p_a(v))^0
\end{aligned}$$

and hence that

$$(\Lambda^{-1}\cdot p_a)^0 = v_a^0 \circ h_{a\Lambda}\ .$$

Further, for $(a_1, \Lambda_1), (a_2, \Lambda_2) \in \mathcal{P}_+^{\uparrow}$ and $f \in L^2_{\mathbb{C}}(\mathbb{R}^3, \varphi_a)$, we have that

$$\begin{aligned}
&\hat{U}_a((a_1, \Lambda_1)\cdot(a_2, \Lambda_2))f = \hat{U}_a(a_1 + \Lambda_1 a_2, \Lambda_1\Lambda_2)f \\
&= T_{\exp(i\,((a_1+\Lambda_1 a_2)_0 v_a^0 - \overrightarrow{a_1+\Lambda_1 a_2}\cdot id_{\mathbb{R}^3}))} U_a(\Lambda_1\Lambda_2)f \\
&= T_{\exp(i\,(a_{10} v_a^0 - \vec{a_1}\cdot id_{\mathbb{R}^3}))} T_{\exp(i\,((\Lambda_1 a_2)_0 v_a^0 - \overrightarrow{\Lambda_1 a_2}\cdot id_{\mathbb{R}^3}))} U_a(\Lambda_1)U_a(\Lambda_2)f \\
&= T_{\exp(i\,(a_{10} v_a^0 - \vec{a_1}\cdot id_{\mathbb{R}^3}))} T_{\exp(i\,((\Lambda_1 a_2)_0 v_a^0 - \overrightarrow{\Lambda_1 a_2}\cdot id_{\mathbb{R}^3}))} [(U_a(\Lambda_2)f)\circ h_{a\Lambda_1}]
\end{aligned}$$

and that

$$\begin{aligned}
&\hat{U}_a(a_1, \Lambda_1)\hat{U}_a(a_2, \Lambda_2)f \\
&= T_{\exp(i\,(a_{10} v_a^0 - \vec{a_1}\cdot id_{\mathbb{R}^3}))} U_a(\Lambda_1) T_{\exp(i\,(a_{20} v_a^0 - \vec{a_2}\cdot id_{\mathbb{R}^3}))} U_a(\Lambda_2)f \\
&= T_{\exp(i\,(a_{10} v_a^0 - \vec{a_1}\cdot id_{\mathbb{R}^3}))} T_{\exp(i\,(a_{20} v_a^0 - \vec{a_2}\cdot id_{\mathbb{R}^3}))\circ h_{a\Lambda_1}} [(U_a(\Lambda_2)f)\circ h_{a\Lambda_1}]\ .
\end{aligned}$$

Since for every $v \in \mathbb{R}^3$,

$$
\begin{aligned}
&\exp(i\,((\Lambda_1 a_2)_0 v_a^0(v) - \overrightarrow{\Lambda_1 a_2}\cdot v)) = \exp(i\,(\Lambda_1 a_2)\cdot p_a(v)) \\
&= \exp(i\,(a_2\cdot\Lambda_1^{-1}p_a(v)) = \exp(i\,[\,a_{20}\,(\Lambda_1^{-1}p_a(v))_0 - \overrightarrow{a_2}\cdot p_a^{-1}(\Lambda_1^{-1}p_a(v))\,]) \\
&= \exp(i\,[\,a_{20}\,v_a^0(h_{a\Lambda_1}(v)) - \overrightarrow{a_2}\cdot h_{a\Lambda_1}(v)\,]) \\
&= \exp(i\,(\,a_{20}\,v_a^0 - \overrightarrow{a_2}\cdot id_{\mathbb{R}^3}\,))(h_{a\Lambda_1}(v))\ ,
\end{aligned}
$$

it follows that

$$
\exp(i\,((\Lambda_1 a_2)_0 v_a^0 - \overrightarrow{\Lambda_1 a_2}\cdot id_{\mathbb{R}^3})) = \exp(i\,(a_{20}v_a^0 - \overrightarrow{a_2}\cdot id_{\mathbb{R}^3}))\circ h_{a\Lambda_1}
$$

and hence that

$$
\hat{U}_a((a_1,\Lambda_1)\cdot(a_2,\Lambda_2))f = \hat{U}_a(a_1,\Lambda_1)\hat{U}_a(a_2,\Lambda_2)f\ .
$$

If $(a_1,\Lambda_1),(a_2,\Lambda_2),\dots$ is a sequence in $\mathscr{P}_+^{\uparrow}$ that is component-wise convergent to $(a,\Lambda)\in\mathscr{P}_+^{\uparrow}$, then it follows that

$$
\begin{aligned}
&\|\hat{U}_a(a_\nu,\Lambda_\nu)f - \hat{U}_a(a,\Lambda)f\|_2 \\
&= \|T_{\exp(i\,(a_{\nu 0}v_a^0 - \overrightarrow{a_\nu}\cdot id_{\mathbb{R}^3}))}U_a(\Lambda_\nu)f - T_{\exp(i\,(a_0 v_a^0 - \vec{a}\cdot id_{\mathbb{R}^3}))}U_a(\Lambda)f\|_2 \\
&= \|\exp(i\,(a_{\nu 0}v_a^0 - \overrightarrow{a_\nu}\cdot id_{\mathbb{R}^3}))U_a(\Lambda_\nu)f - \exp(i\,(a_0 v_a^0 - \vec{a}\cdot id_{\mathbb{R}^3}))U_a(\Lambda)f\|_2 \\
&\leqslant \|\exp(i\,(a_{\nu 0}v_a^0 - \overrightarrow{a_\nu}\cdot id_{\mathbb{R}^3}))[U_a(\Lambda_\nu)f - U_a(\Lambda)f]\|_2 \\
&\quad + \|[\exp(i\,(a_{\nu 0}v_a^0 - \overrightarrow{a_\nu}\cdot id_{\mathbb{R}^3})) - \exp(i\,(a_0 v_a^0 - \vec{a}\cdot id_{\mathbb{R}^3}))]\,U_a(\Lambda)f\|_2 \\
&= \|U_a(\Lambda_\nu)f - U_a(\Lambda)f\|_2 \\
&\quad + \|[\exp(i\,(a_{\nu 0}v_a^0 - \overrightarrow{a_\nu}\cdot id_{\mathbb{R}^3})) - \exp(i\,(a_0 v_a^0 - \vec{a}\cdot id_{\mathbb{R}^3}))]\,U_a(\Lambda)f\|_2\ .
\end{aligned}
$$

Hence, since U_a is strongly continuous and since, with the help of Lebesgue's dominated convergence theorem, it follows that

$$
\lim_{\nu\to\infty}\|[\exp(i\,(a_{\nu 0}v_a^0 - \overrightarrow{a_\nu}\cdot id_{\mathbb{R}^3})) - \exp(i\,(a_0 v_a^0 - \vec{a}\cdot id_{\mathbb{R}^3}))]\,U_a(\Lambda)f\|_2 = 0\ ,
$$

we infer that

$$
\lim_{\nu\to\infty}\hat{U}_a(a_\nu,\Lambda_\nu)f = \hat{U}_a(a,\Lambda)f\ ,
$$

for every $f\in L^2_{\mathbb{C}}(\mathbb{R}^3,\varphi_a)$. □

3.4.1 Generators Associated with Translations

In the following, we analyze the generators of $\hat{U}_a$ for one-parameter subgroups associated with translations.

For $a\in\mathbb{R}^4$, by $M:\mathbb{R}\to\mathscr{P}_+^{\uparrow}$, defined by $M(s) := (sa, E)$, for every $s\in\mathbb{R}$, since

$$M(s_1) \cdot M(s_2) = (s_1 a, E) \cdot (s_2 a, E) = (s_1 a + E s_2 a, E \cdot E)$$
$$= ((s_1 + s_2)\, a, E) = M(s_1 + s_2) \ ,$$

for all $s_1, s_2 \in \mathbb{R}$, there is defined a one-parameter group. In addition, M is component-wise continuous. Hence, $\hat{U}_a \circ M$ is a strongly continuous one-parameter unitary group. According to Stone's theorem, there is a unique densely-defined, linear and self-adjoint operator A_M in $X := L^2_{\mathbb{C}}(\mathbb{R}^3, \varphi_a)$ such that

$$\exp(isA_M) = (\hat{U}_a \circ M)(s) \ ,$$

for every $s \in \mathbb{R}$ and, in particular, that $A_M : D(A_M) \to X$ is given by

$$D(A_M) = \{f \in X : \lim_{s \to 0, s \neq 0} \frac{1}{s} \left[(\hat{U}_a \circ M)(s) - \mathrm{id}_X \right] f \text{ exists}\}$$

and for every $f \in D(A_M)$

$$A_M f = \frac{1}{i} \lim_{s \to 0, s \neq 0} \frac{1}{s} \left[(\hat{U}_a \circ M)(s) - \mathrm{id}_X \right] f \ .$$

In the following, we are going to show that

$$T_{a_0 v_a^0 - \vec{a} \cdot id_{\mathbb{R}^3}} \supset A_M \ , \tag{3.23}$$

where $T_{a_0 v_a^0 - \vec{a} \cdot id_{\mathbb{R}^3}}$ denotes the maximal multiplication operator in $L^2_{\mathbb{C}}(\mathbb{R}^3, \varphi_a)$ corresponding to the function $a_0 v_a^0 - \vec{a} \cdot id_{\mathbb{R}^3}$, $\vec{a} := {}^t(a_1, a_2, a_3)$ and $\vec{a} \cdot id_{\mathbb{R}^3} : \mathbb{R}^3 \to \mathbb{R}^3$ is defined by $(\vec{a} \cdot id_{\mathbb{R}^3})(v) := \vec{a} \cdot v$, for every $v \in \mathbb{R}^3$. For $f \in D(A_M)$, we have that

$$A_M f = \lim_{\nu \to \infty} \frac{\nu}{i} \left[\hat{U}_a(M(1/\nu)) - \mathrm{id}_X \right] f = \lim_{\nu \to \infty} \frac{\nu}{i} \left[\hat{U}_a\left(\frac{1}{\nu} a, E \right) - \mathrm{id}_X \right] f$$
$$= \lim_{\nu \to \infty} \frac{\nu}{i} \left[\exp\left(\frac{i}{\nu} (a_0 v_a^0 - \vec{a} \cdot id_{\mathbb{R}^3}) \right) - 1 \right] f \ .$$

According to the proof of the completeness of L^2-spaces, there is a subsequence of

$$\left(\frac{\nu}{i} \left[\exp\left(\frac{i}{\nu} (a_0 v_a^0 - \vec{a} \cdot id_{\mathbb{R}^3}) \right) - 1 \right] f \right)_{\nu \in \mathbb{N}^*}$$

that is almost everywhere on $\mathbb{R}^3$ pointwise convergent to $A_M f$. Therefore, $A_M f$ is almost everywhere on $\mathbb{R}^3$ equal to $(a_0 v_a^0 - \vec{a} \cdot id_{\mathbb{R}^3}) \cdot f$. As a consequence, f is part of the domain of $T_{a_0 v_a^0 - \vec{a} \cdot id_{\mathbb{R}^3}}$ and $A_M f = T_{a_0 v_a^0 - \vec{a} \cdot id_{\mathbb{R}^3}} f$. Since this is true for every $f \in D(A_M)$, the validity of (3.23) follows. Further, since both operators, A_M and $T_{a_0 v_a^0 - \vec{a} \cdot id_{\mathbb{R}^3}}$, are self-adjoint, it follows from the maximality property of densely-defined, linear and self-adjoint operators in Hilbert spaces that

$$A_M = T_{a_0 v_a^0 - \vec{a} \cdot id_{\mathbb{R}^3}} \ .$$

Hence, we arrive at the following result.

Generators Associated with Translations in $L^2_{\mathbb{C}}(\mathbb{R}^3, \varphi_a)$

For $a \in \mathbb{R}^4$, we have that

$$(\hat{U}_a \circ M)(s) = \exp(is\, T_{a_0 v_a^0 - \vec{a} \cdot id_{\mathbb{R}^3}}) \ ,$$

for every $s \in \mathbb{R}$, where the one-parameter subgroup of $\mathcal{P}_+^\uparrow$, $M : \mathbb{R} \to \mathcal{P}_+^\uparrow$ is defined by $M(s) := (sa, E)$, for every $s \in \mathbb{R}$ and $T_{a_0 v_a^0 - \vec{a} \cdot id_{\mathbb{R}^3}}$ denotes the maximal multiplication operator in $L^2_{\mathbb{C}}(\mathbb{R}^3, \varphi_a)$ corresponding to the function $a_0 v_a^0 - \vec{a} \cdot id_{\mathbb{R}^3}$, $\vec{a} := {}^t(a_1, a_2, a_3)$ and $\vec{a} \cdot id_{\mathbb{R}^3} : \mathbb{R}^3 \to \mathbb{R}^3$ is defined by $(\vec{a} \cdot id_{\mathbb{R}^3})(v) := \vec{a} \cdot v$, for every $v \in \mathbb{R}^3$.

As a side remark, using the Hilbert space isomorphism $\mathcal{V}_a : L^2_{\mathbb{C}}(\mathbb{R}^3, \varphi_a) \to L^2_{\mathbb{C}}(\mathbb{R}^3)$ from Exercise 3.17, defined by

$$\mathcal{V}_a f := (v_a^0)^{-1/2} f \ ,$$

for every $f \in L^2_{\mathbb{C}}(\mathbb{R}^3, \varphi_a)$, with inverse $\mathcal{V}_a^{-1} : L^2_{\mathbb{C}}(\mathbb{R}^3) \to L^2_{\mathbb{C}}(\mathbb{R}^3, \varphi_a)$, given by

$$\mathcal{V}_a^{-1} f := (v_a^0)^{1/2} f \ ,$$

for every $f \in L^2_{\mathbb{C}}(\mathbb{R}^3)$, we note that for every $s \in \mathbb{R}$, it follows from the previous that

$$\mathcal{V}_a \hat{U}_a(M(s)) \mathcal{V}_a^{-1} f = \exp(is\, T_{a_0 v_a^0 - \vec{a} \cdot id_{\mathbb{R}^3}}) f \ ,$$

for every $s \in \mathbb{R}$ and $f \in L^2_{\mathbb{C}}(\mathbb{R}^3)$, where $T_{a_0 v_a^0 - \vec{a} \cdot id_{\mathbb{R}^3}}$ denotes the maximal multiplication operator in $L^2_{\mathbb{C}}(\mathbb{R}^3)$ corresponding to the function $a_0 v_a^0 - \vec{a} \cdot id_{\mathbb{R}^3}$, $\vec{a} := {}^t(a_1, a_2, a_3)$. As a consequence, we also have the following.

Generators Associated with Translations in $L^2_{\mathbb{C}}(\mathbb{R}^3)$

For $a \in \mathbb{R}^4$, we have that

$$\mathcal{V}_a \hat{U}_a(M(s)) \mathcal{V}_a^{-1} = \exp(is\, T_{a_0 v_a^0 - \vec{a} \cdot id_{\mathbb{R}^3}}) \ ,$$

for every $s \in \mathbb{R}$, where the one-parameter subgroup of $\mathcal{P}_+^\uparrow$, $M : \mathbb{R} \to \mathcal{P}_+^\uparrow$ is defined by $M(s) := (sa, E)$, for every $s \in \mathbb{R}$ and $T_{a_0 v_a^0 - \vec{a} \cdot id_{\mathbb{R}^3}}$ denotes the maximal multiplication operator in $L^2_{\mathbb{C}}(\mathbb{R}^3)$ corresponding to the function $a_0 v_a^0 - \vec{a} \cdot id_{\mathbb{R}^3}$, $\vec{a} := {}^t(a_1, a_2, a_3)$ and $\vec{a} \cdot id_{\mathbb{R}^3} : \mathbb{R}^3 \to \mathbb{R}^3$ is defined by $(\vec{a} \cdot id_{\mathbb{R}^3})(v) := \vec{a} \cdot v$, for every $v \in \mathbb{R}^3$.

3.4.2 Observables of a Relativistic Quantum Mechanics of a Scalar Particle

In the following, we connect the generators of $\hat{U}_a$ to relativistic quantum mechanics. In the relativistic quantum mechanical description of the free motion of a scalar particle of mass $m \in [0, \infty)$ in Minkowski space, the dimensionless constant a $(\in [0, \infty))$ is given by

$$a = \frac{mc}{\hbar\kappa} = \frac{1}{\kappa \lambda\!\!\!^{-}_{C}} , \tag{3.24}$$

where $\kappa > 0$ is a scale factor[7] with dimension 1/length and $\lambda\!\!\!^{-}_{C}$ denotes the reduced Compton wavelength of the particle.

Observables of a Relativistic Quantum Mechanics of Scalar Particle I

The generators of the strongly continuous unitary representation $\hat{U}_{mc/(\hbar\kappa)}$ of $\mathcal{P}_+^\uparrow$ are the observables of a relativistic quantum mechanics of scalar particle with mass $m \geqslant 0$ in a momentum representation, where c denotes the speed of light and $\hbar$ is the reduced Planck's constant. In particular, the components of momentum $\hat{p}_k$, $k \in \{0, 1, 2, 3\}$, and Hamiltonian operator $\hat{H}$ are given by maximal multiplication operators $L^2_{\mathbb{C}}(\mathbb{R}^3, \varphi_a)$,

$$\hat{p}_0 = \hbar\kappa \, T_{v_a^0} \, , \quad \hat{p}_\alpha = \hbar\kappa \, T_{v_\alpha} \, , \quad \hat{H} = c\hat{p}_0 = \hbar\kappa c \, T_{v_a^0} \, ,$$

with associated spectra

$$\sigma(\hat{p}_0) = [\hbar\kappa a, \infty) \, , \; \sigma(\hat{p}_\alpha) = \mathbb{R} \, , \; \sigma(\hat{H}) = [\hbar\kappa c \, a, \infty) \, ,$$

where $v_\alpha : \mathbb{R}^3 \to \mathbb{R}$ is the coordinate projection of $\mathbb{R}^3$ onto the α-th component and $\alpha \in \{1, 2, 3\}$.

[7] The quantity κ determines the scale in a position representation, see, Sects. 1.1 and 1.3 in [7].

To this end, we note that

$$e^{-\frac{i}{\hbar}x_0\hat{p}_0} = e^{-\frac{i}{\hbar}x_0\,\hbar\kappa T_{v_a^0}} = e^{-i\,\kappa x_0\,T_{v_a^0}} = (\hat{U}_{mc/(\hbar\kappa)}\circ M_0)(-\kappa x_0)\ ,$$
$$e^{-\frac{i}{\hbar}x_\alpha\hat{p}_\alpha} = e^{-\frac{i}{\hbar}x_\alpha\hbar\kappa T_{v_\alpha}} = e^{-i\kappa x_\alpha T_{v_\alpha}} = (\hat{U}_{mc/(\hbar\kappa)}\circ M_\alpha)(\kappa x_\alpha)\ ,$$
$$e^{-i\frac{t}{\hbar}\hat{H}} = e^{-i\frac{t}{\hbar}\hbar\kappa c\,T_{v_a^0}} = e^{-i\,\kappa ct\,T_{v_a^0}} = (\hat{U}_{mc/(\hbar\kappa)}\circ M_0)(-\kappa ct)\ ,$$

where the parameters x_0, x_1, x_2 and x_3 have the dimension of a length, for $k \in \{0, 1, 2, 3\}$, $M_k : \mathbb{R} \to \mathscr{P}_+^\uparrow$ is defined by $M_k(s) := (se_k, E)$, for every $s \in \mathbb{R}$ and $e_0 := {}^t(1, 0, 0, 0)$, $e_1 := {}^t(0, 1, 0, 0)$, $e_2 := {}^t(0, 0, 1, 0)$ and $e_3 := {}^t(0, 0, 0, 1)$ Since, $M_k(0) = (0, E)$ and

$$\begin{aligned} M_k(s_1)\cdot M_k(s_2) &= (s_1e_k, E)\cdot(s_2e_k, E) = (s_1e_k + E\cdot s_2e_k, E\cdot E) \\ &= ((s_1+s_2)e_k, E) = M_k((s_1+s_2))\ , \end{aligned}$$

for all $s_1, s_2 \in \mathbb{R}$, M_k is a one-parameter subgroup of $\mathscr{P}_+^\uparrow$. Again using the Hilbert space isomorphism U from Exercise 3.17, we also have

Observables of a Relativistic Quantum Mechanics of Scalar Particle II

The generators of the representation $V_a\hat{U}_{mc/(\hbar\kappa)}V_a^{-1}$ of $\mathscr{P}_+^\uparrow$, defined by $(V_a\hat{U}_{mc/(\hbar\kappa)}V_a^{-1})(a, \Lambda) := V_a\hat{U}_{mc/(\hbar\kappa)}(a, \Lambda)V_a^{-1}$, for every $(a, \Lambda) \in \mathscr{P}_+^\uparrow$ are the observables of a relativistic quantum mechanics of scalar particle with mass $m \geqslant 0$ in a momentum representation, where c denotes the speed of light and $\hbar$ is the reduced Planck's constant. In particular, the components of momentum $\hat{p}_k$, $k \in \{0, 1, 2, 3\}$, and Hamiltonian operator $\hat{H}$ are given by maximal multiplication operators in $L^2_{\mathbb{C}}(\mathbb{R}^3)$,

$$\hat{p}_0 = \hbar\kappa\,T_{v_a^0}\ ,\ \hat{p}_\alpha = \hbar\kappa\,T_{v_\alpha}\ ,\ \hat{H} = c\hat{p}_0 = \hbar\kappa c\,T_{v_a^0}\ ,$$

with associated spectra

$$\sigma(\hat{p}_0) = [\hbar\kappa a, \infty)\ ,\ \sigma(\hat{p}_\alpha) = \mathbb{R}\ ,\ \sigma(\hat{H}) = [\hbar\kappa c\,a, \infty)\ ,$$

where $v_\alpha : \mathbb{R}^3 \to \mathbb{R}$ is the coordinate projection of $\mathbb{R}^3$ onto the α-th component and $\alpha \in \{1, 2, 3\}$.

In non-relativistic quantum mechanics, there is a connection between translations in momentum space and the components of the position, see the end of the Sect. 2.6.1 in [7]. Hence, we arrive at the candidate for the α-th component, $\alpha \in \{1, 2, 3\}$, $\hat{q}_\alpha$, of position as the closure of, the densely-defined, linear, symmetric and essentially self-adjoint operator in $L^2_{\mathbb{C}}(\mathbb{R}^3)$ given by

$$\left(\mathscr{S}_{\mathbb{C}}(\mathbb{R}^3) \to L^2_{\mathbb{C}}(\mathbb{R}^3),\ f \mapsto \frac{i}{\kappa}\frac{\partial f}{\partial v_\alpha}\right) ,$$

where we used also a result from Sect. 1.3 of [7]. In particular, we have that

$$\hat{p}_\alpha \hat{q}_\beta f = \hbar\kappa\, v_\alpha \frac{i}{\kappa}\frac{\partial f}{\partial v_\beta} = i\hbar v_\alpha \frac{\partial f}{\partial v_\beta}$$

$$\hat{q}_\beta \hat{p}_\alpha f = \frac{i}{\kappa}\frac{\partial \hbar\kappa\, v_\alpha f}{\partial v_\beta} = i\hbar \frac{\partial v_\alpha f}{\partial v_\beta} = i\hbar\left(v_\alpha \frac{\partial f}{\partial v_\beta} + \delta_{\alpha\beta} f\right) = \hat{p}_\alpha \hat{q}_\beta f + i\hbar\delta_{\alpha\beta} f$$

and hence that

$$[\hat{p}_\alpha, \hat{q}_\beta] f = \frac{\hbar}{i}\,\delta_{\alpha\beta} f ,$$

for $\alpha, \beta \in \{1, 2, 3\}$ and every $f \in \mathscr{S}_{\mathbb{C}}(\mathbb{R}^3)$, where $\delta_{\alpha\beta} := 0$ if $\alpha \neq \beta$ and $\delta_{\alpha\beta} := 1$ if $\alpha = \beta$.

Transfer to a Position Representation

The transfer to a representation that might be called a position representation proceeds with the help of the inverse Fourier transformation F_2^{-1}. In this way, we arrive at

$$F_2^{-1} \hat{p}_\alpha F_2 f = \frac{\hbar\kappa}{i}\frac{\partial f}{\partial u_\alpha} ,$$

for every $f \in \mathscr{S}_{\mathbb{C}}(\mathbb{R}^3)$, and

$$F_2^{-1} \hat{q}_\alpha F_2 = \frac{1}{\kappa}\, T_{u_\alpha} ,$$

with associated spectra

$$\sigma(\hat{q}_\alpha) = \mathbb{R} ,\ \sigma(\hat{p}_\alpha) = \mathbb{R} ,$$

where T_{u_α} is the maximal multiplication operator in $L^2_{\mathbb{C}}(\mathbb{R}^3)$, with the α-th coordinate projection $u_\alpha : \mathbb{R}^3 \to \mathbb{R}$ and $\alpha \in \{1, 2, 3\}$. Hence, the Hamilton operator in this position representation is given by

$$F_2^{-1} \hat{H} F_2 = \hbar\kappa c\, F_2^{-1} T_{v_a^0} F_2 = \hbar\kappa c\, F_2^{-1} T_{(|\ |^2 + a^2)^{1/2}} F_2 ,$$

with associated spectrum

$$\sigma(\hat{H}) = [\hbar\kappa c a, \infty) .$$

As a consequence, by the square $\hat{H}^2$ of $\hat{H}$, defined by

$$\hat{H}^2 f := \hat{H}\hat{H} f ,$$

for every $f \in D(\hat{H}^2) = \{f \in D(\hat{H}) : \hat{H} f \in D(\hat{H})\}$, where $D(\hat{H})$ denotes the domain of $\hat{H}$, there is defined a densely, linear and positive self-adjoint operator $\hat{H}^2$ in $(L^2_{\mathbb{C}}(\mathbb{R}^3, \varphi_a))^2$.

Exercise 3.26 Show that

$$\hat{H}^2 = (\hbar\kappa c)^2 \, T_{|\ |^2+a^2} \ ,$$

where $T_{|\ |^2+a^2}$ denotes the maximal multiplication operator in $L^2_{\mathbb{C}}(\mathbb{R}^3, \varphi_a)$ with $|\ |^2 + a^2$.

Since,

$$\hat{H}^2 = (\hbar\kappa c)^2 \, T_{|\ |^2+a^2} \ , \tag{3.25}$$

where $T_{|\ |^2+a^2}$ denotes the maximal multiplication operator in $L^2_{\mathbb{C}}(\mathbb{R}^3, \varphi_a)$ with $|\ |^2 + a^2$. We note that if $f, g \in D(\hat{H}^2)$, then the paths $u, v : \mathbb{R} \to (L^2_{\mathbb{C}}(\mathbb{R}^3, \varphi_a))^2$, for every $t \in \mathbb{R}$ defined by

$$u(t) := e^{i \frac{t}{\hbar} \hat{H}} f \ , \ v(t) := e^{-i \frac{t}{\hbar} \hat{H}} g \ ,$$

assume their values in $D(\hat{H}^2)$ and are twice differentiable, with derivatives

$$u''(t) = -\frac{1}{\hbar^2} \hat{H}^2 u(t) \ , \ v''(t) = -\frac{1}{\hbar^2} \hat{H}^2 v(t) \ , \tag{3.26}$$

for every $t \in \mathbb{R}$, i.e., essentially, u and v satisfy the Klein-Gordon equation, describing the propagation of a field of mass m in Minkowski space. In this connection, we remind that $L^2_{\mathbb{C}}(\mathbb{R}^3, \varphi_a)$ might be interpreted as momentum space. The observation (3.26) can be used to formulate a well-posed initial value problem for the wave equation

$$u''(t) = -Au(t) \ ,$$

$t \in \mathbb{R}$, for general densely-defined, linear and semi-bounded self-adjoint operators A in Hilbert spaces. For details, we refer to [4], Theorem 2.2.1 and Corollary 2.2.2.

3.5 An Extension to a Unitary/Anti-unitary Representation of the Poincaré Group

In the final step, we extend $\hat{U}_a$ to a representation $\hat{\hat{U}}_a$ of the whole Poincaré group, where the elements of the connected components $\mathcal{P}_+^{\uparrow}$ and $\mathcal{P}_-^{\uparrow}$ of $\mathcal{P}$ are represented by unitary linear operators, whereas the elements of the connected components $\mathcal{P}_+^{\downarrow}$ and $\mathcal{P}_-^{\downarrow}$ of $\mathcal{P}$ are represented by anti-unitary and anti-linear operators.

Theorem 3.11 (A representation of the Poincaré group) *Let $a \geqslant 0$.*

(i) By

$$\hat{\hat{U}}_a(0, \Lambda_P)(f) := f \circ \left(-\mathrm{id}_{\mathbb{R}^3}\right) \text{ and } \hat{\hat{U}}_a(0, \Lambda_T)(f) := f^* \circ \left(-\mathrm{id}_{\mathbb{R}^3}\right) , \tag{3.27}$$

for every $f \in L^2_{\mathbb{C}}(\mathbb{R}^3, \varphi_a)$, there is defined a unitary linear $\hat{\hat{U}}_a(0, \Lambda_P)$ and anti-unitary anti-linear operator $\hat{\hat{U}}_a(0, \Lambda_T)$ on $L^2_{\mathbb{C}}(\mathbb{R}^3, \varphi_a)$, respectively. Further, for $i, j \in \{0, P, T, PT\}$ and $(a, \Lambda) \in \mathscr{P}^{\uparrow}_{+}$, we have that

$$\hat{U}_a((0, \Lambda_i) \cdot (a, \Lambda) \cdot (0, \Lambda_i)^{-1}) = \hat{\hat{U}}_a(0, \Lambda_i)\hat{U}_a(a, \Lambda)[\hat{\hat{U}}_a(0, \Lambda_i)]^{-1} ,$$
$$\hat{\hat{U}}_a((0, \Lambda_i) \cdot (0, \Lambda_j)) = \hat{\hat{U}}_a(0, \Lambda_i)\hat{\hat{U}}_a(0, \Lambda_j) ,$$

where $\hat{\hat{U}}_a(0, \Lambda_0) := \hat{U}_a(0, \Lambda_0)$ and $\hat{\hat{U}}_a(0, \Lambda_{PT}) := \hat{\hat{U}}_a(0, \Lambda_P)\hat{\hat{U}}_a(0, \Lambda_T)$.

(ii) By

$$\hat{\hat{U}}_a((0, \Lambda_i) \cdot (a, \Lambda)) := \hat{\hat{U}}_a(0, \Lambda_i)\hat{U}_a(a, \Lambda) ,$$

for every $i \in \{0, P, T, PT\}$ and $(a, \Lambda) \in \mathscr{P}^{\uparrow}_{+}$, there is defined an extension of $\hat{U}_a$ to a representation of $\mathscr{P}$ by unitary linear or anti-unitary anti-linear operators on $L^2_{\mathbb{C}}(\mathbb{R}^3, \varphi_a)$. In particular, the elements of $\hat{\hat{U}}_a(\mathscr{P}^{\uparrow}_{+}) \cup \hat{\hat{U}}_a(\mathscr{P}^{\uparrow}_{-})$ are unitary linear and the elements of $\hat{\hat{U}}_a(\mathscr{P}^{\downarrow}_{+}) \cup \hat{\hat{U}}_a(\mathscr{P}^{\downarrow}_{-})$ are anti-unitary and anti-linear.

Proof Part (i): We note that $-id_{\mathbb{R}^3} : \mathbb{R}^3 \to \mathbb{R}^3$ is a C^1-diffeomeorphism such that $|\det\left((-id_{\mathbb{R}^3})'\right)| = 1$. Hence, according to the change of variable theorem of Lebesgue, we have that a real-valued function that is defined v^3-a.e. on $\mathbb{R}^3$ is v^3-integrable if and only if $f \circ (-id_{\mathbb{R}^3})$ is defined v^3-a.e. on $\mathbb{R}^3$ as well as v^3-integrable. In addition, in this case, we have that

$$\int_{\mathbb{R}^3} f \, dv^3 = \int_{\mathbb{R}^3} [f \circ (-\mathrm{id}_{\mathbb{R}^3})] \, dv^3 .$$

Also, as a corollary to this theorem, we have that a real-valued function that is defined v^3-a.e. on $\mathbb{R}^3$ is v^3-measurable if and only if $f \circ (-id_{\mathbb{R}^3})$ is defined v^3-a.e. on $\mathbb{R}^3$ as well as v^3-measurable. If $f \in L^2_{\mathbb{C}}(\mathbb{R}^3, \varphi_a)$, then f is φ_a-a.e. defined on $\mathbb{R}^3$ as well as φ_a-measurable and therefore also v^3-a.e. defined on $\mathbb{R}^3$ as well as v^3-measurable. Hence, using this corollary, $f \circ (-id_{\mathbb{R}^3})$ is v^3-a.e. defined on $\mathbb{R}^3$ as well as v^3-measurable and consequently also φ_a-a.e. defined on $\mathbb{R}^3$ as well as φ_a-measurable. Further, we have that

$$\int_{\mathbb{R}^3} |f|^2 \, d\varphi_a = \int_{\mathbb{R}^3} |f|^2 \cdot (v_a^0)^{-1} \, dv^3 = \int_{\mathbb{R}^3} \{[\,|f|^2 \cdot (v_a^0)^{-1}\,] \circ (-id_{\mathbb{R}^3})\} \, dv^3$$
$$= \int_{\mathbb{R}^3} |f \circ (-id_{\mathbb{R}^3})|^2 \, (v_a^0)^{-1} \, dv^3 .$$

Hence, $f \circ (-id_{\mathbb{R}^3}) \in L^2_{\mathbb{C}}(\mathbb{R}^3, \varphi_a)$ and

$$\int_{\mathbb{R}^3} |f|^2 \, d\varphi_a = \int_{\mathbb{R}^3} |f \circ (-id_{\mathbb{R}^3})|^2 \, d\varphi_a$$

as well as $f^* \circ (-id_{\mathbb{R}^3}) \in L^2_{\mathbb{C}}(\mathbb{R}^3, \varphi_a)$ and

$$\int_{\mathbb{R}^3} |f|^2 \, d\varphi_a = \int_{\mathbb{R}^3} |f^*|^2 \, d\varphi_a = \int_{\mathbb{R}^3} |f^* \circ (-id_{\mathbb{R}^3})|^2 \, d\varphi_a \, .$$

It follows that by (3.27) there are defined isometric maps $\hat{\hat{U}}_a(0, \Lambda_P) : L^2_{\mathbb{C}}(\mathbb{R}^3, \varphi_a) \to L^2_{\mathbb{C}}(\mathbb{R}^3, \varphi_a)$ and $\hat{\hat{U}}_a(0, \Lambda_T) : L^2_{\mathbb{C}}(\mathbb{R}^3, \varphi_a) \to L^2_{\mathbb{C}}(\mathbb{R}^3, \varphi_a)$. Further, obviously, $\hat{\hat{U}}_a(0, \Lambda_P)$ is linear and $\hat{\hat{U}}_a(0, \Lambda_T)$ is anti-linear such that $\hat{\hat{U}}_a(0, \Lambda_P) \circ \hat{\hat{U}}_a(0, \Lambda_P)$ and $\hat{\hat{U}}_a(0, \Lambda_T) \circ \hat{\hat{U}}_a(0, \Lambda_T)$ equal the identical map on $L^2_{\mathbb{C}}(\mathbb{R}^3, \varphi_a)$. Therefore, $\hat{\hat{U}}_a(0, \Lambda_P)$ is a bijective linear isometric map, whereas $\hat{\hat{U}}_a(0, \Lambda_T)$ is a bijective anti-linear isometric map. Since we have the polarization identity that

$$\langle f|g\rangle_2 = \frac{1}{4} \left(\|f+g\|_2^2 - \|f-g\|_2^2 - i\,\|f+ig\|_2^2 + i\,\|f-ig\|_2^2 \right) \, ,$$

for all $f, g \in L^2_{\mathbb{C}}(\mathbb{R}^3, \varphi_a)$, where $\langle\,|\,\rangle_2$ denotes the scalar product on $L^2_{\mathbb{C}}(\mathbb{R}^3, \varphi_a)$ and $\|\;\|_2$ the induced norm, it follows that

$$\begin{aligned}
&\left\langle \hat{\hat{U}}_a(0, \Lambda_P) f \middle| \hat{\hat{U}}_a(0, \Lambda_P) g \right\rangle_2 \\
&= \frac{1}{4} \Big(\|\hat{\hat{U}}_a(0, \Lambda_P) f + \hat{\hat{U}}_a(0, \Lambda_P) g\|_2^2 - \|\hat{\hat{U}}_a(0, \Lambda_P) f - \hat{\hat{U}}_a(0, \Lambda_P) g\|_2^2 \\
&\qquad - i\,\|\hat{\hat{U}}_a(0, \Lambda_P) f + i\hat{\hat{U}}_a(0, \Lambda_P) g\|_2^2 + i\,\|\hat{\hat{U}}_a(0, \Lambda_P) f - i\hat{\hat{U}}_a(0, \Lambda_P) g\|_2^2 \Big) \\
&= \frac{1}{4} \Big(\|\hat{\hat{U}}_a(0, \Lambda_P)(f+g)\|_2^2 - \|\hat{\hat{U}}_a(0, \Lambda_P)(f-g)\|_2^2 \\
&\qquad - i\,\|\hat{\hat{U}}_a(0, \Lambda_P)(f+ig)\|_2^2 + i\,\|\hat{\hat{U}}_a(0, \Lambda_P)(f-ig)\|_2^2 \Big) \\
&= \frac{1}{4} \left(\|f+g\|_2^2 - \|f-g\|_2^2 - i\,\|f+ig\|_2^2 + i\,\|f-ig\|_2^2 \right) = \langle f|g\rangle_2 \, ,
\end{aligned}$$

and that

$$\begin{aligned}
&\left\langle \hat{\hat{U}}_a(0,\Lambda_T) f | \hat{\hat{U}}_a(0,\Lambda_T) g \right\rangle_2 \\
&= \frac{1}{4}\Big(\|\hat{\hat{U}}_a(0,\Lambda_T) f + \hat{\hat{U}}_a(0,\Lambda_T) g\|_2^2 - \|\hat{\hat{U}}_a(0,\Lambda_T) f - \hat{\hat{U}}_a(0,\Lambda_T) g\|_2^2 \\
&\qquad - i\,\|\hat{\hat{U}}_a(0,\Lambda_T) f + i\hat{\hat{U}}_a(0,\Lambda_T) g\|_2^2 + i\,\|\hat{\hat{U}}_a(0,\Lambda_T) f - i\hat{\hat{U}}_a(0,\Lambda_T) g\|_2^2 \Big) \\
&= \frac{1}{4}\Big(\|\hat{\hat{U}}_a(0,\Lambda_T)(f+g)\|_2^2 - \|\hat{\hat{U}}_a(0,\Lambda_T)(f-g)\|_2^2 \\
&\qquad - i\,\|\hat{\hat{U}}_a(0,\Lambda_T)(f-ig)\|_2^2 + i\,\|\hat{\hat{U}}_a(0,\Lambda_T)(f+ig)\|_2^2 \Big) \\
&= \frac{1}{4}\left(\|f+g\|_2^2 - \|f-g\|_2^2 - i\,\|f-ig\|_2^2 + i\,\|f+ig\|_2^2 \right) \\
&= \frac{1}{4}\left(\|f+g\|_2^2 - \|f-g\|_2^2 - i\,\|f+ig\|_2^2 + i\,\|f-ig\|_2^2 \right)^* = \langle f|g\rangle_2^* \ ,
\end{aligned}$$

for all $f, g \in L^2_{\mathbb{C}}(\mathbb{R}^3, \varphi_a)$. Hence, $\hat{\hat{U}}_a(0, \Lambda_P)$ is an unitary linear operator on $L^2_{\mathbb{C}}(\mathbb{R}^3, \varphi_a)$, and $\hat{\hat{U}}_a(0, \Lambda_T)$ is an anti-unitary anti-linear operator on $L^2_{\mathbb{C}}(\mathbb{R}^3, \varphi_a)$. Further, for $(a, \Lambda) \in \mathscr{P}$, it follows that

$$\begin{aligned}
&(0, \Lambda_i) \cdot (a, \Lambda) \cdot (0, \Lambda_i)^{-1} = (\Lambda_i a, \Lambda_i \Lambda) \cdot (0, \Lambda_i^{-1}) \\
&= (\Lambda_i a, \Lambda_i \Lambda) \cdot (0, \Lambda_i) = (\Lambda_i a, \Lambda_i \Lambda \Lambda_i) \ ,
\end{aligned}$$

$$\begin{aligned}
\Lambda_P \Lambda &= \begin{pmatrix} 1 & 0 & 0 & 0 \\ 0 & -1 & 0 & 0 \\ 0 & 0 & -1 & 0 \\ 0 & 0 & 0 & -1 \end{pmatrix} \cdot \begin{pmatrix} \Lambda_{00} & \Lambda_{10} & \Lambda_{20} & \Lambda_{30} \\ \Lambda_{01} & \Lambda_{11} & \Lambda_{21} & \Lambda_{31} \\ \Lambda_{02} & \Lambda_{12} & \Lambda_{22} & \Lambda_{32} \\ \Lambda_{03} & \Lambda_{13} & \Lambda_{23} & \Lambda_{33} \end{pmatrix} \\
&= \begin{pmatrix} \Lambda_{00} & \Lambda_{10} & \Lambda_{20} & \Lambda_{30} \\ -\Lambda_{01} & -\Lambda_{11} & -\Lambda_{21} & -\Lambda_{31} \\ -\Lambda_{02} & -\Lambda_{12} & -\Lambda_{22} & -\Lambda_{32} \\ -\Lambda_{03} & -\Lambda_{13} & -\Lambda_{23} & -\Lambda_{33} \end{pmatrix} , \\
\Lambda_P \Lambda \Lambda_P &= \begin{pmatrix} \Lambda_{00} & \Lambda_{10} & \Lambda_{20} & \Lambda_{30} \\ -\Lambda_{01} & -\Lambda_{11} & -\Lambda_{21} & -\Lambda_{31} \\ -\Lambda_{02} & -\Lambda_{12} & -\Lambda_{22} & -\Lambda_{32} \\ -\Lambda_{03} & -\Lambda_{13} & -\Lambda_{23} & -\Lambda_{33} \end{pmatrix} \cdot \begin{pmatrix} 1 & 0 & 0 & 0 \\ 0 & -1 & 0 & 0 \\ 0 & 0 & -1 & 0 \\ 0 & 0 & 0 & -1 \end{pmatrix} \\
&= \begin{pmatrix} \Lambda_{00} & -\Lambda_{10} & -\Lambda_{20} & -\Lambda_{30} \\ -\Lambda_{01} & \Lambda_{11} & \Lambda_{21} & \Lambda_{31} \\ -\Lambda_{02} & \Lambda_{12} & \Lambda_{22} & \Lambda_{32} \\ -\Lambda_{03} & \Lambda_{13} & \Lambda_{23} & \Lambda_{33} \end{pmatrix} = \Lambda^{-1} ,
\end{aligned}$$

where we used (3.4), and

$$\Lambda_T \Lambda = -\Lambda_P \Lambda = \begin{pmatrix} -\Lambda_{00} & -\Lambda_{10} & -\Lambda_{20} & -\Lambda_{30} \\ \Lambda_{01} & \Lambda_{11} & \Lambda_{21} & \Lambda_{31} \\ \Lambda_{02} & \Lambda_{12} & \Lambda_{22} & \Lambda_{32} \\ \Lambda_{03} & \Lambda_{13} & \Lambda_{23} & \Lambda_{33} \end{pmatrix} ,$$

$$\Lambda_T \Lambda \Lambda_T = (-\Lambda_P)\Lambda(-\Lambda_P) = \Lambda_P \Lambda \Lambda_P = \Lambda^{-1} .$$

Hence, if $(a, \Lambda) \in \mathscr{P}_+^\uparrow$, then

$$(0, \Lambda_i) \cdot (a, \Lambda) \cdot (0, \Lambda_i)^{-1} = (\Lambda_i a, \Lambda^{-1}) \in \mathscr{P}_+^\uparrow ,$$

for every $i \in \{P, T\}$. In this case, it follows for $f \in L^2_{\mathbb{C}}(\mathbb{R}^3, \varphi_a)$ further that

$$\begin{aligned}
&\hat{U}_a((0, \Lambda_P) \cdot (a, \Lambda) \cdot (0, \Lambda_P)^{-1}) f = \hat{U}_a(\Lambda_P \cdot a, \Lambda^{-1}) f \\
&= \exp(i\,((\Lambda_P \cdot a)_0 v_a^0 - \overrightarrow{\Lambda_P \cdot a} \cdot \mathrm{id}_{\mathbb{R}^3}))\,(f \circ h_{a\Lambda^{-1}}) \\
&= \exp(i\,(a_0 v_a^0 + \vec{a} \cdot \mathrm{id}_{\mathbb{R}^3}))\,(f \circ h_{a\Lambda^{-1}}) , \\
&\hat{\hat{U}}_a(0, \Lambda_P)\hat{U}_a(a, \Lambda)[\hat{\hat{U}}_a(0, \Lambda_P)]^{-1} f = \hat{\hat{U}}_a(0, \Lambda_P)\hat{U}_a(a, \Lambda)\hat{\hat{U}}_a(0, \Lambda_P) f \\
&= \hat{\hat{U}}_a(0, \Lambda_P)\hat{U}_a(a, \Lambda)[f \circ (-\mathrm{id}_{\mathbb{R}^3})] \\
&= \hat{\hat{U}}_a(0, \Lambda_P)[\exp(i\,(a_0 v_a^0 - \vec{a} \cdot \mathrm{id}_{\mathbb{R}^3}))\,(f \circ (-\mathrm{id}_{\mathbb{R}^3}) \circ h_{a\Lambda})] \\
&= \exp(i\,(a_0 v_a^0 + \vec{a} \cdot \mathrm{id}_{\mathbb{R}^3}))\,(f \circ (-\mathrm{id}_{\mathbb{R}^3}) \circ h_{a\Lambda} \circ (-\mathrm{id}_{\mathbb{R}^3}))
\end{aligned}$$

and that

$$\begin{aligned}
&\hat{U}_a((0, \Lambda_T) \cdot (a, \Lambda) \cdot (0, \Lambda_T)^{-1}) f = \hat{U}_a(\Lambda_T \cdot a, \Lambda^{-1}) f \\
&= \exp(i\,((\Lambda_T \cdot a)_0 v_a^0 - \overrightarrow{\Lambda_T \cdot a} \cdot \mathrm{id}_{\mathbb{R}^3}))\,(f \circ h_{a\Lambda^{-1}}) \\
&= \exp(-i\,(a_0 v_a^0 + \vec{a} \cdot \mathrm{id}_{\mathbb{R}^3}))\,(f \circ h_{a\Lambda^{-1}}) , \\
&\hat{\hat{U}}_a(0, \Lambda_T)\hat{U}_a(a, \Lambda)[\hat{\hat{U}}_a(0, \Lambda_T)]^{-1} f = \hat{\hat{U}}_a(0, \Lambda_T)\hat{U}_a(a, \Lambda)\hat{\hat{U}}_a(0, \Lambda_T) f \\
&= \hat{\hat{U}}_a(0, \Lambda_T)\hat{U}_a(a, \Lambda)[f^* \circ (-\mathrm{id}_{\mathbb{R}^3})] \\
&= \hat{\hat{U}}_a(0, \Lambda_T)[\exp(i\,(a_0 v_a^0 - \vec{a} \cdot \mathrm{id}_{\mathbb{R}^3}))\,(f^* \circ (-\mathrm{id}_{\mathbb{R}^3}) \circ h_{a\Lambda})] \\
&= \exp(-i\,(a_0 v_a^0 + \vec{a} \cdot \mathrm{id}_{\mathbb{R}^3}))\,(f \circ (-\mathrm{id}_{\mathbb{R}^3}) \circ h_{a\Lambda} \circ (-\mathrm{id}_{\mathbb{R}^3})) .
\end{aligned}$$

Since we have for $v \in \mathbb{R}^3$ that

$$p_a(-v) = \begin{pmatrix} (a^2+|v|^2)^{1/2} \\ -v_1 \\ -v_2 \\ -v_3 \end{pmatrix} ,$$

$$\Lambda^{-1} p_a(-v)$$

$$= \begin{pmatrix} (\Lambda^{-1})_{00} & (\Lambda^{-1})_{10} & (\Lambda^{-1})_{20} & (\Lambda^{-1})_{30} \\ (\Lambda^{-1})_{01} & (\Lambda^{-1})_{11} & (\Lambda^{-1})_{21} & (\Lambda^{-1})_{31} \\ (\Lambda^{-1})_{02} & (\Lambda^{-1})_{12} & (\Lambda^{-1})_{22} & (\Lambda^{-1})_{32} \\ (\Lambda^{-1})_{03} & (\Lambda^{-1})_{13} & (\Lambda^{-1})_{23} & (\Lambda^{-1})_{33} \end{pmatrix} \cdot \begin{pmatrix} (a^2+|v|^2)^{1/2} \\ -v_1 \\ -v_2 \\ -v_3 \end{pmatrix}$$

$$= \begin{pmatrix} (\Lambda^{-1})_{00}(a^2+|v|^2)^{1/2} - (\Lambda^{-1})_{10}v_1 - (\Lambda^{-1})_{20}v_2 - (\Lambda^{-1})_{30}v_3 \\ (\Lambda^{-1})_{01}(a^2+|v|^2)^{1/2} - (\Lambda^{-1})_{11}v_1 - (\Lambda^{-1})_{21}v_2 - (\Lambda^{-1})_{31}v_3 \\ (\Lambda^{-1})_{02}(a^2+|v|^2)^{1/2} - (\Lambda^{-1})_{12}v_1 - (\Lambda^{-1})_{22}v_2 - (\Lambda^{-1})_{32}v_3 \\ (\Lambda^{-1})_{03}(a^2+|v|^2)^{1/2} - (\Lambda^{-1})_{13}v_1 - (\Lambda^{-1})_{23}v_2 - (\Lambda^{-1})_{33}v_3 \end{pmatrix} ,$$

$$h_{a\Lambda}(-v)$$

$$= \begin{pmatrix} (\Lambda^{-1})_{01}(a^2+|v|^2)^{1/2} - (\Lambda^{-1})_{11}v_1 - (\Lambda^{-1})_{21}v_2 - (\Lambda^{-1})_{31}v_3 \\ (\Lambda^{-1})_{02}(a^2+|v|^2)^{1/2} - (\Lambda^{-1})_{12}v_1 - (\Lambda^{-1})_{22}v_2 - (\Lambda^{-1})_{32}v_3 \\ (\Lambda^{-1})_{03}(a^2+|v|^2)^{1/2} - (\Lambda^{-1})_{13}v_1 - (\Lambda^{-1})_{23}v_2 - (\Lambda^{-1})_{33}v_3 \end{pmatrix}$$

$$= -h_{a\Lambda}(-v)$$

$$= \begin{pmatrix} -(\Lambda^{-1})_{01}(a^2+|v|^2)^{1/2} + (\Lambda^{-1})_{11}v_1 + (\Lambda^{-1})_{21}v_2 + (\Lambda^{-1})_{31}v_3 \\ -(\Lambda^{-1})_{02}(a^2+|v|^2)^{1/2} + (\Lambda^{-1})_{12}v_1 + (\Lambda^{-1})_{22}v_2 + (\Lambda^{-1})_{32}v_3 \\ -(\Lambda^{-1})_{03}(a^2+|v|^2)^{1/2} + (\Lambda^{-1})_{13}v_1 + (\Lambda^{-1})_{23}v_2 + (\Lambda^{-1})_{33}v_3 \end{pmatrix}$$

$$= \begin{pmatrix} \Lambda_{01}\,(a^2+|v|^2)^{1/2} + \Lambda_{11}v_1 + \Lambda_{21}v_2 + \Lambda_{31}v_3 \\ \Lambda_{02}\,(a^2+|v|^2)^{1/2} + \Lambda_{12}v_1 + \Lambda_{22}v_2 + \Lambda_{32}v_3 \\ \Lambda_{03}\,(a^2+|v|^2)^{1/2} + \Lambda_{13}v_1 + \Lambda_{23}v_2 + \Lambda_{33}v_3 \end{pmatrix} = h_{a\Lambda^{-1}}(v) ,$$

where we used (3.4), it follows that

$$\hat{U}_a((0,\Lambda_P)\cdot(a,\Lambda)\cdot(0,\Lambda_P)^{-1})f = \hat{\hat{U}}_a(0,\Lambda_P)\hat{U}_a(a,\Lambda)[\hat{\hat{U}}_a(0,\Lambda_P)]^{-1}f ,$$

$$\hat{U}_a((0,\Lambda_T)\cdot(a,\Lambda)\cdot(0,\Lambda_T)^{-1})f = \hat{\hat{U}}_a(0,\Lambda_T)\hat{U}_a(a,\Lambda)[\hat{\hat{U}}_a(0,\Lambda_T)]^{-1}f .$$

Further,

$$
\begin{aligned}
&\hat{U}_a((0,\Lambda_0)\cdot(a,\Lambda)\cdot(0,\Lambda_0)^{-1}) = \hat{U}_a(a,\Lambda)\\
&= \hat{U}_a(0,\Lambda_0)\hat{U}_a(a,\Lambda)\hat{U}_a((0,\Lambda_0)^{-1}) = \hat{U}_a(0,\Lambda_0)\hat{U}_a(a,\Lambda)[\hat{U}_a(0,\Lambda_0)]^{-1}\\
&= \hat{\hat{U}}_a(0,\Lambda_0)\hat{U}_a(a,\Lambda)[\hat{\hat{U}}_a(0,\Lambda_0)]^{-1}\ ,\\
&\hat{U}_a((0,\Lambda_{PT})\cdot(a,\Lambda)\cdot(0,\Lambda_{PT})^{-1})\\
&= \hat{U}_a((0,\Lambda_P)\cdot(0,\Lambda_T)\cdot(a,\Lambda)\cdot((0,\Lambda_P)\cdot(0,\Lambda_T))^{-1})\\
&= \hat{U}_a((0,\Lambda_P)\cdot(0,\Lambda_T)\cdot(a,\Lambda)\cdot(0,\Lambda_T)^{-1}\cdot(0,\Lambda_P)^{-1})\\
&= \hat{\hat{U}}_a(0,\Lambda_P)\hat{U}_a((0,\Lambda_T)\cdot(a,\Lambda)\cdot(0,\Lambda_T)^{-1})[\hat{\hat{U}}_a(0,\Lambda_P)]^{-1}\\
&= \hat{\hat{U}}_a(0,\Lambda_P)\hat{\hat{U}}_a(0,\Lambda_T)\hat{U}_a(a,\Lambda)[\hat{\hat{U}}_a(0,\Lambda_T)]^{-1}[\hat{\hat{U}}_a(0,\Lambda_P)]^{-1}\\
&= \hat{\hat{U}}_a(0,\Lambda_P)\hat{\hat{U}}_a(0,\Lambda_T)\hat{U}_a(a,\Lambda)[\hat{\hat{U}}_a(0,\Lambda_P)\hat{\hat{U}}_a(0,\Lambda_T)]^{-1}\\
&= \hat{\hat{U}}_a(0,\Lambda_{PT})\hat{U}_a(a,\Lambda)[\hat{\hat{U}}_a(0,\Lambda_{PT})]^{-1}\ .
\end{aligned}
$$

Hence, it follows that

$$
\hat{U}_a((0,\Lambda_i)\cdot(a,\Lambda)\cdot(0,\Lambda_i)^{-1}) = \hat{\hat{U}}_a(0,\Lambda_i)\hat{U}_a(a,\Lambda)[\hat{\hat{U}}_a(0,\Lambda_i)]^{-1}\ ,
$$

for every $i \in \{0, P, T, PT\}$. Moreover, for $f \in L^2_{\mathbb{C}}(\mathbb{R}^3, \varphi_a)$, we have that

$$
\begin{aligned}
\hat{\hat{U}}_a((0,\Lambda_0)\cdot(0,\Lambda_0))f &= \hat{\hat{U}}_a(0,\Lambda_0)f = \hat{U}_a(0,\Lambda_0)f = f\\
&= \hat{U}_a(0,\Lambda_0)\hat{U}_a(0,\Lambda_0)f = \hat{\hat{U}}_a(0,\Lambda_0)\hat{\hat{U}}_a(0,\Lambda_0)f\ ,\\
\hat{\hat{U}}_a((0,\Lambda_0)\cdot(0,\Lambda_P))f &= \hat{\hat{U}}_a(0,\Lambda_P)f = \hat{U}_a(0,\Lambda_0)\hat{\hat{U}}_a(0,\Lambda_P)f\\
&= \hat{\hat{U}}_a(0,\Lambda_0)\hat{\hat{U}}_a(0,\Lambda_P)f\ ,\\
\hat{\hat{U}}_a((0,\Lambda_0)\cdot(0,\Lambda_T))f &= \hat{\hat{U}}_a(0,\Lambda_T)f = \hat{U}_a(0,\Lambda_0)\hat{\hat{U}}_a(0,\Lambda_T)f\\
&= \hat{\hat{U}}_a(0,\Lambda_0)\hat{\hat{U}}_a(0,\Lambda_T)f\ ,\\
\hat{\hat{U}}_a((0,\Lambda_0)\cdot(0,\Lambda_{PT}))f &= \hat{\hat{U}}_a(0,\Lambda_{PT})f = \hat{U}_a(0,\Lambda_0)\hat{\hat{U}}_a(0,\Lambda_{PT})f\\
&= \hat{\hat{U}}_a(0,\Lambda_0)\hat{\hat{U}}_a(0,\Lambda_{PT})f\ ,\\
\hat{\hat{U}}_a((0,\Lambda_P)\cdot(0,\Lambda_0))f &= \hat{\hat{U}}_a(0,\Lambda_P)f = \hat{\hat{U}}_a(0,\Lambda_P)\hat{U}_a(0,\Lambda_0)f\\
&= \hat{\hat{U}}_a(0,\Lambda_P)\hat{\hat{U}}_a(0,\Lambda_0)f\ ,\\
\hat{\hat{U}}_a((0,\Lambda_P)\cdot(0,\Lambda_P))f &= \hat{\hat{U}}_a(0,\Lambda_0)f = [f\circ(-\mathrm{id}_{\mathbb{R}^3})]\circ(-\mathrm{id}_{\mathbb{R}^3})\\
&= \hat{\hat{U}}_a(0,\Lambda_P)\hat{\hat{U}}_a(0,\Lambda_P)f\ ,\\
\hat{\hat{U}}_a((0,\Lambda_P)\cdot(0,\Lambda_T))f &= \hat{\hat{U}}_a(0,\Lambda_{PT})f = \hat{\hat{U}}_a(0,\Lambda_P)\hat{\hat{U}}_a(0,\Lambda_T)f\ ,\\
\hat{\hat{U}}_a((0,\Lambda_P)\cdot(0,\Lambda_{PT}))f &= \hat{\hat{U}}_a(0,\Lambda_P\cdot\Lambda_{PT})f = \hat{\hat{U}}_a(0,\Lambda_T)f
\end{aligned}
$$

$$
\begin{aligned}
&= f^* \circ (-id_{\mathbb{R}^3}) = \hat{\hat{U}}_a(0, \Lambda_P)\hat{\hat{U}}_a(0, \Lambda_{PT})f \ , \\
\hat{\hat{U}}_a((0, \Lambda_T) \cdot (0, \Lambda_0))f &= \hat{\hat{U}}_a(0, \Lambda_T)f = \hat{\hat{U}}_a(0, \Lambda_T)\hat{\hat{U}}_a(0, \Lambda_0)f \ , \\
\hat{\hat{U}}_a((0, \Lambda_T) \cdot (0, \Lambda_P))f &= \hat{\hat{U}}_a(0, \Lambda_{PT})f = f^* = [f \circ (-id_{\mathbb{R}^3})]^* \circ (-id_{\mathbb{R}^3}) \\
&= \hat{\hat{U}}_a(0, \Lambda_T)\hat{\hat{U}}_a(0, \Lambda_P)f \ , \\
\hat{\hat{U}}_a((0, \Lambda_T) \cdot (0, \Lambda_T))f &= \hat{\hat{U}}_a(0, \Lambda_0)f = [f^* \circ (-id_{\mathbb{R}^3})]^* \circ (-id_{\mathbb{R}^3}) \\
&= \hat{\hat{U}}_a(0, \Lambda_T)\hat{\hat{U}}_a(0, \Lambda_T)f \ , \\
\hat{\hat{U}}_a((0, \Lambda_T) \cdot (0, \Lambda_{PT}))f &= \hat{\hat{U}}_a(0, \Lambda_P)f = f \circ (-id_{\mathbb{R}^3}) \\
&= (f^*)^* \circ (-id_{\mathbb{R}^3}) = \hat{\hat{U}}_a(0, \Lambda_T)\hat{\hat{U}}_a(0, \Lambda_{PT})f \ , \\
\hat{\hat{U}}_a((0, \Lambda_{PT}) \cdot (0, \Lambda_0))f &= \hat{\hat{U}}_a(0, \Lambda_{PT})f = \hat{\hat{U}}_a(0, \Lambda_{PT})\hat{\hat{U}}_a(0, \Lambda_0)f \ , \\
\hat{\hat{U}}_a((0, \Lambda_{PT}) \cdot (0, \Lambda_P))f &= \hat{\hat{U}}_a(0, \Lambda_T)f = f^* \circ (-id_{\mathbb{R}^3}) = [f \circ (-id_{\mathbb{R}^3})]^* \\
&= \hat{\hat{U}}_a(0, \Lambda_{PT})\hat{\hat{U}}_a(0, \Lambda_P)f \ , \\
\hat{\hat{U}}_a((0, \Lambda_{PT}) \cdot (0, \Lambda_T))f &= \hat{\hat{U}}_a(0, \Lambda_P)f = f \circ (-id_{\mathbb{R}^3}) = [f^* \circ (-id_{\mathbb{R}^3})]^* \\
&= \hat{\hat{U}}_a(0, \Lambda_{PT})\hat{\hat{U}}_a(0, \Lambda_T)f \ , \\
\hat{\hat{U}}_a((0, \Lambda_{PT}) \cdot (0, \Lambda_{PT}))f &= \hat{\hat{U}}_a(0, \Lambda_0)f = f = (f^*)^* \\
&= \hat{\hat{U}}_a(0, \Lambda_{PT})\hat{\hat{U}}_a(0, \Lambda_{PT})f \ .
\end{aligned}
$$

Part (ii): Since $\mathscr{P} = \mathscr{P}_+^\uparrow \cup \mathscr{P}_+^\downarrow \cup \mathscr{P}_-^\uparrow \cup \mathscr{P}_-^\downarrow$ and as a consequence of the uniqueness of the representations in Exercise 3.25, by

$$
\hat{\hat{U}}_a((0, \Lambda_i) \cdot (a, \Lambda)) := \hat{\hat{U}}_a(0, \Lambda_i)\hat{U}_a(a, \Lambda) \ ,
$$

for every $i \in \{0, P, T, PT\}$ and $(a, \Lambda) \in \mathscr{P}_+^\uparrow$, there is defined a map from $\mathscr{P}$ into the set of bijective continuous maps, from $L^2_{\mathbb{C}}(\mathbb{R}^3, \varphi_a)$ to $L^2_{\mathbb{C}}(\mathbb{R}^3, \varphi_a)$. Further, since the composition of unitary linear with anti-unitary anti-linear operators on a complex Hilbert space is anti-linear and anti-unitary, it follows that the elements of $U(\mathscr{P}_+^\uparrow) \cup U(\mathscr{P}_-^\uparrow)$ are unitary linear and that the elements of $U(\mathscr{P}_+^\downarrow) \cup U(\mathscr{P}_-^\downarrow)$ are anti-unitary and anti-linear. Moreover, since

$$
\begin{aligned}
\hat{\hat{U}}_a(a, \Lambda) = \hat{\hat{U}}_a((0, \Lambda_0) \cdot (a, \Lambda)) &:= \hat{\hat{U}}_a(0, \Lambda_0)\hat{U}_a(a, \Lambda) \\
&= \hat{U}_a(0, \Lambda_0)\hat{U}_a(a, \Lambda) = \hat{U}_a(a, \Lambda) \ ,
\end{aligned}
$$

for every $(a, \Lambda) \in \mathscr{P}_+^\uparrow$, it follows that $\hat{\hat{U}}_a$ is an extension of $\hat{U}_a$. Furthermore, if $i, j \in \{0, P, T, PT\}$ and $(a, \Lambda), (\bar{a}, \bar{\Lambda}) \in \mathscr{P}_+^\uparrow$, using Part (i) and the fact that $\hat{U}_a$ is a representation of $\mathscr{P}_+^\uparrow$, we infer as follows.

$$
\begin{aligned}
&\hat{\hat{U}}_a((0, \Lambda_i) \cdot (a, \Lambda) \cdot (0, \Lambda_j) \cdot (\bar{a}, \bar{\Lambda})) \\
&= \hat{\hat{U}}_a((0, \Lambda_i) \cdot (0, \Lambda_j) \cdot (0, \Lambda_j)^{-1} \cdot (a, \Lambda) \cdot (0, \Lambda_j) \cdot (\bar{a}, \bar{\Lambda})) \\
&= \hat{\hat{U}}_a((0, \Lambda_i) \cdot (0, \Lambda_j))\, \hat{U}_a((0, \Lambda_j)^{-1} \cdot (a, \Lambda) \cdot (0, \Lambda_j) \cdot (\bar{a}, \bar{\Lambda})) \\
&= \hat{\hat{U}}_a(0, \Lambda_i)\, \hat{\hat{U}}_a(0, \Lambda_j)\, \hat{U}_a((0, \Lambda_j) \cdot (a, \Lambda) \cdot (0, \Lambda_j)^{-1} \cdot (\bar{a}, \bar{\Lambda})) \\
&= \hat{\hat{U}}_a(0, \Lambda_i)\, \hat{\hat{U}}_a(0, \Lambda_j)\, \hat{U}_a((0, \Lambda_j) \cdot (a, \Lambda) \cdot (0, \Lambda_j)^{-1})\, \hat{U}_a(\bar{a}, \bar{\Lambda}) \\
&= \hat{\hat{U}}_a(0, \Lambda_i)\, \hat{\hat{U}}_a(0, \Lambda_j)\, \hat{\hat{U}}_a(0, \Lambda_j)\, \hat{U}_a(a, \Lambda)\, [\hat{\hat{U}}_a(0, \Lambda_j)]^{-1}\, \hat{U}_a(\bar{a}, \bar{\Lambda}) \\
&= \hat{\hat{U}}_a(0, \Lambda_i)\, \hat{U}_a(a, \Lambda)\, [\hat{\hat{U}}_a(0, \Lambda_j)]^{-1}\, \hat{U}_a(\bar{a}, \bar{\Lambda}) \\
&= \hat{\hat{U}}_a(0, \Lambda_i)\, \hat{U}_a(a, \Lambda)\, \hat{\hat{U}}_a(0, \Lambda_j)\, \hat{U}_a(\bar{a}, \bar{\Lambda}) \\
&= \hat{\hat{U}}_a((0, \Lambda_i) \cdot (a, \Lambda))\, \hat{\hat{U}}_a((0, \Lambda_j) \cdot (\bar{a}, \bar{\Lambda}))\ .
\end{aligned}
$$

Hence, $\hat{\hat{U}}_a$ is a representation of $\mathscr{P}$. □

Appendix

A.1 Polar Decomposition

Lemma A.1 *Let* $(X, \langle\,|\,\rangle)$ *be a non-trivial Hilbert space over* $\mathbb{K} \in \{\mathbb{R}, \mathbb{C}\}$ *and* A *a densely-defined, linear and closed operator in* X. *Then*

$$\overline{Ran(A^*)} = (kerA)^{\perp} \ .$$

Proof If $g \in \text{Ran}(A^*)$ and $f \in D(A^*)$ is such that $g = A^* f$, then

$$\langle g | f' \rangle = \langle A^* f | f' \rangle = \langle f | A f' \rangle = 0 \ ,$$

for every $f' \in \ker A$ and hence

$$g \in (\ker A)^{\perp} \ .$$

As a consequence,

$$\text{Ran}(A^*) \subset (\ker A)^{\perp} \ ,$$

implying that

$$\overline{\text{Ran}(A^*)} \subset (\ker A)^{\perp} \ .$$

On the other hand, using that

$$A^{**} = \bar{A} = A \ ,$$

it follows that if $f' \in [\text{Ran}(A^*)]^{\perp}$, then

$$\langle f' | A^* f \rangle = 0 \ ,$$

for every $f \in D(A^*)$ and hence that

$$f' \in D(A^{**}) = D(\bar{A}) = D(A)$$

H. R. Beyer, *Quantum Spin and Representations of the Poincaré Group, Part I*, Synthesis Lectures on Engineering, Science, and Technology,
https://doi.org/10.1007/978-3-031-84140-8

as well as that

$$Af' = 0 \ .$$

Hence

$$f' \in \ker A \ .$$

As a consequence,

$$[\mathrm{Ran}(A^*)]^{\perp} \subset \ker A \ ,$$

implying that

$$\overline{\mathrm{Ran}(A^*)} = [\mathrm{Ran}(A^*)]^{\perp\perp} \supset (\ker A)^{\perp} \ .$$

□

Definition A.2 (*Partial isometries*) Let $(X, \langle\,|\,\rangle)$ be a non-trivial Hilbert space over $\mathbb{K} \in \{\mathbb{R}, \mathbb{C}\}$ and $V \in L(X, X)$. We call V a partial isometry if

$$V\big|_{(\ker V)^{\perp}} : (\ker V)^{\perp} \to X$$

is isometric. Obviously, if V is a partial isometry, $\mathrm{Ran} V$ is a closed subspace of X and, as a consequence of the polarization identities for the scalar product, $\langle\,|\,\rangle$,

$$\begin{pmatrix} (\ker V)^{\perp} & \to & \mathrm{Ran} V \\ f & \mapsto & Vf \end{pmatrix}$$

is a Hilbert space isomorphism, and we call $(\ker V)^{\perp}$ the initial space of V, $\mathrm{Ran} V$ the final space of V, the orthogonal projection on $(\ker V)^{\perp}$ the initial projection of V and the orthogonal projection on $\mathrm{Ran} V$ the final projection of V.

Proposition A.3 (Elementary properties of partial isometries) *Let* $(X, \langle\,|\,\rangle)$ *be a non-trivial Hilbert space over* $\mathbb{K} \in \{\mathbb{R}, \mathbb{C}\}$ *and* $V \in L(X, X)$. *Then*

(i) *V is a partial isometry iff V^*V is an orthogonal projection.*
(ii) *If V is a partial isometry, then V^*V is the initial projection of V, VV^* is the final projection of V, and V^* is a partial isometry with initial projection VV^* and final projection V^*V.*

Proof First, we assume that V is a partial isometry. In this case,we show the auxiliar statement that V^* is a partial isometry with initial space $\mathrm{Ran} V$ and the final space $(\ker V)^{\perp}$. In addition,

$$V^*V\big|_{(\ker V)^{\perp}} = \mathrm{id}_X\big|_{(\ker V)^{\perp}} \ .$$

For the proof, we show that

$$(\ker V^*)^{\perp} = \mathrm{Ran} V \ , \quad V^*\mathrm{Ran} V \subset (\ker V)^{\perp} \ ,$$
$$(V^*V - \mathrm{id}_X)((\ker V)^{\perp}) \subset \ker V \ . \tag{A.1}$$

Indeed,

$$(\ker V^*)^{\perp} = \mathrm{Ran} V \ ,$$

since if $f \in \ker V^*$, then

$$\langle f|Vg\rangle = \langle V^* f|g\rangle = 0 \ ,$$

for every $g \in X$ and hence $f \in (\mathrm{Ran} V)^{\perp}$; for $f \in (\mathrm{Ran} V)^{\perp}$ it follows that

$$\langle V^* f|g\rangle = \langle f|Vg\rangle = 0 \ ,$$

for every $g \in X$ and hence $f \in \ker V^*$. As a consequence,

$$\ker V^* = (\mathrm{Ran} V)^{\perp} \ ,$$

implying that

$$(\ker V^*)^{\perp} = (\mathrm{Ran} V)^{\perp\perp} = \mathrm{Ran} V \ .$$

Further,

$$V^*\mathrm{Ran} V \subset (\ker V)^{\perp} \ ,$$

since for $f \in X$, $g \in \ker V$, it follows that

$$\langle g|V^*Vf\rangle = \langle Vg|Vf\rangle = 0$$

and hence that $V^*Vf \in (\ker V)^{\perp}$. Finally,

$$(V^*V - \mathrm{id}_X)((\ker V)^{\perp}) \subset \ker V \ ,$$

since for $f, g \in (\ker V)^{\perp}$, it follows that

$$\langle V^*Vf - f|g\rangle = \langle V^*Vf|g\rangle - \langle f|g\rangle = \langle Vf|Vg\rangle - \langle f|g\rangle = 0$$

and hence that $V^*Vf - f \in (\ker V)^{\perp\perp} = \ker V$. Hence, (A.1) has been proved. As a consequence of (A.1), it follows that

$$(V^*V - \mathrm{id}_X)((\ker V)^{\perp}) \subset \ker V \cap (\ker V)^{\perp} = \{0\}$$

and hence that

$$V^*V\big|_{(\ker V)^{\perp}} = \mathrm{id}_X\big|_{(\ker V)^{\perp}} \ .$$

The latter implies that

$$V^*\mathrm{Ran} V = (\ker V)^{\perp} \ .$$

Also, for $f \in (\ker V)^{\perp}$, it follows that

$$\|V^*Vf\| = \|f\| = \|Vf\|$$

and hence that

$$V^*\big|_{\text{Ran}\,V} : \text{Ran}\,V \to X$$

is isometric. As a consequence, the auxiliary statement follows. Further, let P_i be the initial projection of V and P_f be the final projection of V, i.e., P_i is the orthogonal projection on $(\ker V)^\perp$ and P_f is the orthogonal projection on $\text{Ran}\,V$. Then, it follows for $f \in X$ that $P_i f \in (\ker V)^\perp$, $f - P_i f \in (\ker V)^{\perp\perp} = \ker V$ and hence that

$$V^*Vf = V^*VP_i f + V^*V(f - P_i f) = P_i f$$

as well as for $f \in X$ that $P_f f \in \text{Ran}\,V$, $f - P_f f \in (\text{Ran}\,V)^\perp = (\ker V^*)^{\perp\perp} = \ker V^*$ and hence that

$$VV^*f = VV^*P_f f + VV^*(f - P_f f) = VV^*P_f f = VV^*Vg = Vg = P_f f\ ,$$

where $g \in (\ker V)^\perp$ is such that $Vg = P_f f$. As a consequence,

$$V^*V = P_i\ ,\ \ VV^* = P_f\ .$$

Second, it remains to be proved that if V^*V is an orthogonal projection that V is a partial isometry. If V^*V is an orthogonal projection, it follows that

$$\text{Ran}\,V^*V = \overline{\text{Ran}\,V^*V} = \overline{\text{Ran}(V^*V)^*} = (\ker V^*V)^\perp = (\ker V)^\perp\ .$$

Hence it follows for $f \in (\ker V)^\perp$ that

$$\|Vf\|^2 = \langle Vf|Vf\rangle = \big\langle f|V^*Vf\big\rangle = \langle f|f\rangle = \|f\|^2$$

and hence that

$$V\big|_{(\ker V)^\perp} : (\ker V)^\perp \to X$$

is an isometry as well as that V is a partial isometry. □

Theorem A.4 (Polar decomposition) *Let $(X, \langle\,|\,\rangle)$ be a non-trivial Hilbert space over $\mathbb{K} \in \{\mathbb{R}, \mathbb{C}\}$ and $A : D(A) \to X$ a densely-defined, linear and closed operator in X. Then there is a partial isometry $V \in L(X, X)$ with initial space $\overline{\text{Ran}\ A^*}$ and end space $\overline{\text{Ran}\ A}$ such that*

$$A = V(A^*A)^{1/2}\ .$$

If, in addition, A is injective, then V is an isometry,

$$V^*V = id_X$$

and

$$V^*A = V^*V(A^*A)^{1/2} = (A^*A)^{1/2} .$$

If A is injective and $Ran A$ is dense in X, then the final space of V, too, is given by the whole space, X, and hence V is orthogonal / unitary, implying that,

$$V^* = V^{-1}$$

as well as that

$$VV^* = id_X .$$

In addition, we have in this case that

$$V^*|_{RanA} = (A^*A)^{1/2}A^{-1}|_{RanA} .$$

Further, if B is a densely-defined, linear and positive self-adjoint operator in X, $W \in L(X, X)$ a partial isometry with initial space $\overline{Ran(B)}$ such that

$$WB = A ,$$

then $B = (A^*A)^{1/2}$ and $W = V$.

Proof We are going to use that for every densely-defined, linear and closed operator B in X, B^*B is a densely-defined, linear and positive self-adjoint operator in X, $D(B^*B)$ is a core for B, and

$$\ker B = \ker(B^*B) .$$

In particular, for the cases $B = (A^*A)^{1/2}$ and $B = A$, this implies that

$$\ker(A^*A)^{1/2} = \ker((A^*A)^{1/2*}(A^*A)^{1/2}) = \ker((A^*A)^{1/2}(A^*A)^{1/2})$$
$$= \ker(A^*A) = \ker A$$

and hence that

$$\ker(A^*A)^{1/2} = \ker A . \tag{A.2}$$

In the next step, we are going to show that

$$D((A^*A)^{1/2}) = D(A) . \tag{A.3}$$

First, we note that, since $D(A^*A)$ is a core for $(A^*A)^{1/2}$, for $f \in D((A^*A)^{1/2})$, there is a sequence $f_1, f_2, \dots$ in $D(A^*A)$ such that

$$\lim_{\nu\to\infty} f_\nu = f \wedge \lim_{\nu\to\infty} (A^*A)^{1/2} f_\nu = (A^*A)^{1/2} f .$$

Further,

$$\|Af_\mu - Af_\nu\|^2 = \|A(f_\mu - f_\nu)\|^2 = \langle A(f_\mu - f_\nu)|A(f_\mu - f_\nu)\rangle$$
$$= \langle f_\mu - f_\nu|A^*A(f_\mu - f_\nu)\rangle = \langle f_\mu - f_\nu|(A^*A)^{1/2}(A^*A)^{1/2}(f_\mu - f_\nu)\rangle$$
$$= \|(A^*A)^{1/2}(f_\mu - f_\nu)\|^2 = \|(A^*A)^{1/2}f_\mu - (A^*A)^{1/2}f_\nu\|^2 ,$$

for all $\mu, \nu \in \mathbb{N}^*$. Hence $Af_1, Af_2, \ldots$ is a Cauchy sequence in X and therefore also convergent. Since A is closed, this implies that $f \in D(A)$. Hence it follows that

$$D((A^*A)^{1/2}) \subset D(A) .$$

Second, since $D(A^*A)$ is a core for $D(A)$, for $f \in D(A)$, there is a sequence $f_1, f_2, \ldots$ in $D(A^*A)$ such that

$$\lim_{\nu\to\infty} f_\nu = f \wedge \lim_{\nu\to\infty} Af_\nu = Af .$$

Further,

$$\|Af_\mu - Af_\nu\|^2 = \|A(f_\mu - f_\nu)\|^2 = \langle A(f_\mu - f_\nu)|A(f_\mu - f_\nu)\rangle$$
$$= \langle f_\mu - f_\nu|A^*A(f_\mu - f_\nu)\rangle = \langle f_\mu - f_\nu|(A^*A)^{1/2}(A^*A)^{1/2}(f_\mu - f_\nu)\rangle$$
$$= \|(A^*A)^{1/2}(f_\mu - f_\nu)\|^2 = \|(A^*A)^{1/2}f_\mu - (A^*A)^{1/2}f_\nu\|^2 ,$$

for all $\mu, \nu \in \mathbb{N}^*$. Hence $(A^*A)^{1/2}f_1, (A^*A)^{1/2}f_2, \ldots$ is a Cauchy sequence in X and therefore also convergent. Since $(A^*A)^{1/2}$ is closed, this implies that $f \in D((A^*A)^{1/2})$. Hence it follows that

$$D(A) \subset D((A^*A)^{1/2})$$

and finally (A.3). In the next step, we define,

$$V_0 : \mathrm{Ran}(A^*A)^{1/2} \to X$$

by

$$V_0(A^*A)^{1/2}f := Af ,$$

for every $f \in D((A^*A)^{1/2})$. As a consequence of (A.2), it follows that V_0 is well-defined. Further, if $f, g \in D((A^*A)^{1/2})$, $\lambda \in \mathbb{K}$, then

$$V_0(A^*A)^{1/2}(f+g) = A(f+g) = Af + Ag$$
$$= V_0(A^*A)^{1/2}f + V_0(A^*A)^{1/2}g ,$$
$$V_0(A^*A)^{1/2}\lambda f = A\lambda f = \lambda Af = \lambda V_0(A^*A)^{1/2}f .$$

Hence V_0 is linear. In addition, V_0 is isometric, since for $f \in D((A^*A)^{1/2})$, it follows that

$$\|V_0(A^*A)^{1/2}f\|^2 = \|Af\|^2 = \langle Af|Af\rangle = \langle f|A^*Af\rangle$$
$$= \langle f|(A^*A)^{1/2}(A^*A)^{1/2}f\rangle = \|(A^*A)^{1/2}f\|^2 .$$

Hence $V_0 \in L(\operatorname{Ran}(A^*A)^{1/2}, X)$ and according to the bounded linear extension theorem, there is a unique extension of V_0 to an element

$$\hat{V}_0 \in L(\overline{\operatorname{Ran}(A^*A)^{1/2}}, X) .$$

In particular, this $\hat{V}_0$ is isometric, too. Further, we conclude that

$$\overline{\operatorname{Ran}(A^*A)^{1/2}} = [\ker(A^*A)^{1/2}]^{\perp} = (\ker A)^{\perp} = \overline{\operatorname{Ran} A^*} .$$

Hence,

$$\hat{V}_0 \in L(\overline{\operatorname{Ran} A^*}, X) .$$

We note that $\operatorname{Ran}\hat{V}_0$ is a closed subspace of X, since for $g \in \overline{\operatorname{Ran}\hat{V}_0}$, there is a sequence $f_1, f_2, \ldots$ in $\overline{\operatorname{Ran} A^*}$ such that

$$\lim_{\nu\to\infty} \hat{V}_0 f_\nu = g .$$

Since $\hat{V}_0$ is isometric, this implies that $f_1, f_2, \ldots$ is a Cauchy-sequence in $\overline{\operatorname{Ran} A^*}$ and hence convergent to some $f \in \overline{\operatorname{Ran} A^*}$. As a consequence,

$$g = \lim_{\nu\to\infty} \hat{V}_0 f_\nu = \hat{V}_0 \lim_{\nu\to\infty} f_\nu = \hat{V}_0 f ,$$

i.e., $g \in \operatorname{Ran}\hat{V}_0$. For the next step, let $P \in L(X, X)$ be the orthogonal projection onto $\overline{\operatorname{Ran} A^*}$ and

$$V := \hat{V}_0 \circ P \in L(X, X) .$$

Then,

$$\begin{aligned} &Vf = 0 \Leftrightarrow \hat{V}_0 Pf = 0 \Leftrightarrow Pf = 0 \Leftrightarrow \\ &f \in (\overline{\operatorname{Ran} A^*})^{\perp} = (\ker A)^{\perp\perp} = \ker A . \end{aligned}$$

Hence,

$$\ker V = \ker A$$

and

$$(\ker V)^{\perp} = (\ker A)^{\perp} = \overline{\operatorname{Ran} A^*} .$$

Hence, V is a partial isometry with initial space

$$(\ker V)^{\perp} = \overline{\operatorname{Ran} A^*} .$$

Further, $\overline{\operatorname{Ran} A}$ is the end space of V. This can be seen as follows. First, according to definition,

$$\operatorname{Ran} V_0 \subset \operatorname{Ran} A$$

and hence

$$\mathrm{Ran}\,\hat{V}_0 \subset \overline{\mathrm{Ran}\,A}\ .$$

On the other hand,

$$\mathrm{Ran}A \subset \mathrm{Ran}\,\hat{V}_0\ .$$

This can be seen as follows. Since $D(A^*A)$ is a core for A, for $f \in D(A)$, there is a sequence $f_1, f_2, \ldots$ in $D(A^*A)$ such that

$$\lim_{\nu\to\infty} f_\nu = f \wedge Af = \lim_{\nu\to\infty} Af_\nu = \lim_{\nu\to\infty} V_0(A^*A)^{1/2} f_\nu\ .$$

Since V_0 is isometric, this implies that

$$(A^*A)^{1/2} f_1, (A^*A)^{1/2} f_2, \ldots$$

is a Cauchy-sequence in $\overline{\mathrm{Ran}A^*}$ and hence convergent to some $g \in \overline{\mathrm{Ran}A^*}$,

$$\lim_{\nu\to\infty} (A^*A)^{1/2} f_\nu = g\ ,$$

implying that

$$\hat{V}_0 g = Af\ .$$

Further, since $\mathrm{Ran}\,\hat{V}_0$ is a closed subspace of X, we conclude

$$\overline{\mathrm{Ran}A} \subset \mathrm{Ran}\,\hat{V}_0$$

and, summarizing, that

$$\mathrm{Ran}\,\hat{V}_0 = \overline{\mathrm{Ran}A}\ .$$

The latter implies that

$$\mathrm{Ran}\,V = \overline{\mathrm{Ran}A}\ .$$

Further, for $f \in D((A^*A)^{1/2})$, it follows that

$$V(A^*A)^{1/2} f = \hat{V}_0 P(A^*A)^{1/2} f = \hat{V}_0(A^*A)^{1/2} f = V_0(A^*A)^{1/2} f = Af\ .$$

If, in addition, A is injective, then the initial space of V coincides with the whole space, X. Hence, V is an isometry. Further, with the help of the polarization identities for the scalar product, $\langle\,|\,\rangle$, it follows for $f, g \in X$ that

$$\big\langle f | (V^*V - \mathrm{id}_X) g \big\rangle = \langle Vf | Vg\rangle - \langle f|g\rangle = \langle f|g\rangle - \langle f|g\rangle = 0$$

and hence that

$$V^*V = \mathrm{id}_X\ .$$

In particular, in this case, we have

$$V^*A = V^*V(A^*A)^{1/2} = (A^*A)^{1/2} \ .$$

If A is injective and $\text{Ran}A$ is dense in X, then the final space of V, too, is given by the whole space, X, and hence V is orthogonal / unitary, implying that,

$$V^* = V^{-1}$$

as well as that

$$VV^* = \text{id}_X \ .$$

In addition, we have in this case

$$V^*Af = (A^*A)^{1/2}A^{-1}Af$$

and hence that

$$V^*|_{\text{Ran}A} = (A^*A)^{1/2}A^{-1}|_{\text{Ran}A} \ .$$

Further, if B is a densely-defined, linear and postive self-adjoint operator in X, $W \in L(X, X)$ a partial isometry with initial space $\overline{\text{Ran}(B)}$ such that

$$WB = A \ ,$$

then $D(B) = D(A)$ and, it follows for $f \in D(A^*)$, $g \in D(A)$ that

$$\left\langle A^*f|g\right\rangle = \langle f|Ag\rangle = \langle f|WBg\rangle = \left\langle W^*f|Bg\right\rangle$$

and hence, since B is self-adjoint, for every $f \in D(A^*)$ that $W^*f \in D(B)$ as well as that

$$A^*f = BW^*f \ .$$

In particular, this implies for $f \in D(A^*A)$ that

$$A^*Af = BW^*Af = BW^*WBf = B^2f \ ,$$

where we used that W^*W is the projection onto the initial space $\overline{\text{Ran}B}$ of W. As a consequence, B^2 is a symmetric extension of the densely-defined, linear and positive self-adjoint operator A^*A in X and hence

$$B^2 = A^*A \ .$$

Since B is a densely-defined, linear and positive self-adjoint operator in X, the latter implies that

$$B = (A^*A)^{1/2} \ .$$

Since,

$$A = V(A^*A)^{1/2} = W(A^*A)^{1/2} \ ,$$

we conclude further that

$$W\big|_{\mathrm{Ran}(A^*A)^{1/2}} = V\big|_{\mathrm{Ran}(A^*A)^{1/2}}$$

and hence from the bounded linear extension theorem that

$$W\big|_{\overline{\mathrm{Ran}(A^*A)^{1/2}}} = V\big|_{\overline{\mathrm{Ran}(A^*A)^{1/2}}} \ .$$

and hence also that $W = V$. □

A.2 The Minkowski Model of the 3-Dimensional Hyperbolic Space of Constant Negative Curvature

Equipped with the induced topology, $H_{a+} \subset \mathbb{R}^4$ is a Hausdorff topological space, with a countable basis. Further, we define

$$\Omega_a := \begin{cases} \mathbb{R}^3 & \text{if } a > 0 \\ \mathbb{R}^3 \backslash \{0\} & \text{if } a = 0 \end{cases}$$

and $u^{-1} : \Omega_a \to H_{a+}$ by

$$u^{-1}(v) := {}^t((a^2 + |v|^2)^{1/2}, v_1, v_2, v_3) \ ,$$

for every $v \in \Omega_a$. Then u^{-1} is a homeomorphism with inverse $u : H_{a+} \to \Omega_a$ given by

$$u(v) := {}^t(v_1, v_2, v_3) \ ,$$

for every $v \in H_{a+}$. Hence, u is a global 3-dimensional chart for H_{a+} and $\mathcal{A} := \{u\}$ is 3-dimensional atlas for H_{a+}. Therefore, H_{a+} is a 3-dimensional topological manifold. Further, $\mathcal{A}$ is also a 3-dimensional C^∞-atlas for H_{a+}. Hence, the C^∞-atlas $\mathcal{A}_{\max}$, consisting of all homeomorphisms $\bar{u}$ from an subsets of H_{a+} to open subsets of $\mathbb{R}^3$ such that $\bar{u} \circ u^{-1}$ is C^∞, provides a C^∞-structure for H_{a+}. Equipped with this structure, H_{a+} becomes a C^∞-manifold.

In the following way, the embedding of H_{a+} into Minkowki space leads to a Riemannian metric g for H_{a+}. If $v = {}^t((a^2 + |v|^2)^{1/2}, v_1, v_2, v_3) \in H_{a+}$, $w \in \mathbb{R}^3$ and

$$\begin{aligned} c_{v,w}(s) &:= u^{-1}(u(v) + sw) \\ &= {}^t((a^2 + |u(v) + sw|^2)^{1/2}, [u(v) + sw]_1, [u(v) + sw]_2, [u(v) + sw]_3) \ , \end{aligned}$$

for every $s \in \mathbb{R}$, then

$$(u_k \circ c_{v,w})(s) = v_k + s\, w_k \ , \quad (u_k \circ c_{v,w})'(s) = w_k \ ,$$

for every $k \in \{1, 2, 3\}$ and hence

$$\dot{c}_{v,w}(s) = \sum_{k=1}^{3} (\dot{c}_{v,w}(s) \cdot u_k) \left.\frac{\partial}{\partial u_k}\right|_{c_{v,w}(s)}$$
$$= \sum_{k=1}^{3} (u_k \circ c_{v,w})'(s) \left.\frac{\partial}{\partial u_k}\right|_{c_{v,w}(s)} = \sum_{k=1}^{3} w_k \left.\frac{\partial}{\partial u_k}\right|_{c_{v,w}(s)} ,$$

for every $s \in \mathbb{R}$. In particular, we have that

$$\dot{c}_{v,w}(0) = \sum_{k=1}^{3} w_k \left.\frac{\partial}{\partial u_k}\right|_{v} . \tag{A.4}$$

Hence, if $g_v : T_v H_{a+} \times T_v H_{a+} \to \mathbb{R}$ is a bilinear form, then

$$g_v(\dot{c}_{v,w_1}(0), \dot{c}_{v,w_2}(0)) = \sum_{k,l=1}^{3} g_v\left(\left.\frac{\partial}{\partial u_k}\right|_{v}, \left.\frac{\partial}{\partial u_l}\right|_{v} \right) (w_1)_k\,(w_2)_l ,$$

for $w_1, w_2 \in \mathbb{R}^3$. Further,

$$c'_{v,w}(s) = {}^t((a^2 + |u(v) + sw|^2)^{-1/2}\, w \cdot [u(v) + sw], w_1, w_2, w_3) ,$$

for every $s \in \mathbb{R}$, and in particular

$$c'_{v,w}(0) = {}^t((a^2 + |\vec{v}\,|^2)^{-1/2}\, w \cdot \vec{v}, w_1, w_2, w_3) ,$$

where $\vec{v} := {}^t(v_1, v_2, v_3)$. Hence, it follows for $w_1, w_2 \in \mathbb{R}^3$ that

$$c'_{v,w_1}(0) \cdot c'_{v,w_2}(0) = (a^2 + |\vec{v}\,|^2)^{-1}(\vec{v} \cdot w_1)(\vec{v} \cdot w_2) - w_1 \cdot w_2$$
$$= \sum_{k,l=1}^{3} \left(\frac{v_k v_l}{a^2 + |\vec{v}\,|^2} - E_{kl} \right) (w_1)_k\,(w_2)_l .$$

The requirement that

$$c'_{v,w_1}(0) \cdot c'_{v,w_2}(0) = -g_v(\dot{c}_{v,w_1}(0), \dot{c}_{v,w_2}(0)) , \tag{A.5}$$

for very $v \in H_{a+}$ and $w_1, w_2 \in \mathbb{R}^3$, defines in a unique way the C^∞-tensor field of type $(2, 0)$ on H_{a+} given by

$$g = \sum_{j,k=1}^{3} g_{jk}\, du_j \otimes du_k ,$$

where

$$(g_{jk})_{j,k\in\{1,2,3\}}$$
$$= \frac{1}{a^2+u_1^2+u_2^2+u_3^2}\begin{pmatrix} a^2+u_2^2+u_3^2 & -u_1u_2 & -u_1u_3 \\ -u_1u_2 & a^2+u_1^2+u_3^2 & -u_2u_3 \\ -u_1u_3 & -u_2u_3 & a^2+u_1^2+u_2^2 \end{pmatrix} .$$

Since for every $v \in H_a$, we have that

$$(g_{jk}(v))_{j,k\in\{1,2,3\}}$$
$$= \frac{1}{a^2+v_1^2+v_2^2+v_3^2}\begin{pmatrix} a^2+v_2^2+v_3^2 & -v_1v_2 & -v_1v_3 \\ -v_1v_2 & a^2+v_1^2+v_3^2 & -v_2v_3 \\ -v_1v_3 & -v_2v_3 & a^2+v_1^2+v_2^2 \end{pmatrix} ,$$

is symmetric with principal minors given by

$$\frac{a^2+v_2^2+v_3^2}{a^2+v_1^2+v_2^2+v_3^2} > 0 \ , \quad \frac{a^2+v_3^2}{a^2+v_1^2+v_2^2+v_3^2} > 0 \ , \quad \frac{a^2}{a^2+v_1^2+v_2^2+v_3^2} > 0 \ ,$$

it follows from Sylvester's criterion, e.g., see Lemma 3.40 of [8], that

$$(g_{jk}(v))_{j,k\in\{1,2,3\}}$$

is positive definite if and only if $a > 0$.

Hence, if $a > 0$, then

$$((H_{a+}, \mathcal{A}_{\max}), g)$$

is a Riemannian C^∞-manifold. From a standard, but lengthy, calculation, it follows that this Riemannian manifold has *constant Riemann (or sectional) curvature* $-1/a^2$.

For this calculation, the free mathematica application "Riemannian Geometry & Tensor Calculus" (RGTC) might be used [13]. The volume form that is associated to g is given by

$$\sqrt{|g|}\, du^1 \wedge du^2 \wedge du^3 = \frac{a}{\sqrt{a^2+u_1^2+u_2^2+u_3^2}}\, du^1 \wedge du^2 \wedge du^3 \ .$$

Further, for $\Lambda \in \mathcal{L}_+^\uparrow$ and $\mathfrak{v} \in \Omega_a$, we have that

$$\begin{aligned}\Lambda \cdot u^{-1}(\mathfrak{v}) &= \Lambda \cdot {}^t((a^2+|\mathfrak{v}|^2)^{1/2}, \mathfrak{v}_1, \mathfrak{v}_2, \mathfrak{v}_3) \\ &= {}^t((a^2+|\bar{\mathfrak{v}}|^2)^{1/2}, \bar{\mathfrak{v}}_1, \bar{\mathfrak{v}}_2, \bar{\mathfrak{v}}_3) = u^{-1}(\bar{\mathfrak{v}}) \ ,\end{aligned}$$

where

$$\bar{\mathfrak{v}} = \begin{pmatrix} \Lambda_{10}\,(a^2 + |\mathfrak{v}|^2)^{1/2} + \sum_{k=1}^3 \Lambda_{1k}\mathfrak{v}_k \\ \Lambda_{20}\,(a^2 + |\mathfrak{v}|^2)^{1/2} + \sum_{k=1}^3 \Lambda_{2k}\mathfrak{v}_k \\ \Lambda_{30}\,(a^2 + |\mathfrak{v}|^2)^{1/2} + \sum_{k=1}^3 \Lambda_{3k}\mathfrak{v}_k \end{pmatrix} ,$$

and that

$$(u^{-1})'(\mathfrak{v}) = \begin{pmatrix} \mathfrak{v}_1\,(a^2 + |\mathfrak{v}|^2)^{-1/2} & \mathfrak{v}_2\,(a^2 + |\mathfrak{v}|^2)^{-1/2} & \mathfrak{v}_3\,(a^2 + |\mathfrak{v}|^2)^{-1/2} \\ 1 & 0 & 0 \\ 0 & 1 & 0 \\ 0 & 0 & 1 \end{pmatrix} .$$

Hence, it follows for $v \in H_{a+}$, $\vec{v} := {}^t(v_1, v_2, v_3)$, $w \in \mathbb{R}^3$ and $s \in (-\delta, \delta)$, where $\delta > 0$ is sufficiently small, that

$$\begin{aligned} \Lambda \cdot c_{v,w}(s) &= \Lambda \cdot u^{-1}(u(v) + sw) = \Lambda \cdot u^{-1}(\vec{v} + sw) \\ &= u^{-1}\left(\begin{pmatrix} \Lambda_{10}\,(a^2 + |\vec{v} + sw|^2)^{1/2} + \sum_{k=1}^3 \Lambda_{1k}\,(\vec{v} + sw)_k \\ \Lambda_{20}\,(a^2 + |\vec{v} + sw|^2)^{1/2} + \sum_{k=1}^3 \Lambda_{2k}\,(\vec{v} + sw)_k \\ \Lambda_{30}\,(a^2 + |\vec{v} + sw|^2)^{1/2} + \sum_{k=1}^3 \Lambda_{3k}\,(\vec{v} + sw)_k \end{pmatrix}\right) . \end{aligned}$$

In particular,

$$\Lambda \cdot c_{v,w}(0) = u^{-1}\left(\begin{pmatrix} \Lambda_{10}\,(a^2 + |\vec{v}\,|^2)^{1/2} + \sum_{k=1}^3 \Lambda_{1k}\,\vec{v}_k \\ \Lambda_{20}\,(a^2 + |\vec{v}\,|^2)^{1/2} + \sum_{k=1}^3 \Lambda_{2k}\,\vec{v}_k \\ \Lambda_{30}\,(a^2 + |\vec{v}\,|^2)^{1/2} + \sum_{k=1}^3 \Lambda_{3k}\,\vec{v}_k \end{pmatrix}\right) = u^{-1}(\vec{\bar{v}}) ,$$

where

$$\begin{aligned} \bar{v} &:= (\Lambda \cdot c_{v,w})(0) = \Lambda \cdot v = \Lambda \cdot {}^t((a^2 + |\vec{v}\,|^2)^{1/2}, v_1, v_2, v_3) \\ &= \begin{pmatrix} \Lambda_{00}\,(a^2 + |\vec{v}\,|^2)^{1/2} + \sum_{k=1}^3 \Lambda_{0k}\vec{v}_k \\ \Lambda_{10}\,(a^2 + |\vec{v}\,|^2)^{1/2} + \sum_{k=1}^3 \Lambda_{1k}\vec{v}_k \\ \Lambda_{20}\,(a^2 + |\vec{v}\,|^2)^{1/2} + \sum_{k=1}^3 \Lambda_{2k}\vec{v}_k \\ \Lambda_{30}\,(a^2 + |\vec{v}\,|^2)^{1/2} + \sum_{k=1}^3 \Lambda_{3k}\vec{v}_k \end{pmatrix} , \end{aligned}$$

and $\vec{\bar{v}} := (\bar{v}_1, \bar{v}_2, \bar{v}_3)$. Further,

$$
\begin{aligned}
&(\Lambda \cdot c_{v,w})'(0) \\
&= \begin{pmatrix} \bar{v}_1\,(a^2+|\vec{\bar{v}}\,|^2)^{-1/2} & \bar{v}_2\,(a^2+|\vec{\bar{v}}\,|^2)^{-1/2} & \bar{v}_3\,(a^2+|\vec{\bar{v}}\,|^2)^{-1/2} \\ 1 & 0 & 0 \\ 0 & 1 & 0 \\ 0 & 0 & 1 \end{pmatrix} \\
&\quad \cdot \begin{pmatrix} \Lambda_{10}\, v_0^{-1}\, \vec{v}\cdot w + \sum_{k=1}^{3} \Lambda_{1k}\, w_k \\ \Lambda_{20}\, v_0^{-1}\, \vec{v}\cdot w + \sum_{k=1}^{3} \Lambda_{2k}\, w_k \\ \Lambda_{30}\, v_0^{-1}\, \vec{v}\cdot w + \sum_{k=1}^{3} \Lambda_{3k}\, w_k \end{pmatrix} \\
&= \begin{pmatrix} (a^2+|\vec{\bar{v}}\,|^2)^{-1/2}\, \vec{\bar{v}}\cdot \bar{w} \\ \bar{w}_1 \\ \bar{w}_2 \\ \bar{w}_3 \end{pmatrix} ,
\end{aligned}
$$

where

$$
\bar{w} := \begin{pmatrix} \sum_{k=1}^{3} (\Lambda_{1k} + \Lambda_{10}\, \frac{v_k}{v_0})\, w_k \\ \sum_{k=1}^{3} (\Lambda_{2k} + \Lambda_{20}\, \frac{v_k}{v_0})\, w_k \\ \sum_{k=1}^{3} (\Lambda_{3k} + \Lambda_{30}\, \frac{v_k}{v_0})\, w_k \end{pmatrix} = (f_\Lambda)'_v(w) \ , \tag{A.6}
$$

and we define $(f_\Lambda)'_v \in \mathrm{L}(\mathbb{R}^3, \mathbb{R}^3)$ by

$$
(f_\Lambda)'_v(w) := \begin{pmatrix} \Lambda_{11} + \Lambda_{10}\, \frac{v_1}{v_0} & \Lambda_{12} + \Lambda_{10}\, \frac{v_2}{v_0} & \Lambda_{13} + \Lambda_{10}\, \frac{v_3}{v_0} \\ \Lambda_{21} + \Lambda_{20}\, \frac{v_1}{v_0} & \Lambda_{22} + \Lambda_{20}\, \frac{v_2}{v_0} & \Lambda_{23} + \Lambda_{20}\, \frac{v_3}{v_0} \\ \Lambda_{31} + \Lambda_{30}\, \frac{v_1}{v_0} & \Lambda_{32} + \Lambda_{30}\, \frac{v_2}{v_0} & \Lambda_{33} + \Lambda_{30}\, \frac{v_3}{v_0} \end{pmatrix} \cdot w \ ,
$$

for every $w \in \mathbb{R}^3$. We note that

$$
\begin{aligned}
&c_{\bar{v},\bar{w}}(0) = \bar{v} = \Lambda \cdot c_{v,w}(0) \ , \\
&c'_{\bar{v},\bar{w}}(0) = {}^t((a^2+|\vec{\bar{v}}\,|^2)^{-1/2}\, \bar{w}\cdot\vec{\bar{v}}\,, \bar{w}_1, \bar{w}_2, \bar{w}_3) = \Lambda \cdot c'_{v,w}(0) \ .
\end{aligned}
$$

Hence, it follows from (A.5) that

$$
\begin{aligned}
&g_v(\dot{c}_{v,w_1}(0), \dot{c}_{v,w_2}(0)) = -c'_{v,w_1}(0)\cdot c'_{v,w_2}(0) \\
&= -(\Lambda\cdot c_{v,w_1})'(0)\cdot(\Lambda\cdot c_{v,w_2})'(0) = -c'_{\bar{v},\bar{w}_1}(0)\cdot c'_{\bar{v},\bar{w}_2}(0) \\
&= g_{f_\Lambda(v)}(\dot{c}_{f_\Lambda(v),(f_\Lambda)'_v(w_1)}(0), \dot{c}_{f_\Lambda(v),(f_\Lambda)'_v(w_2)}(0)) \ ,
\end{aligned} \tag{A.7}
$$

for very $v \in H_{a+}$ and $w_1, w_2 \in \mathbb{R}^3$, where we define $f_\Lambda : H_{a+} \to H_{a+}$ by

$$
f_\Lambda(v) := \Lambda \cdot v \ ,
$$

for every $v \in H_{a+}$. Since

$$(u \circ f_\Lambda \circ u^{-1})(\mathfrak{v}) = \begin{pmatrix} \Lambda_{10}\,(a^2 + |\mathfrak{v}|^2)^{1/2} + \sum_{k=1}^{3} \Lambda_{1k}\mathfrak{v}_k \\ \Lambda_{20}\,(a^2 + |\mathfrak{v}|^2)^{1/2} + \sum_{k=1}^{3} \Lambda_{2k}\mathfrak{v}_k \\ \Lambda_{30}\,(a^2 + |\mathfrak{v}|^2)^{1/2} + \sum_{k=1}^{3} \Lambda_{3k}\mathfrak{v}_k \end{pmatrix} ,$$

for every $\mathfrak{v} \in \Omega_a$, f_Λ is C^∞. In addition, we note that f_Λ is bijective. Further, for $v \in H_{a+}$, the representation matrix $M(f_{v*}) \in \mathrm{M}(3, \mathbb{R})$ of f_{v*} with respect to the bases

$$\left.\frac{\partial}{\partial u_1}\right|_v , \left.\frac{\partial}{\partial u_2}\right|_v , \left.\frac{\partial}{\partial u_3}\right|_v$$

of $T_v H_{a+}$ and

$$\left.\frac{\partial}{\partial u_1}\right|_{f(v)} , \left.\frac{\partial}{\partial u_2}\right|_{f(v)} , \left.\frac{\partial}{\partial u_3}\right|_{f(v)}$$

of $T_{f(v)} H_{a+}$ is given by

$$\begin{aligned} [M(f_{v*})]_{kl} &= \left(\frac{\partial (u_k \circ f_\Lambda \circ u^{-1})}{\partial \mathfrak{v}_l} (\vec{v}\,) \right)_{(k,l) \in \{1,2,3\} \times \{1,2,3\}} \\ &= \begin{pmatrix} \Lambda_{11} + \Lambda_{10}\,\frac{v_1}{v_0} & \Lambda_{12} + \Lambda_{10}\,\frac{v_2}{v_0} & \Lambda_{13} + \Lambda_{10}\,\frac{v_3}{v_0} \\ \Lambda_{21} + \Lambda_{20}\,\frac{v_1}{v_0} & \Lambda_{22} + \Lambda_{20}\,\frac{v_2}{v_0} & \Lambda_{23} + \Lambda_{20}\,\frac{v_3}{v_0} \\ \Lambda_{31} + \Lambda_{30}\,\frac{v_1}{v_0} & \Lambda_{32} + \Lambda_{30}\,\frac{v_2}{v_0} & \Lambda_{33} + \Lambda_{30}\,\frac{v_3}{v_0} \end{pmatrix} \end{aligned}$$

Hence, using (A.4), it follows for $w \in \mathbb{R}^3$ that

$$\begin{aligned} f_{*v}\,\dot{c}_{v,w}(0) &= \sum_{l=1}^{3} w_l\, f_{*v} \left.\frac{\partial}{\partial u_l}\right|_v = \sum_{l=1}^{3} w_l \sum_{k=1}^{3} [M(f_{v*})]_{kl} \left.\frac{\partial}{\partial u_k}\right|_{f_\Lambda(v)} \\ &= \sum_{k=1}^{3} \left(\sum_{l=1}^{3} [M(f_{v*})]_{kl} w_l \right) \left.\frac{\partial}{\partial u_k}\right|_{f_\Lambda(v)} = \sum_{k=1}^{3} \bar{w}_k \left.\frac{\partial}{\partial u_k}\right|_{f_\Lambda(v)} = \dot{c}_{f_\Lambda(v),(f_\Lambda)'_v(w)} , \end{aligned}$$

where $\bar{w}$ is defined by (A.6). Therefore, we infer from (A.7) that

$$g_v(\dot{c}_{v,w_1}(0), \dot{c}_{v,w_2}(0)) = g_{f_\Lambda(v)}(f_{*v}\,\dot{c}_{v,w_1}(0), f_{*v}\,\dot{c}_{v,w_2}(0)) ,$$

for all $w_1, w_2 \in \mathbb{R}^3$ and hence that f_Λ is an isometry.

> The map $f_\Lambda : H_{a+} \to H_{a+}$, defined by
>
> $$f_\Lambda(v) := \Lambda \cdot v ,$$
>
> for every $v \in H_{a+}$, is an isometry.

A.3 Solutions

Solution 1.1 Equipped with the composition of maps as an additional operation, L(X, X) is an associative Banach algebra with unit element. Further, if $A, B \in \mathrm{GL}(X)$, then $A \circ B$ is bijective, as a composition of bijective maps, and hence $A \circ B \in \mathrm{GL}(X)$. Also $\mathrm{id}_X \in \mathrm{GL}(X)$ and, if $A \in \mathrm{GL}(X)$, then A, A^{-1} are bijective and hence, since according to the bounded inverse theorem $A^{-1} \in \mathrm{L(X, X)}$, $A^{-1} \in \mathrm{GL}(X)$. Hence, $\mathrm{GL}(X)$ equipped with the operations of composition and inversion is a group. In addition, completely analogous to case of $\mathrm{GL}(n, \mathbb{K})$, it follows that operations of composition and inversion are continuous and hence that $\mathrm{GL}(X)$ equipped with the operations of composition and inversion is a topological group. Finally, $\mathrm{GL}(X) \subset \mathrm{L(X, X)}$ is open. For the proof, let $A \in \mathrm{GL}(X)$ and $B \in U_{\|A^{-1}\|_{\mathrm{op}}^{-1}}(0)$. Then

$$A - B = (\mathrm{id}_X - BA^{-1}) \circ A \; .$$

Since

$$\|BA^{-1}\|_{\mathrm{op}} \leqslant \|B\|_{\mathrm{op}} \|A^{-1}\|_{\mathrm{op}} < 1 \; ,$$

$\mathrm{id}_X - BA^{-1} \in \mathrm{L(X, X)}$ is bijective and therefore $\mathrm{id}_X - BA^{-1} \in \mathrm{GL}(X)$. As a consequence, as a composition of bijective maps, $A - B$ is bijective, and therefore $A - B \in \mathrm{GL}(X)$. Hence $U_{\|A^{-1}\|_{\mathrm{op}}^{-1}}(A) \subset \mathrm{GL}(X)$. We note that the fact, that $\mathrm{id}_X - BA^{-1} \in \mathrm{L(X, X)}$ is bijective, follows from the facts that $\sum_{k=0}^{\infty} (BA^{-1})^k \in \mathrm{L(X, X)}$ and

$$\left[\sum_{k=0}^{\infty} (BA^{-1})^k\right] (\mathrm{id}_X - BA^{-1}) = (\mathrm{id}_X - BA^{-1}) \left[\sum_{k=0}^{\infty} (BA^{-1})^k\right] = \mathrm{id}_X \; .$$

Solution 1.2 "Part (a)": According to the Cayley-Hamilton theorem, we have that

$$A^2 - \mathrm{Tr}(A)A + \det(A)\, E = 0 \; ,$$

for every $A \in \mathrm{M}(2, \mathbb{C})$. Hence, if $A \in \mathrm{M}(2, \mathbb{C})$ is trace-free, then

$$A^2 = -\det(A)\, E \; ,$$

and it follows by induction on $k \in \mathbb{N}$ that

$$A^{2k} = (-1)^k \, [\det(A)]^k \, E \; ,$$

for every $k \in \mathbb{N}$ and hence that

$$\begin{aligned}
\exp(A) &= \sum_{k=0}^{\infty} \frac{1}{k!} \,.\, A^k = \sum_{k=0}^{\infty} \frac{1}{(2k)!} \,.\, A^{2k} + \sum_{k=0}^{\infty} \frac{1}{(2k+1)!} \,.\, A^{2k+1} \\
&= \sum_{k=0}^{\infty} \frac{1}{(2k)!} \,.\, (-1)^k \, [\det(A)]^k \, E + \sum_{k=0}^{\infty} \frac{1}{(2k+1)!} \,.\, (-1)^k \, [\det(A)]^k \, A \\
&= \sum_{k=0}^{\infty} \frac{1}{(2k)!} \,.\, (-1)^k \, [\sqrt{\det(A)}\,]^{2k} \, E \\
&\quad + \frac{1}{\sqrt{\det(A)}} \sum_{k=0}^{\infty} \frac{1}{(2k+1)!} \,.\, (-1)^k \, [\sqrt{\det(A)}\,]^{2k+1} \, A \\
&= \cos(\sqrt{\det(A)}\,) E + \frac{\sin(\sqrt{\det(A)}\,)}{\sqrt{\det(A)}} \, A \ ,
\end{aligned}$$

where $\sqrt{\det(A)}$ denotes some complex number such that $\sqrt{\det(A)}^{\,2} = \sqrt{\det(A)}$. "Part (b)": If $A \in \mathrm{M}(2, \mathbb{C})$ is trace-free and such that

$$\exp(A) = \begin{pmatrix} -1 & 1 \\ 0 & -1 \end{pmatrix} ,$$

then

$$\mathrm{Tr}(\exp(A)) = 2\cos(\sqrt{\det(A)}\,) = -2$$

and hence

$$\sqrt{\det(A)} = (2n+1)\,\pi \ ,$$

for some $n \in \mathbb{N}$. As a consequence,

$$\begin{aligned}
\cos(\sqrt{\det(A)}\,) &= \cos((2n+1)\,\pi) = \cos(\pi) = -1 \ , \\
\sin(\sqrt{\det(A)}\,) &= \sin((2n+1)\,\pi) = \sin(\pi) = 0 \ ,
\end{aligned}$$

and

$$\exp(A) = \begin{pmatrix} -1 & 0 \\ 0 & -1 \end{pmatrix} . \lightning$$

"Part (c)": Since,

$$\begin{pmatrix} -1 & 0 \\ 0 & -1 \end{pmatrix} \in \mathrm{SL}(2, \mathbb{R}) \ ,$$

it follows that

$$\exp(T_E \mathrm{SL}(2, \mathbb{R})) \subsetneqq \mathrm{SL}(2, \mathbb{R}) \ , \ \exp(T_E \mathrm{SL}(2, \mathbb{C})) \subsetneqq \mathrm{SL}(2, \mathbb{C}) \ .$$

Solution 1.3 "Part (a)": Let

$$A := \begin{pmatrix} -1 & 1 \\ 0 & -1 \end{pmatrix} .$$

If $B \in \mathrm{M}(2, \mathbb{R})$ is such that $B^2 = A$, then it follows with the help of the Cayley-Hamilton theorem that

$$\begin{aligned} \mathrm{Tr}(B)B - \det(B)E &= A \ , \\ [\mathrm{Tr}(B)]^2 - 2\det(B) &= \mathrm{Tr}(A) = -2 \ , \\ [\det(B)]^2 &= \det(B^2) = \det(A) = 1 \ . \end{aligned}$$

As a consequence, we have that $\det(B) \in \{-1, 1\}$. If $\det(B) = 1$, it follows that $\mathrm{Tr}(B) = 0$ and hence that $A = -E$. ↯ If $\det(B) = -1$, it follows that $[\mathrm{Tr}(B)]^2 = -4$. ↯

"Part (b)": If $B \in \mathrm{M}(2, \mathbb{R})$ is such such that $\exp(B) = A$, then

$$A = \exp(B) = \exp((1/2)B)\exp((1/2)B) = [\exp((1/2)B)]^2 \ . ↯$$

"Part (c)": Since $A \in \mathrm{GL}_+(2, \mathbb{R})$ and $T_E\mathrm{GL}_+(2, \mathbb{R}) = \mathrm{M}(2, \mathbb{R})$, we have that $\exp(T_E \mathrm{GL}_+(2, \mathbb{R})) \subsetneqq \mathrm{GL}_+(2, \mathbb{R})$.

Solution 2.1 Since

$$|z_1 \cdot z_2| = |z_1| \cdot |z_2| = 1 \cdot 1 = 1 \ ,$$

for all $z_1, z_2 \in S^1$, S^1 is closed under complex multiplication. Further, since complex multiplication is associative, $1 \in S^1$ and for every $z \in S^1$ we have that

$$z^* \cdot z = z \cdot z^* = 1$$

as well as that $z^* \in S^1$, it follows that S^1 equipped matrix multiplication is a group. Further, the operations of complex multiplication and complex conjugation are continuous. Hence, $S^1 \subset \mathbb{C}$ equipped with the induced topology and the operations of complex multiplication and inversion is a topological group. Further, it follows from (2.2) that f is well-defined. Also, f is a group homomorphism, since

$$\begin{aligned} f(z_1 \cdot z_2) &= f((x_1 + iy_1) \cdot (x_2 + iy_2)) = f(x_1x_2 - y_1y_2 + i(x_1y_2 + y_1x_2)) \\ &= \begin{pmatrix} x_1x_2 - y_1y_2 & x_1y_2 + y_1x_2 \\ -(x_1y_2 + y_1x_2) & x_1x_2 - y_1y_2 \end{pmatrix} = \begin{pmatrix} x_1 & y_1 \\ -y_1 & x_1 \end{pmatrix} \cdot \begin{pmatrix} x_2 & y_2 \\ -y_2 & x_2 \end{pmatrix} \\ &= f(z_1) \cdot f(z_2) \ , \end{aligned}$$

for all $z_1 = (x_1, y_1), z_2 = (x_2, y_2) \in S^1$. If

$$f(z_1) = \begin{pmatrix} x_1 & y_1 \\ -y_1 & x_1 \end{pmatrix} = f(z_2) = \begin{pmatrix} x_2 & y_2 \\ -y_2 & x_2 \end{pmatrix} \ ,$$

for $z_1 = (x_1, y_1), z_2 = (x_2, y_2) \in S^1$, then $z_1 = z_2$ and hence f is injective. Also, as a consequence of (2.2), f is surjective. If $z = (x, y) \in \mathbb{C}$, then

$$\left\| \begin{pmatrix} x & y \\ -y & x \end{pmatrix} \right\|_{op} = \sup_{(a,b)\in S^1} |(xa+yb, -ya+xb)|$$
$$= \sup_{(a,b)\in S^1} |(x-iy)(a+ib)| = \sup_{(a,b)\in S^1} |x-iy| \cdot |a+ib| = |z| \, .$$

$$\|f(z_1) - f(z_2)\|_{op} = \left\| \begin{pmatrix} x_1 - x_2 & y_1 - y_2 \\ -(y_1 - y_2) & x_1 - x_2 \end{pmatrix} \right\|_{op} = |z_1 - z_2| \, ,$$

for $z_1 = (x_1, y_1), z_2 = (x_2, y_2) \in S^1$. As a consequence, f is continuous. The inverse f^{-1} : $\mathrm{SO}(2) \to S^1$ of f is given by

$$f^{-1}\left(\begin{pmatrix} x & y \\ -y & x \end{pmatrix}\right) = x + iy \, ,$$

for every

$$\begin{pmatrix} x & y \\ -y & x \end{pmatrix} \in \mathrm{SO}(2) \, .$$

In particular,

$$\left| f^{-1}\left(\begin{pmatrix} x_1 & y_1 \\ -y_1 & x_1 \end{pmatrix}\right) - f^{-1}\left(\begin{pmatrix} x_2 & y_2 \\ -y_2 & x_2 \end{pmatrix}\right) \right| = |x_1 + iy_1 - (x_2 + iy_2)|$$
$$= \left\| \begin{pmatrix} x_1 - x_2 & y_1 - y_2 \\ -(y_1 - y_2) & x_1 - x_2 \end{pmatrix} \right\|_{op} = \left\| \begin{pmatrix} x_1 & y_1 \\ -y_1 & x_1 \end{pmatrix} - \begin{pmatrix} x_2 & y_2 \\ -y_2 & x_2 \end{pmatrix} \right\|_{op} ,$$

for all

$$\begin{pmatrix} x_1 & y_1 \\ -y_1 & x_1 \end{pmatrix} , \begin{pmatrix} x_2 & y_2 \\ -y_2 & x_2 \end{pmatrix} \in \mathrm{SO}(2) \, .$$

As a consequence, f^{-1} is continuous.

Solution 2.2 From (2.8) and (2.7), it follows that

$$\begin{aligned}
& R^* \cdot R \\
& = (E - \sin(\varphi)\, A + [1 - \cos(\varphi)]\, A^2) \cdot (E + \sin(\varphi)\, A + [1 - \cos(\varphi)]\, A^2) \\
& = E + \sin(\varphi)\, A + [1 - \cos(\varphi)]\, A^2 \\
& \quad - \sin(\varphi)\, A \cdot (E + \sin(\varphi)\, A + [1 - \cos(\varphi)]\, A^2) \\
& \quad + [1 - \cos(\varphi)]\, A^2 \cdot (E + \sin(\varphi)\, A + [1 - \cos(\varphi)]\, A^2) \\
& = E + \sin(\varphi)\, A + [1 - \cos(\varphi)]\, A^2 \\
& \quad - \sin(\varphi)\, (A + \sin(\varphi)\, A^2 + [1 - \cos(\varphi)]\, A^3) \\
& \quad + [1 - \cos(\varphi)]\, (A^2 + \sin(\varphi)\, A^3 + [1 - \cos(\varphi)]\, A^4) \\
& = E + \sin(\varphi)\, A + [1 - \cos(\varphi)]\, A^2
\end{aligned}$$

$$
\begin{aligned}
&- \sin(\varphi)\,(A + \sin(\varphi)\,A^2 - [1 - \cos(\varphi)]\,A) \\
&+ [1 - \cos(\varphi)]\,(A^2 - \sin(\varphi)\,A - [1 - \cos(\varphi)]\,A^2) \\
= {}& E + \sin(\varphi)\,A + [1 - \cos(\varphi)]\,A^2 \\
&- \sin(\varphi)\,(\sin(\varphi)\,A^2 + \cos(\varphi)\,A) \\
&+ [1 - \cos(\varphi)]\,(-\sin(\varphi)\,A + \cos(\varphi)\,A^2) \\
= {}& E + \sin(\varphi)\,A + [1 - \cos(\varphi)]\,A^2 \\
&- \sin^2(\varphi)\,A^2 - \sin(\varphi)\cos(\varphi)\,A \\
&- \sin(\varphi)[1 - \cos(\varphi)]\,A + \cos(\varphi)[1 - \cos(\varphi)]\,A^2 = E\ .
\end{aligned}
$$

Solution 2.3

$$
\sigma_1 := \begin{pmatrix} 0 & 1 \\ 1 & 0 \end{pmatrix}\ , \quad \sigma_2 := \begin{pmatrix} 0 & -i \\ i & 0 \end{pmatrix}\ , \quad \sigma_3 := \begin{pmatrix} 1 & 0 \\ 0 & -1 \end{pmatrix}
$$

$$
\begin{aligned}
\sigma_1 \cdot \sigma_1 &= \begin{pmatrix} 0 & 1 \\ 1 & 0 \end{pmatrix} \cdot \begin{pmatrix} 0 & 1 \\ 1 & 0 \end{pmatrix} = E\ , \quad \sigma_2 \cdot \sigma_2 = \begin{pmatrix} 0 & -i \\ i & 0 \end{pmatrix} \cdot \begin{pmatrix} 0 & -i \\ i & 0 \end{pmatrix} = E\ , \\
\sigma_3 \cdot \sigma_3 &= \begin{pmatrix} 1 & 0 \\ 0 & -1 \end{pmatrix} \cdot \begin{pmatrix} 1 & 0 \\ 0 & -1 \end{pmatrix} = E\ , \\
\sigma_1 \cdot \sigma_2 &= \begin{pmatrix} 0 & 1 \\ 1 & 0 \end{pmatrix} \cdot \begin{pmatrix} 0 & -i \\ i & 0 \end{pmatrix} = \begin{pmatrix} i & 0 \\ 0 & -i \end{pmatrix} = i\sigma_3\ , \\
\sigma_2 \cdot \sigma_1 &= \begin{pmatrix} 0 & -i \\ i & 0 \end{pmatrix} \cdot \begin{pmatrix} 0 & 1 \\ 1 & 0 \end{pmatrix} = \begin{pmatrix} -i & 0 \\ 0 & i \end{pmatrix} = -i\sigma_3\ , \\
\sigma_1 \cdot \sigma_3 &= \begin{pmatrix} 0 & 1 \\ 1 & 0 \end{pmatrix} \cdot \begin{pmatrix} 1 & 0 \\ 0 & -1 \end{pmatrix} = \begin{pmatrix} 0 & -1 \\ 1 & 0 \end{pmatrix} = -i\sigma_2\ , \\
\sigma_3 \cdot \sigma_1 &= \begin{pmatrix} 1 & 0 \\ 0 & -1 \end{pmatrix} \cdot \begin{pmatrix} 0 & 1 \\ 1 & 0 \end{pmatrix} = \begin{pmatrix} 0 & 1 \\ -1 & 0 \end{pmatrix} = i\sigma_2\ , \\
\sigma_2 \cdot \sigma_3 &= \begin{pmatrix} 0 & -i \\ i & 0 \end{pmatrix} \cdot \begin{pmatrix} 1 & 0 \\ 0 & -1 \end{pmatrix} = \begin{pmatrix} 0 & i \\ i & 0 \end{pmatrix} = i\sigma_1\ , \\
\sigma_3 \cdot \sigma_2 &= \begin{pmatrix} 1 & 0 \\ 0 & -1 \end{pmatrix} \cdot \begin{pmatrix} 0 & -i \\ i & 0 \end{pmatrix} = \begin{pmatrix} 0 & -i \\ -i & 0 \end{pmatrix} = -i\sigma_1\ .
\end{aligned}
$$

Solution 2.4 It follows from the Definitions (2.11), (2.12), (2.13) and for $G \in \mathrm{SU}(2)$ and $x \in \mathbb{R}^3$ that

$$
\begin{aligned}
\Phi_1(G)(x) &= (\sigma^{-1} \circ R_G \circ \sigma)(x) = \sigma^{-1}(R_G(\sigma(x))) \\
&= \sigma^{-1}(G \cdot \sigma(x) \cdot G^*)
\end{aligned}
$$

and hence that

$$
\sigma(\Phi_1(G)(x)) = G \cdot \sigma(x) \cdot G^*\ . \tag{A.8}
$$

Therefore,

$$i\sum_{k=1}^{3}(\Phi_1(G)\cdot x)_k\,\sigma_k = G\cdot\left(i\sum_{l=1}^{3}x_l\,\sigma_l\right)\cdot G^* \ .$$

From the latter, we infer that

$$\begin{aligned}\sum_{k=1}^{3}\sum_{l=1}^{3}[\Phi_1(G)]_{kl}\cdot x_l\,\sigma_k &= \sum_{l=1}^{3}\sum_{k=1}^{3}[\Phi_1(G)]_{kl}\cdot x_l\,\sigma_k\\ &= G\cdot\left(\sum_{l=1}^{3}x_l\,\sigma_l\right)\cdot G^* = \sum_{l=1}^{3}x_l\,G\cdot\sigma_l\cdot G^*\end{aligned}$$

and hence that

$$\sum_{k=1}^{3}[\Phi_1(G)]_{kl}\,\sigma_k = G\cdot\sigma_l\cdot G^* \ ,$$

for all $l \in \{1,2,3\}$.

Solution 2.5 According to the proof of Theorem 2.13, for $G \in \mathrm{SU}(2)$, we have that

$$\Phi_1(G) = \begin{pmatrix} \mathrm{Re}(g_{11}^2 - g_{12}^2) & \mathrm{Im}(g_{11}^2 + g_{12}^2) & -2\,\mathrm{Re}(g_{11}g_{12}) \\ -\,\mathrm{Im}(g_{11}^2 - g_{12}^2) & \mathrm{Re}(g_{11}^2 + g_{12}^2) & 2\,\mathrm{Im}(g_{11}g_{12}) \\ 2\,\mathrm{Re}(g_{11}g_{12}^*) & 2\,\mathrm{Im}(g_{11}g_{12}^*) & |g_{11}|^2 - |g_{12}|^2 \end{pmatrix} \ .$$

Hence,

$$\begin{aligned}[\Phi_1(G^*)] &= \begin{pmatrix} \mathrm{Re}((g_{11}^*)^2 - g_{12}^2) & \mathrm{Im}((g_{11}^*)^2 + g_{12}^2) & 2\,\mathrm{Re}((g_{11}^*)g_{12}) \\ -\,\mathrm{Im}((g_{11}^*)^2 - g_{12}^2) & \mathrm{Re}((g_{11}^*)^2 + g_{12}^2) & -2\,\mathrm{Im}((g_{11}^*)g_{12}) \\ -2\,\mathrm{Re}((g_{11}^*)g_{12}^*) & -2\,\mathrm{Im}((g_{11}^*)g_{12}^*) & |(g_{11}^*)|^2 - |g_{12}|^2 \end{pmatrix}\\ &= \begin{pmatrix} \mathrm{Re}(g_{11}^2 - g_{12}^2) & -\,\mathrm{Im}(g_{11}^2 - g_{12}^2) & 2\,\mathrm{Re}(g_{11}g_{12}^*) \\ \mathrm{Im}(g_{11}^2 + g_{12}^2) & \mathrm{Re}(g_{11}^2 + g_{12}^2) & 2\,\mathrm{Im}(g_{11}g_{12}^*) \\ -2\,\mathrm{Re}(g_{11}g_{12}) & 2\,\mathrm{Im}(g_{11}g_{12}) & |g_{11}|^2 - |g_{12}|^2 \end{pmatrix} = [\Phi_1(G)]^t \ .\end{aligned}$$

Further, according to (2.14), for $l \in \{1,2,3\}$, we have that

$$\sum_{k=1}^{3}[\Phi_1(G)]_{kl}\,\sigma_k = G\cdot\sigma_l\cdot G^* \ .$$

Hence,

$$\sum_{k=1}^{3}[\Phi_1(G)]_{lk}\,\sigma_k = \sum_{k=1}^{3}[[\Phi_1(G)]^t]_{kl}\,\sigma_k = \sum_{k=1}^{3}[\Phi_1(G^*)]_{kl}\,\sigma_k = G^*\cdot\sigma_l\cdot G \ .$$

Solution 3.1 According to proof of Lemma 3.2, if $\Lambda \in \mathcal{L}$, then

$$\Lambda^{-1} = \eta \, \Lambda^t \eta \ ,$$

where

$$\eta := \begin{pmatrix} 1 & 0 & 0 & 0 \\ 0 & -1 & 0 & 0 \\ 0 & 0 & -1 & 0 \\ 0 & 0 & 0 & -1 \end{pmatrix} .$$

Hence,

$$\Lambda^t = \eta \, \Lambda^{-1} \eta \ .$$

Since,

$$\begin{aligned} (\eta \, x) \cdot (\eta \, x) &= {}^t(x_0, -x_1, -x_2, -x_3) \cdot {}^t(x_0, -x_1, -x_2, -x_3) \\ &= (x_0)^2 - (-x_1)^2 - (-x_2)^2 - (-x_3)^2 \\ &= (x_0)^2 - (x_1)^2 - (x_2)^2 - (x_3)^2 = x \cdot x \ , \end{aligned}$$

for every $x = {}^t(x_0, x_1, x_2, x_3) \in \mathbb{R}^4$, it follows that $\eta \in \mathcal{L}$ and hence, since $\mathcal{L}$, equipped with operation of matrix multiplication, is a group, that $\Lambda^t \in \mathcal{L}$.

Solution 3.2 If $\Lambda_1, \lambda_2, \dots$ is a sequence in $\mathcal{L}$ that is component-wise convergent to $\Lambda \in$ M$(4, \mathbb{R})$, it follows for every $x \in \mathbb{R}^4$ that

$$\begin{aligned} x \cdot x &= \lim_{\nu \to \infty} (\Lambda_\nu \cdot x) \cdot (\Lambda_\nu \cdot x) \\ &= \lim_{\nu \to \infty} \sum_{k,l=0}^{3} \eta_{kl} \left(\sum_{m=0}^{3} \Lambda_{\nu km} x_m \right) \left(\sum_{n=0}^{3} \Lambda_{\nu ln} x_n \right) \\ &= \lim_{\nu \to \infty} \sum_{k,l,m,n=0}^{3} \eta_{kl} \Lambda_{\nu km} \Lambda_{\nu ln} x_m x_n = \sum_{k,l,m,n=0}^{3} \eta_{kl} \Lambda_{km} \Lambda_{ln} x_m x_n \\ &= \sum_{k,l=0}^{3} \eta_{kl} \left(\sum_{m=0}^{3} \Lambda_{km} x_m \right) \left(\sum_{n=0}^{3} \Lambda_{ln} x_n \right) = (\Lambda \cdot x) \cdot (\Lambda \cdot x) \end{aligned}$$

and hence that $\Lambda \in \mathcal{L}$. Further, since according to Exercise 3.5, we have that

$$M_{01}(s) := \begin{pmatrix} \cosh(s) & \sinh(s) & 0 & 0 \\ \sinh(s) & \cosh(s) & 0 & 0 \\ 0 & 0 & 1 & 0 \\ 0 & 0 & 0 & 1 \end{pmatrix} \in \mathcal{L} \ ,$$

for every $s \in \mathbb{R}$, it follows that $\mathcal{L}$ is unbounded.

Solution 3.3 If $\Lambda_1, \Lambda_2, \ldots$ is a sequence in $\mathcal{L}_+^\uparrow$ that is component-wise convergent to $\Lambda \in \mathrm{M}(4, \mathbb{R})$, then $(\Lambda_1)_{00}, (\Lambda_2)_{00}, \ldots$ is a sequence of positive real numbers that is convergent to Λ_{00}. Hence, $\Lambda_{00} \geqslant 0$. Further, since $\mathcal{L} \subset \mathrm{M}(4, \mathbb{R})$ is closed, we have that $\Lambda \in \mathcal{L}$ and hence that $\Lambda \in \mathcal{L}_+^\uparrow$. Also, since according to Exercise 3.5, we have that

$$M_{01}(s) := \begin{pmatrix} \cosh(s) & \sinh(s) & 0 & 0 \\ \sinh(s) & \cosh(s) & 0 & 0 \\ 0 & 0 & 1 & 0 \\ 0 & 0 & 0 & 1 \end{pmatrix} \in \mathcal{L}$$

and $\cosh(s) \geqslant 0$, for every $s \in \mathbb{R}$, it follows that $M_{01}(s) \in \mathcal{L}_+^\uparrow$, for every $s \in \mathbb{R}$, and hence that $\mathcal{L}_+^\uparrow$ is unbounded.

Solution 3.4 For every $s \in \mathbb{R}$ and $x \in \mathbb{R}^4$, we have that

$$\begin{aligned}
&(M_1(s)\,x) \cdot (M_1(s)\,x) \\
&= x_0^2 - x_1^2 - (\cos(s)\,x_2 + \sin(s)\,x_3)^2 - (-\sin(s)\,x_2 + \cos(s)\,x_3)^2 \\
&= x_0^2 - x_1^2 - (\cos^2(s) + \sin^2(s))\,x_2^2 - (\sin^2(s) + \cos^2(s))\,x_3^2 = x \cdot x\ , \\
&(M_2(s)\,x) \cdot (M_2(s)\,x) \\
&= x_0^2 - (\cos(s)\,x_1 - \sin(s)\,x_3)^2 - x_2^2 - (\sin(s)\,x_1 + \cos(s)\,x_3)^2 \\
&= x_0^2 - (\cos^2(s) + \sin^2(s))\,x_1^2 - x_2^2 - (\sin^2(s) + \cos^2(s))\,x_3^2 = x \cdot x\ , \\
&(M_3(s)\,x) \cdot (M_3(s)\,x) \\
&= x_0^2 - (\cos(s)\,x_1 + \sin(s)\,x_2)^2 - (-\sin(s)\,x_1 + \cos(s)\,x_2)^2 - x_3^2 \\
&= x_0^2 - (\cos^2(s) + \sin^2(s))\,x_1^2 - (\sin^2(s) + \cos^2(s))\,x_2^2 - x_3^2 = x \cdot x\ .
\end{aligned}$$

Further, it follows that

$$\det(M_1(s)) = 1 \cdot 1 \cdot \begin{vmatrix} \cos(s) & \sin(s) \\ -\sin(s) & \cos(s) \end{vmatrix} = \cos^2(s) + \sin^2(s) = 1\ ,$$

$$\det(M_2(s)) = 1 \cdot \begin{vmatrix} \cos(s) & 0 & -\sin(s) \\ 0 & 1 & 0 \\ \sin(s) & 0 & \cos(s) \end{vmatrix} = 1 \cdot \begin{vmatrix} \cos(s) & -\sin(s) \\ \sin(s) & \cos(s) \end{vmatrix}$$

$$= \cos^2(s) + \sin^2(s) = 1\ ,$$

$$\det(M_3(s)) = 1 \cdot \begin{vmatrix} \cos(s) & \sin(s) & 0 \\ -\sin(s) & \cos(s) & 0 \\ 0 & 0 & 1 \end{vmatrix} = 1 \cdot \begin{vmatrix} \cos(s) & \sin(s) \\ -\sin(s) & \cos(s) \end{vmatrix}$$

$$= \cos^2(s) + \sin^2(s) = 1$$

and that $(M_1(s))_{00} = (M_2(s))_{00} = (M_3(s))_{00} = 1 \geqslant 0$. Therefore $M_1(s), M_2(s), M_3(s) \in \mathcal{L}_+^{\uparrow}$, for every $s \in \mathbb{R}$. Further,

$$M_1(s) \cdot M_1(\bar{s}) = \begin{pmatrix} 1 & 0 & 0 & 0 \\ 0 & 1 & 0 & 0 \\ 0 & 0 & \cos(s) & \sin(s) \\ 0 & 0 & -\sin(s) & \cos(s) \end{pmatrix} \cdot \begin{pmatrix} 1 & 0 & 0 & 0 \\ 0 & 1 & 0 & 0 \\ 0 & 0 & \cos(\bar{s}) & \sin(\bar{s}) \\ 0 & 0 & -\sin(\bar{s}) & \cos(\bar{s}) \end{pmatrix}$$

$$= \begin{pmatrix} 1 & 0 & 0 & 0 \\ 0 & 1 & 0 & 0 \\ 0 & 0 & \cos(s)\cos(\bar{s}) - \sin(s)\sin(\bar{s}) & \cos(s)\sin(\bar{s}) + \sin(s)\cos(\bar{s}) \\ 0 & 0 & -\sin(s)\cos(\bar{s}) - \cos(s)\sin(\bar{s}) & -\sin(s)\sin(\bar{s}) + \cos(s)\cos(\bar{s}) \end{pmatrix}$$

$$= \begin{pmatrix} 1 & 0 & 0 & 0 \\ 0 & 1 & 0 & 0 \\ 0 & 0 & \cos(s+\bar{s}) & \sin(s+\bar{s}) \\ 0 & 0 & -\sin(s+\bar{s}) & \cos(s+\bar{s}) \end{pmatrix} = M_1(s+\bar{s}) \ ,$$

$$M_2(s) \cdot M_2(\bar{s}) = \begin{pmatrix} 1 & 0 & 0 & 0 \\ 0 & \cos(s) & 0 & -\sin(s) \\ 0 & 0 & 1 & 0 \\ 0 & \sin(s) & 0 & \cos(s) \end{pmatrix} \cdot \begin{pmatrix} 1 & 0 & 0 & 0 \\ 0 & \cos(\bar{s}) & 0 & -\sin(\bar{s}) \\ 0 & 0 & 1 & 0 \\ 0 & \sin(\bar{s}) & 0 & \cos(\bar{s}) \end{pmatrix}$$

$$= \begin{pmatrix} 1 & 0 & 0 & 0 \\ 0 & \cos(s)\cos(\bar{s}) - \sin(s)\sin(\bar{s}) & 0 & -\cos(s)\sin(\bar{s}) - \sin(s)\cos(\bar{s}) \\ 0 & 0 & 1 & 0 \\ 0 & \sin(s)\cos(\bar{s}) + \cos(s)\sin(\bar{s}) & 0 & -\sin(s)\sin(\bar{s}) + \cos(s)\cos(\bar{s}) \end{pmatrix}$$

$$= \begin{pmatrix} 1 & 0 & 0 & 0 \\ 0 & \cos(s+\bar{s}) & 0 & -\sin(s+\bar{s}) \\ 0 & 0 & 1 & 0 \\ 0 & \sin(s+\bar{s}) & 0 & \cos(s+\bar{s}) \end{pmatrix} = M_2(s+\bar{s}) \ ,$$

$$M_3(s) \cdot M_3(\bar{s}) = \begin{pmatrix} 1 & 0 & 0 & 0 \\ 0 & \cos(s) & \sin(s) & 0 \\ 0 & -\sin(s) & \cos(s) & 0 \\ 0 & 0 & 0 & 1 \end{pmatrix} \cdot \begin{pmatrix} 1 & 0 & 0 & 0 \\ 0 & \cos(\bar{s}) & \sin(\bar{s}) & 0 \\ 0 & -\sin(\bar{s}) & \cos(\bar{s}) & 0 \\ 0 & 0 & 0 & 1 \end{pmatrix}$$

$$= \begin{pmatrix} 1 & 0 & 0 & 0 \\ 0 & \cos(s)\cos(\bar{s}) - \sin(s)\sin(\bar{s}) & \cos(s)\sin(\bar{s}) + \sin(s)\cos(\bar{s}) & 0 \\ 0 & -\sin(s)\cos(\bar{s}) - \cos(s)\sin(\bar{s}) & -\sin(s)\sin(\bar{s}) + \cos(s)\cos(\bar{s}) & 0 \\ 0 & 0 & 0 & 1 \end{pmatrix}$$

$$= \begin{pmatrix} 1 & 0 & 0 & 0 \\ 0 & \cos(s+\bar{s}) & \sin(s+\bar{s}) & 0 \\ 0 & -\sin(s+\bar{s}) & \cos(s+\bar{s}) & 0 \\ 0 & 0 & 0 & 1 \end{pmatrix} = M_3(s+\bar{s}) \ ,$$

for all $s, \bar{s} \in \mathbb{R}$ and $M_1(0) = M_2(0) = M_3(0) = E$. Hence, by (3.8), for every $s \in \mathbb{R}$, there are defined one-parameter groups $M_j : \mathbb{R} \to \mathcal{L}_+^\uparrow$, $j \in \{1, 2, 3\}$.

Solution 3.5 For every $s \in \mathbb{R}$ and $x \in \mathbb{R}^4$, we have that

$$
\begin{aligned}
&(M_{01}(s)\,x) \cdot (M_{01}(s)\,x) \\
&= (\cosh(s)\,x_0 + \sinh(s)\,x_1)^2 - (\sinh(s)\,x_0 + \cosh(s)\,x_1)^2 - x_2^2 - x_3^2 \\
&= (\cosh^2(s) - \sinh^2(s))\,x_0^2 + (\sinh^2(s) - \cosh^2(s))\,x_1^2 - x_2^2 - x_3^2 \\
&= x_0^2 - x_1^2 - x_2^2 - x_3^2 = x \cdot x \ , \\
&(M_{02}(s)\,x) \cdot (M_{02}(s)\,x) \\
&= (\cosh(s)\,x_0 + \sinh(s)\,x_2)^2 - x_1^2 - (\sinh(s)\,x_0 + \cosh(s)\,x_2)^2 - x_3^2 \\
&= (\cosh^2(s) - \sinh^2(s))\,x_0^2 + (\sinh^2(s) - \cosh^2(s))\,x_2^2 - x_1^2 - x_3^2 \\
&= x_0^2 - x_1^2 - x_2^2 - x_3^2 = x \cdot x \ , \\
&(M_{03}(s)\,x) \cdot (M_{03}(s)\,x) \\
&= (\cosh(s)\,x_0 + \sinh(s)\,x_3)^2 - (\sinh(s)\,x_0 + \cosh(s)\,x_3)^2 - x_1^2 - x_2^2 \\
&= (\cosh^2(s) - \sinh^2(s))\,x_0^2 + (\sinh^2(s) - \cosh^2(s))\,x_3^2 - x_1^2 - x_2^2 \\
&= x_0^2 - x_1^2 - x_2^2 - x_3^2 = x \cdot x \ .
\end{aligned}
$$

Further,

$$
\det(M_{01}(s)) = \begin{vmatrix} \cosh(s) & \sinh(s) \\ \sinh(s) & \cosh(s) \end{vmatrix} \cdot 1 \cdot 1 = \cosh^2(s) - \sinh^2(s) = 1 \ ,
$$

$$
\det(M_{02}(s)) = \begin{vmatrix} \cosh(s) & 0 & \sinh(s) \\ 0 & 1 & 0 \\ \sinh(s) & 0 & \cosh(s) \end{vmatrix} \cdot 1 = 1 \cdot \begin{vmatrix} \cosh(s) & \sinh(s) \\ \sinh(s) & \cosh(s) \end{vmatrix}
$$
$$
= \cosh^2(s) - \sinh^2(s) = 1 \ ,
$$

$$
\det(M_{03}(s)) = 1 \cdot \begin{vmatrix} \cosh(s) & 0 & \sinh(s) \\ 0 & 1 & 0 \\ \sinh(s) & 0 & \cosh(s) \end{vmatrix} = 1 \cdot \begin{vmatrix} \cosh(s) & \sinh(s) \\ \sinh(s) & \cosh(s) \end{vmatrix}
$$
$$
= \cosh^2(s) - \sinh^2(s) = 1
$$

and that $(M_{01}(s))_{00}{=}(M_{02}(s))_{00} = (M_{03}(s))_{00} = \cosh(s) \geqslant 0$. Therefore $M_{01}(s), M_{02}(s), M_{03}(s) \in \mathcal{L}_+^\uparrow$, for every $s \in \mathbb{R}$. Further,

$$
M_{01}(s)\cdot M_{01}(\bar{s}) = \begin{pmatrix} \cosh(s) & \sinh(s) & 0 & 0 \\ \sinh(s) & \cosh(s) & 0 & 0 \\ 0 & 0 & 1 & 0 \\ 0 & 0 & 0 & 1 \end{pmatrix} \cdot \begin{pmatrix} \cosh(\bar{s}) & \sinh(\bar{s}) & 0 & 0 \\ \sinh(\bar{s}) & \cosh(\bar{s}) & 0 & 0 \\ 0 & 0 & 1 & 0 \\ 0 & 0 & 0 & 1 \end{pmatrix}
$$

$$
= \begin{pmatrix} \cosh(s)\cosh(\bar{s}) + \sinh(s)\sinh(\bar{s}) & \cosh(s)\sinh(\bar{s}) + \sinh(s)\cosh(\bar{s}) & 0 & 0 \\ \sinh(s)\cosh(\bar{s}) + \cosh(s)\sinh(\bar{s}) & \sinh(s)\sinh(\bar{s}) + \cosh(s)\cosh(\bar{s}) & 0 & 0 \\ 0 & 0 & 1 & 0 \\ 0 & 0 & 0 & 1 \end{pmatrix}
$$

$$
= \begin{pmatrix} \cosh(s+\bar{s}) & \sinh(s+\bar{s}) & 0 & 0 \\ \sinh(s+\bar{s}) & \cosh(s+\bar{s}) & 0 & 0 \\ 0 & 0 & 1 & 0 \\ 0 & 0 & 0 & 1 \end{pmatrix} = M_{01}(s+\bar{s}) \ ,
$$

$$
M_{02}(s)\cdot M_{02}(\bar{s}) = \begin{pmatrix} \cosh(s) & 0 & \sinh(s) & 0 \\ 0 & 1 & 0 & 0 \\ \sinh(s) & 0 & \cosh(s) & 0 \\ 0 & 0 & 0 & 1 \end{pmatrix} \cdot \begin{pmatrix} \cosh(\bar{s}) & 0 & \sinh(\bar{s}) & 0 \\ 0 & 1 & 0 & 0 \\ \sinh(\bar{s}) & 0 & \cosh(\bar{s}) & 0 \\ 0 & 0 & 0 & 1 \end{pmatrix}
$$

$$
= \begin{pmatrix} \cosh(s)\cosh(\bar{s}) + \sinh(s)\sinh(\bar{s}) & 0 & \cosh(s)\sinh(\bar{s}) + \sinh(s)\cosh(\bar{s}) & 0 \\ 0 & 1 & 0 & 0 \\ \sinh(s)\cosh(\bar{s}) + \cosh(s)\sinh(\bar{s}) & 0 & \sinh(s)\sinh(\bar{s}) + \cosh(s)\cosh(\bar{s}) & 0 \\ 0 & 0 & 0 & 1 \end{pmatrix}
$$

$$
= \begin{pmatrix} \cosh(s+\bar{s}) & 0 & \sinh(s+\bar{s}) & 0 \\ 0 & 1 & 0 & 0 \\ \sinh(s+\bar{s}) & 0 & \cosh(s+\bar{s}) & 0 \\ 0 & 0 & 0 & 1 \end{pmatrix} = M_{02}(s+\bar{s}) \ ,
$$

$$M_{03}(s) \cdot M_{03}(\bar{s}) = \begin{pmatrix} \cosh(s) & 0 & 0 & \sinh(s) \\ 0 & 1 & 0 & 0 \\ 0 & 0 & 1 & 0 \\ \sinh(s) & 0 & 0 & \cosh(s) \end{pmatrix} \cdot \begin{pmatrix} \cosh(\bar{s}) & 0 & 0 & \sinh(\bar{s}) \\ 0 & 1 & 0 & 0 \\ 0 & 0 & 1 & 0 \\ \sinh(\bar{s}) & 0 & 0 & \cosh(\bar{s}) \end{pmatrix}$$

$$= \begin{pmatrix} \cosh(s)\cosh(\bar{s}) + \sinh(s)\sinh(\bar{s}) & 0 & 0 & \cosh(s)\sinh(\bar{s}) + \sinh(s)\cosh(\bar{s}) \\ 0 & 1 & 0 & 0 \\ 0 & 0 & 1 & 0 \\ \sinh(s)\cosh(\bar{s}) + \cosh(s)\sinh(\bar{s}) & 0 & 0 & \sinh(s)\sinh(\bar{s}) + \cosh(s)\cosh(\bar{s}) \end{pmatrix}$$

$$= \begin{pmatrix} \cosh(s+\bar{s}) & 0 & 0 & \sinh(s+\bar{s}) \\ 0 & 1 & 0 & 0 \\ 0 & 0 & 1 & 0 \\ \sinh(s+\bar{s}) & 0 & 0 & \cosh(s+\bar{s}) \end{pmatrix} = M_{03}(s+\bar{s}) \ ,$$

for all $s, \bar{s} \in \mathbb{R}$ and $M_{01}(0) = M_{02}(0) = M_{03}(0) = E$. Hence, by (3.9), for every $s \in \mathbb{R}$, there are defined one-parameter groups $M_{0j} : \mathbb{R} \to \mathcal{L}_+^\uparrow$, $j \in \{1, 2, 3\}$.

Solution 3.6 If $c : I \to \mathrm{M}(4, \mathbb{R})$ is differentiable with $\mathrm{Ran}(c) \subset \mathcal{L}$ and $c(0) = E$, where I is an open interval of $\mathbb{R}$ around 0, then

$$c(s) \cdot \eta \cdot (c(s))^t \cdot \eta = E$$

and hence

$$c'(s) \cdot \eta \cdot (c(s))^t \cdot \eta + c(s) \cdot \eta \cdot (c'(s))^t \cdot \eta = 0 \ ,$$

for every $s \in \mathbb{R}$. In particular, the latter implies that

$$0 = c'(0) \cdot \eta \cdot (c(0))^t \cdot \eta + c(0) \cdot \eta \cdot (c'(0))^t \cdot \eta = c'(0) + \eta \cdot (c'(0))^t \cdot \eta$$

and hence that

$$(c'(0))^t = -\eta \cdot c'(0) \cdot \eta \ .$$

Defining $M := c'(0)$, it follows that

$$M = \begin{pmatrix} M_{00} & M_{01} & M_{02} & M_{03} \\ M_{10} & M_{11} & M_{12} & M_{13} \\ M_{20} & M_{21} & M_{22} & M_{23} \\ M_{30} & M_{31} & M_{32} & M_{33} \end{pmatrix}$$

$$\begin{pmatrix} M_{00} & M_{10} & M_{20} & M_{30} \\ M_{01} & M_{11} & M_{21} & M_{31} \\ M_{02} & M_{12} & M_{22} & M_{32} \\ M_{03} & M_{13} & M_{23} & M_{33} \end{pmatrix} = M^t = -\eta \cdot M \cdot \eta$$

$$= \begin{pmatrix} -M_{00} & M_{01} & M_{02} & M_{03} \\ M_{10} & -M_{11} & -M_{12} & -M_{13} \\ M_{20} & -M_{21} & -M_{22} & -M_{23} \\ M_{30} & -M_{31} & -M_{32} & -M_{33} \end{pmatrix}.$$

Hence,

$$M = \begin{pmatrix} 0 & M_{01} & M_{02} & M_{03} \\ M_{01} & 0 & M_{12} & M_{13} \\ M_{02} & -M_{12} & 0 & M_{23} \\ M_{03} & -M_{13} & -M_{23} & 0 \end{pmatrix}$$

$$\in \left\{ \begin{pmatrix} 0 & a_1 & a_2 & a_3 \\ a_1 & 0 & -b_3 & b_2 \\ a_2 & b_3 & 0 & -b_1 \\ a_3 & -b_2 & b_1 & 0 \end{pmatrix} : a, b \in \mathbb{R}^3 \right\}.$$

From Exercises 3.4 and 3.5, it follows that $A_1, A_2, A_3, B_1, B_2, B_3 \in T_E\mathcal{L}$, where

$$A_1 := \begin{pmatrix} 0&1&0&0 \\ 1&0&0&0 \\ 0&0&0&0 \\ 0&0&0&0 \end{pmatrix}, \quad A_2 := \begin{pmatrix} 0&0&1&0 \\ 0&0&0&0 \\ 1&0&0&0 \\ 0&0&0&0 \end{pmatrix},$$

$$A_3 := \begin{pmatrix} 0&0&0&1 \\ 0&0&0&0 \\ 0&0&0&0 \\ 1&0&0&0 \end{pmatrix}, \quad B_1 := \begin{pmatrix} 0&0&0&0 \\ 0&0&0&0 \\ 0&0&0&1 \\ 0&0&-1&0 \end{pmatrix},$$

$$B_2 := \begin{pmatrix} 0&0&0&0 \\ 0&0&0&-1 \\ 0&0&0&0 \\ 0&1&0&0 \end{pmatrix}, \quad B_3 := \begin{pmatrix} 0&0&0&0 \\ 0&0&1&0 \\ 0&-1&0&0 \\ 0&0&0&0 \end{pmatrix}.$$

Hence

$$a_1 A_1 + a_2 A_2 - b_1 B_1 - b_2 B_2 - b_3 B_3 = \begin{pmatrix} 0 & a_1 & a_2 & a_3 \\ a_1 & 0 & -b_3 & b_2 \\ a_2 & b_3 & 0 & -b_1 \\ a_3 & -b_2 & b_1 & 0 \end{pmatrix} \in T_E\mathcal{L},$$

for $a, b \in \mathbb{R}^3$. Therefore,

$$T_E\mathcal{L} \supset \left\{ \begin{pmatrix} 0 & a_1 & a_2 & a_3 \\ a_1 & 0 & -b_3 & b_2 \\ a_2 & b_3 & 0 & -b_1 \\ a_3 & -b_2 & b_1 & 0 \end{pmatrix} : a, b \in \mathbb{R}^3 \right\} .$$

Solution 3.7 For $a, b \in \mathbb{R}^3$ and

$$M := \begin{pmatrix} 0 & a_1 & a_2 & a_3 \\ a_1 & 0 & -b_3 & b_2 \\ a_2 & b_3 & 0 & -b_1 \\ a_3 & -b_2 & b_1 & 0 \end{pmatrix},$$

we have that

$$\begin{aligned}
&-\eta \cdot \begin{pmatrix} 0 & a_1 & a_2 & a_3 \\ a_1 & 0 & -b_3 & b_2 \\ a_2 & b_3 & 0 & -b_1 \\ a_3 & -b_2 & b_1 & 0 \end{pmatrix}^t \\
&= \begin{pmatrix} -1 & 0 & 0 & 0 \\ 0 & 1 & 0 & 0 \\ 0 & 0 & 1 & 0 \\ 0 & 0 & 0 & 1 \end{pmatrix} \cdot \begin{pmatrix} 0 & a_1 & a_2 & a_3 \\ a_1 & 0 & b_3 & -b_2 \\ a_2 & -b_3 & 0 & b_1 \\ a_3 & b_2 & -b_1 & 0 \end{pmatrix} \\
&= \begin{pmatrix} 0 & -a_1 & -a_2 & -a_3 \\ a_1 & 0 & b_3 & -b_2 \\ a_2 & -b_3 & 0 & b_1 \\ a_3 & b_2 & -b_1 & 0 \end{pmatrix} \\
&= \begin{pmatrix} 0 & a_1 & a_2 & a_3 \\ a_1 & 0 & -b_3 & b_2 \\ a_2 & b_3 & 0 & -b_1 \\ a_3 & -b_2 & b_1 & 0 \end{pmatrix} \cdot \begin{pmatrix} 1 & 0 & 0 & 0 \\ 0 & -1 & 0 & 0 \\ 0 & 0 & -1 & 0 \\ 0 & 0 & 0 & -1 \end{pmatrix} .
\end{aligned}$$

Hence, $M^t = -\eta \cdot M \cdot \eta$. Further, for $M \in \mathrm{M}(4, \mathbb{R})$ such that $M^t = -\eta \cdot M \cdot \eta$, it follows that

$$-\eta \cdot M^t = \begin{pmatrix} -1 & 0 & 0 & 0 \\ 0 & 1 & 0 & 0 \\ 0 & 0 & 1 & 0 \\ 0 & 0 & 0 & 1 \end{pmatrix} \cdot \begin{pmatrix} M_{00} & M_{10} & M_{20} & M_{30} \\ M_{01} & M_{11} & M_{21} & M_{31} \\ M_{02} & M_{12} & M_{22} & M_{32} \\ M_{03} & M_{13} & M_{23} & M_{33} \end{pmatrix}$$

$$= \begin{pmatrix} -M_{00} & -M_{10} & -M_{20} & -M_{30} \\ M_{01} & M_{11} & M_{21} & M_{31} \\ M_{02} & M_{12} & M_{22} & M_{32} \\ M_{03} & M_{13} & M_{23} & M_{33} \end{pmatrix}$$

$$= \begin{pmatrix} M_{00} & M_{01} & M_{02} & M_{03} \\ M_{10} & M_{11} & M_{12} & M_{13} \\ M_{20} & M_{21} & M_{22} & M_{23} \\ M_{30} & M_{31} & M_{32} & M_{33} \end{pmatrix} \cdot \begin{pmatrix} 1 & 0 & 0 & 0 \\ 0 & -1 & 0 & 0 \\ 0 & 0 & -1 & 0 \\ 0 & 0 & 0 & -1 \end{pmatrix}$$

$$= \begin{pmatrix} M_{00} & -M_{01} & -M_{02} & -M_{03} \\ M_{10} & -M_{11} & -M_{12} & -M_{13} \\ M_{20} & -M_{21} & -M_{22} & -M_{23} \\ M_{30} & -M_{31} & -M_{32} & -M_{33} \end{pmatrix}$$

and hence that

$$M = \begin{pmatrix} 0 & M_{01} & M_{02} & M_{03} \\ M_{01} & 0 & M_{12} & M_{13} \\ M_{02} & -M_{12} & 0 & M_{23} \\ M_{03} & -M_{13} & -M_{23} & 0 \end{pmatrix} \in T_E\mathcal{L} \ .$$

Further, if $M \in \mathbb{R}^4$ such that $M^t = -\eta \cdot M \cdot \eta$, then

$$(\eta \cdot M \cdot \eta)^k = \eta \cdot M^k \cdot \eta \ .$$

The proof proceeds by induction over $k \in \mathbb{N}$. Since

$$(\eta \cdot M \cdot \eta)^0 = E = \eta \cdot \eta = \eta \cdot E \cdot \eta = \eta \cdot M^0 \cdot \eta \ ,$$

the statement is true for $k = 0$. If the statement is true for $k \in \mathbb{N}$, then

$$\begin{aligned}(\eta \cdot M \cdot \eta)^{k+1} &= (\eta \cdot M \cdot \eta)^k \cdot (\eta \cdot M \cdot \eta) = \eta \cdot M^k \cdot \eta \cdot \eta \cdot M \cdot \eta \\ &= \eta \cdot M^k \cdot M \cdot \eta = \eta \cdot M^{k+1} \cdot \eta\end{aligned}$$

and hence the statement is true for the subsequent case. Further, for $x, y \in \mathbb{R}^4$, it follows that

$$
\begin{aligned}
x \cdot [\exp(M)\, y] &= \sum_{k=0}^{\infty} \frac{1}{k!}\, x \cdot M^k y = \sum_{k=0}^{\infty} \frac{1}{k!} \left\langle x | \eta M^k y \right\rangle = \sum_{k=0}^{\infty} \frac{1}{k!} \left\langle \eta x | M^k y \right\rangle \\
&= \sum_{k=0}^{\infty} \frac{1}{k!} (-1)^k \left\langle (\eta \cdot M \cdot \eta)^k \eta x | y \right\rangle = \sum_{k=0}^{\infty} \frac{1}{k!} (-1)^k \left\langle \eta M^k \eta \eta x | y \right\rangle \\
&= \sum_{k=0}^{\infty} \frac{1}{k!} (-1)^k \left\langle \eta M^k x | y \right\rangle = \langle \eta \exp(-M) x | y \rangle = \langle \exp(-M) x | \eta y \rangle \\
&= [\exp(-M)\, x] \cdot y \ ,
\end{aligned}
$$

where $\langle \, | \, \rangle$ denotes the Euclidean scalar product on $\mathbb{R}^3$, and hence that

$$
[\exp(M)\, x] \cdot [\exp(M)\, y] = [\exp(-M) \exp(M)\, x] \cdot y = x \cdot y \ .
$$

Therefore, $\exp(A) \in \mathcal{L}$. In addition, we have that

$$
\det(\exp(A)) = \exp(\mathrm{Tr}(A)) = \exp(0) = 1 \ .
$$

Further, we note that from (3.5), it follows for $\Lambda \in \mathcal{L}$ that either $\Lambda_{00} \geqslant 1$ or $\Lambda_{00} \leqslant -1$. Hence, $h := ([0, 1] \to \mathbb{R}, [\exp(tA)]_{00})$ is a continuous function such that $h(0) = 1$. If $s \in (0, 1]$ is such that $f(s) < 0$, then it follows from the Intermediate Value Theorem the existence of $\tau \in (0, s)$ such that $f(0) = 0$. $\lightning$ Hence, it follows that $f(t) \geqslant 0$, for every $t \in [0, 1]$ and therefore that $[\exp(tA)]_{00} \geqslant 0$. Thus, $\exp(A) \in \mathcal{L}_+^{\uparrow}$.

Solution 3.8 If $M \in \mathrm{M}(4, \mathbb{R})$ is such that $M^t = \pm \eta \cdot M \cdot \eta$, it follows that

$$
\begin{aligned}
(\Lambda M \Lambda^{-1})^t &= (\Lambda^{-1})^t M^t \Lambda^t = \pm (\Lambda^{-1})^t \, \eta \, M \, \eta \, \Lambda^t \\
&= \pm (\eta \, \Lambda^t \eta)^t \, \eta \, M \, \eta \, \eta \, \Lambda^{-1} \, \eta = \pm \eta \, \Lambda \eta \, \eta \, M \, \eta \, \eta \, \Lambda^{-1} \, \eta = \pm \eta \, \Lambda M \Lambda^{-1} \eta \ .
\end{aligned}
$$

Solution 3.9 If $\lambda \notin \{0, -a_1, a_1\}$, it follows that

$$\det(A-\lambda E)=\begin{vmatrix}-\lambda & a_1 & a_2 & a_3\\ a_1 & -\lambda & -b_3 & b_2\\ a_2 & b_3 & -\lambda & -b_1\\ a_3 & -b_2 & b_1 & -\lambda\end{vmatrix}=\begin{vmatrix}-\lambda & a_1 & a_2 & a_3\\ a_1-\frac{a_1}{\lambda}\lambda & \frac{a_1}{\lambda}a_1-\lambda & \frac{a_1}{\lambda}a_2-b_3 & \frac{a_1}{\lambda}a_3+b_2\\ a_2-\frac{a_2}{\lambda}\lambda & \frac{a_2}{\lambda}a_1+b_3 & \frac{a_2}{\lambda}a_2-\lambda & \frac{a_2}{\lambda}a_3-b_1\\ a_3-\frac{a_3}{\lambda}\lambda & \frac{a_3}{\lambda}a_1-b_2 & \frac{a_3}{\lambda}a_2+b_1 & \frac{a_3}{\lambda}a_3-\lambda\end{vmatrix}$$

$$=-\lambda\begin{vmatrix}\frac{a_1}{\lambda}a_1-\lambda & \frac{a_1}{\lambda}a_2-b_3 & \frac{a_1}{\lambda}a_3+b_2\\ \frac{a_2}{\lambda}a_1+b_3 & \frac{a_2}{\lambda}a_2-\lambda & \frac{a_2}{\lambda}a_3-b_1\\ \frac{a_3}{\lambda}a_1-b_2 & \frac{a_3}{\lambda}a_2+b_1 & \frac{a_3}{\lambda}a_3-\lambda\end{vmatrix}=-\frac{1}{\lambda^2}\begin{vmatrix}a_1^2-\lambda^2 & a_1a_2-b_3\lambda & a_1a_3+b_2\lambda\\ a_1a_2+b_3\lambda & a_2^2-\lambda^2 & a_2a_3-b_1\lambda\\ a_1a_3-b_2\lambda & a_2a_3+b_1\lambda & a_3^2-\lambda^2\end{vmatrix}$$

$$=-\frac{1}{\lambda^2}\begin{vmatrix}a_1^2-\lambda^2 & a_1a_2-b_3\lambda & a_1a_3+b_2\lambda\\ a_1a_2+b_3\lambda-\frac{a_1a_2+b_3\lambda}{a_1^2-\lambda^2}(a_1^2-\lambda^2) & a_2^2-\lambda^2-\frac{(a_1a_2+b_3\lambda)(a_1a_2-b_3\lambda)}{a_1^2-\lambda^2} & a_2a_3-b_1\lambda-\frac{(a_1a_2+b_3\lambda)(a_1a_3+b_2\lambda)}{a_1^2-\lambda^2}\\ a_1a_3-b_2\lambda-\frac{a_1a_3-b_2\lambda}{a_1^2-\lambda^2}(a_1^2-\lambda^2) & a_2a_3+b_1\lambda-\frac{a_1a_3-b_2\lambda}{a_1^2-\lambda^2}(a_1a_2-b_3\lambda) & a_3^2-\lambda^2-\frac{a_1a_3-b_2\lambda}{a_1^2-\lambda^2}(a_1a_3+b_2\lambda)\end{vmatrix}$$

$$=-\frac{a_1^2-\lambda^2}{\lambda^2}\begin{vmatrix}a_2^2-\lambda^2-\frac{(a_1a_2+b_3\lambda)(a_1a_2-b_3\lambda)}{a_1^2-\lambda^2} & a_2a_3-b_1\lambda-\frac{(a_1a_2+b_3\lambda)(a_1a_3+b_2\lambda)}{a_1^2-\lambda^2}\\ a_2a_3+b_1\lambda-\frac{a_1a_3-b_2\lambda}{a_1^2-\lambda^2}(a_1a_2-b_3\lambda) & a_3^2-\lambda^2-\frac{a_1a_3-b_2\lambda}{a_1^2-\lambda^2}(a_1a_3+b_2\lambda)\end{vmatrix}$$

$$=-\frac{1}{\lambda^2(a_1^2-\lambda^2)}\begin{vmatrix}(a_1^2-\lambda^2)(a_2^2-\lambda^2)-(a_1a_2+b_3\lambda)(a_1a_2-b_3\lambda) & (a_1^2-\lambda^2)(a_2a_3-b_1\lambda)-(a_1a_2+b_3\lambda)(a_1a_3+b_2\lambda)\\ (a_1^2-\lambda^2)(a_2a_3+b_1\lambda)-(a_1a_3-b_2\lambda)(a_1a_2-b_3\lambda) & (a_1^2-\lambda^2)(a_3^2-\lambda^2)-(a_1a_3-b_2\lambda)(a_1a_3+b_2\lambda)\end{vmatrix}$$

$$=-\frac{1}{\lambda^2(a_1^2-\lambda^2)}\begin{vmatrix}(a_1^2-\lambda^2)(a_2^2-\lambda^2)-(a_1^2a_2^2-b_3^2\lambda^2) & (a_1^2-\lambda^2)(a_2a_3-b_1\lambda)-(a_1a_2+b_3\lambda)(a_1a_3+b_2\lambda)\\ (a_1^2-\lambda^2)(a_2a_3+b_1\lambda)-(a_1a_2-b_3\lambda)(a_1a_3-b_2\lambda) & (a_1^2-\lambda^2)(a_3^2-\lambda^2)-(a_1^2a_3^2-b_2^2\lambda^2)\end{vmatrix}$$

$$=-\frac{1}{\lambda^2(a_1^2-\lambda^2)}\begin{vmatrix}\lambda^4-(a_1^2+a_2^2-b_3^2)\,\lambda^2 & b_1\lambda^3-(a_2a_3+b_2b_3)\lambda^2-a_1\,(a\cdot b)\,\lambda\\ -b_1\lambda^3-(a_2a_3+b_2b_3)\lambda^2+a_1\,(a\cdot b)\,\lambda & \lambda^4-(a_1^2+a_3^2-b_2^2)\,\lambda^2\end{vmatrix}$$

$$=-\frac{1}{(a_1^2-\lambda^2)}\begin{vmatrix}\lambda^3-(a_1^2+a_2^2-b_3^2)\,\lambda & b_1\lambda^2-(a_2a_3+b_2b_3)\lambda-a_1\,(a\cdot b)\\ -b_1\lambda^2-(a_2a_3+b_2b_3)\lambda+a_1\,(a\cdot b) & \lambda^3-(a_1^2+a_3^2-b_2^2)\,\lambda\end{vmatrix}$$

$$=-\frac{1}{(a_1^2-\lambda^2)}\{\lambda^2\cdot[\lambda^2-(a_1^2+a_2^2-b_3^2)]\cdot[\lambda^2-(a_1^2+a_3^2-b_2^2)]-[\,b_1\lambda^2-(a_2a_3+b_2b_3)\,\lambda-a_1\,(a\cdot b)]\cdot[-b_1\lambda^2-(a_2a_3+b_2b_3)\,\lambda+a_1\,(a\cdot b)]\}$$

$$=-\frac{1}{(a_1^2-\lambda^2)}\{\lambda^2\cdot[\lambda^2-a_1^2-(a_2^2-b_3^2)]\cdot[\lambda^2-a_1^2-(a_3^2-b_2^2)]+[\,b_1\lambda^2-a_1\,(a\cdot b)]^2-(a_2a_3+b_2b_3)^2\,\lambda^2\}$$

$$=-\frac{1}{(a_1^2-\lambda^2)}\{\lambda^2\cdot[(\lambda^2-a_1^2)^2-(a_2^2+a_3^2-b_2^2-b_3^2)(\lambda^2-a_1^2)+(a_2^2-b_3^2)(a_3^2-b_2^2)]+[\,b_1\lambda^2-a_1\,(a\cdot b)]^2-(a_2a_3+b_2b_3)^2\,\lambda^2\}$$

$$=-\frac{1}{(a_1^2-\lambda^2)}\{\lambda^2\cdot(\lambda^2-a_1^2)\,[\lambda^2-a_1^2-(a_2^2+a_3^2-b_2^2-b_3^2)]+[\,b_1\lambda^2-a_1\,(a\cdot b)]^2+[(a_2^2-b_3^2)(a_3^2-b_2^2)-(a_2a_3+b_2b_3)^2]\,\lambda^2\}$$

$$=-\frac{1}{(a_1^2-\lambda^2)}\{\lambda^2\cdot(\lambda^2-a_1^2)\,[\lambda^2-a_1^2-(a_2^2+a_3^2-b_2^2-b_3^2)]+[\,b_1\lambda^2-a_1\,(a\cdot b)]^2-(a_2^2b_2^2+a_3^2b_3^2+2a_2a_3b_2b_3)\,\lambda^2\}$$

$$=-\frac{1}{(a_1^2-\lambda^2)}\{\lambda^2\cdot(\lambda^2-a_1^2)\,[\lambda^2-a_1^2-(a_2^2+a_3^2-b_2^2-b_3^2)]+[\,b_1\lambda^2-a_1\,(a\cdot b)]^2-(a_2b_2+a_3b_3)^2\,\lambda^2\}$$

$$=-\frac{1}{(a_1^2-\lambda^2)}\{\lambda^2\cdot(\lambda^2-a_1^2)\,[\lambda^2-a_1^2-(a_2^2+a_3^2-b_2^2-b_3^2)]+[\,b_1^2\lambda^4-2a_1b_1\,(a\cdot b)\,\lambda^2+a_1^2\,(a\cdot b)^2]-(a\cdot b-a_1b_1)^2\,\lambda^2\}$$

$$=-\frac{1}{(a_1^2-\lambda^2)}\{\lambda^2\cdot(\lambda^2-a_1^2)\,[\lambda^2-a_1^2-(a_2^2+a_3^2-b_2^2-b_3^2)]+b_1^2\lambda^4-[(a\cdot b)^2+a_1^2b_1^2]\,\lambda^2+a_1^2\,(a\cdot b)^2\}$$

$$=-\frac{1}{(a_1^2-\lambda^2)}\{\lambda^2\cdot(\lambda^2-a_1^2)\,[\lambda^2-a_1^2-(a_2^2+a_3^2-b_2^2-b_3^2)]+[\,b_1^2\lambda^4-2a_1b_1\,(a\cdot b)\,\lambda^2+a_1^2\,(a\cdot b)^2]-(a\cdot b-a_1b_1)^2\,\lambda^2\}$$

$$=-\frac{1}{(a_1^2-\lambda^2)}\{\lambda^2\cdot(\lambda^2-a_1^2)\,[\lambda^2-a_1^2-(a_2^2+a_3^2-b_2^2-b_3^2)]+b_1^2\lambda^4-[(a\cdot b)^2+a_1^2b_1^2]\,\lambda^2+a_1^2\,(a\cdot b)^2-(b_1^2a_1^4-[(a\cdot b)^2+a_1^2b_1^2]\,a_1^2+a_1^2\,(a\cdot b)^2)\}$$

$$=-\frac{1}{(a_1^2-\lambda^2)}\{\lambda^2\cdot(\lambda^2-a_1^2)\,[\lambda^2-a_1^2-(a_2^2+a_3^2-b_2^2-b_3^2)]+b_1^2\,(\lambda^4-a_1^4)-[(a\cdot b)^2+a_1^2b_1^2]\,(\lambda^2-a_1^2)\}$$

$$=\lambda^2\,[\lambda^2-a_1^2-(a_2^2+a_3^2-b_2^2-b_3^2)]+b_1^2\,(\lambda^2+a_1^2)-[(a\cdot b)^2+a_1^2b_1^2]$$

$$=\lambda^2\,(\lambda^2-|a|^2+b_2^2+b_3^2)+b_1^2\,(\lambda^2+a_1^2)-[(a\cdot b)^2+a_1^2b_1^2]$$

$$=\lambda^4+(-|a|^2+b_2^2+b_3^2)\,\lambda^2+b_1^2\,(\lambda^2+a_1^2)-[(a\cdot b)^2+a_1^2b_1^2]$$

$$=\lambda^4+(\,|b|^2-|a|^2)\,\lambda^2-(a\cdot b)^2\ .$$

Since $(\mathbb{C}\to\mathbb{C}, \lambda\mapsto\det(A-\lambda E))$, as a polynomial of degree 4, is holomorphic, it follows that

$$\det(A-\lambda E)=\lambda^4+(\,|b|^2-|a|^2)\,\lambda^2-(a\cdot b)^2\ ,$$

for every $\lambda\in\mathbb{C}$. Hence, it follows from the Cayley-Hamilton theorem that

$$A^4+(\,|b|^2-|a|^2)\,A^2-(a\cdot b)^2\,E=0\ .$$

Solution 3.10 If $\Lambda\in\mathscr{L}_+^{\uparrow}$, then

$$\det(\Lambda_0\Lambda) = \det(\Lambda) = 1\ ,\quad (\Lambda_0\Lambda)_{00} = \Lambda_{00} \geqslant 0\ ,$$

and hence $\bar{\Lambda} := \Lambda_0\Lambda \in \mathcal{L}_+^{\uparrow}$ as well as $\Lambda = \Lambda_0\bar{\Lambda}$; if $\Lambda \in \mathcal{L}_+^{\downarrow}$, then

$$\det(\Lambda_{PT}\Lambda) = \det(\Lambda) = 1\ ,\quad (\Lambda_{PT}\Lambda)_{00} = -\Lambda_{00} \geqslant 0\ ,$$

and hence $\bar{\Lambda} := \Lambda_{PT}\Lambda \in \mathcal{L}_+^{\uparrow}$ as well as $\Lambda = \Lambda_{PT}\bar{\Lambda}$; if $\Lambda \in \mathcal{L}_-^{\uparrow}$, then

$$\det(\Lambda_P\Lambda) = -\det(\Lambda) = 1\ ,\quad (\Lambda_P\Lambda)_{00} = \Lambda_{00} \geqslant 0\ ,$$

and hence $\bar{\Lambda} := \Lambda_P\Lambda \in \mathcal{L}_+^{\uparrow}$ as well as $\Lambda = \Lambda_P\bar{\Lambda}$; if $\Lambda \in \mathcal{L}_-^{\downarrow}$, then

$$\det(\Lambda_T\Lambda) = -\det(\Lambda) = 1\ ,\quad (\Lambda_T\Lambda)_{00} = -\Lambda_{00} \geqslant 0\ ,$$

and hence $\bar{\Lambda} := \Lambda_T\Lambda \in \mathcal{L}_+^{\uparrow}$ as well as $\Lambda = \Lambda_T\bar{\Lambda}$. Moreover, such $\bar{\Lambda}$, with the prescribed properties, are obviously uniquely determined.

Solution 3.11 If $a > 0$ and $w \in \mathbb{R}^4$ is a non-zero tangent vector of H_{a+} at ${}^t((a^2 + |v|^2)^{1/2}, v_1, v_2, v_3) \in H_{a+}$ such that $w_0 \geqslant 0$, where $v \in \mathbb{R}^3$, then there is differentiable path $\gamma : I \to \mathbb{R}^4$, where I is an open interval of $\mathbb{R}$ around 0, such that $\mathrm{Ran}(\gamma) \subset H_{a+}$, $\gamma(0) = {}^t((a^2 + |v|^2)^{1/2}, v_1, v_2, v_3)$ and $\gamma'(0) = w$. Since $\gamma(s)\cdot\gamma(s) = a^2$, for every $s \in I$, it follows from differentiation that $\gamma(0)\cdot\gamma'(0) = 0$ and hence that

$$\begin{aligned} 0 &= w \cdot {}^t((a^2 + v_1^2 + v_2^2 + v_3^2)^{1/2}, v_1, v_2, v_3) = w_0\,(a^2 + v_1^2 + v_2^2 + v_3^2)^{1/2} - \vec{w}\cdot v \\ &\geqslant w_0\,(a^2 + |v|^2)^{1/2} - |\vec{w}|\cdot|v|\ , \end{aligned}$$

where $\vec{w} := {}^t(w_1, w_2, w_3)$, $\vec{w}\cdot v$ denotes the canonical scalar product of $\vec{w}$ and v and $|\ |$ denotes the canonical norm on $\mathbb{R}^3$, and hence that

$$0 \leqslant w_0 \leqslant |\vec{w}|\cdot\frac{|v|}{(a^2 + |v|^2)^{1/2}}\ ,\quad w_0^2 \leqslant |\vec{w}|^2\cdot\frac{|v|^2}{a^2 + |v|^2}\ ,$$

resulting

$$w\cdot w \leqslant -\frac{a^2|\vec{w}|^2}{a^2 + |v|^2}\ .$$

We consider cases, if $\vec{w} \neq 0$, the latter implies that $w\cdot w < 0$. If $\vec{w} = 0$, then $w\cdot w = w_0^2 \leqslant 0$, hence $w_0 = 0$ and therefore we arrive at $w = 0$. The latter is in contradiction to our assumption that $w \neq 0$. $\lightning$ If $w \in \mathbb{R}^4$ is a non-zero tangent vector of H_{a+} at $\{{}^t((a^2 + |v|^2)^{1/2}, v_1, v_2, v_3) \in H_{a+}$ such that $w_0 \leqslant 0$, then $-w \in \mathbb{R}^4$ is a non-zero tangent vector of H_{a+} at $\{{}^t((a^2 + |v|^2)^{1/2}, v_1, v_2, v_3) \in H_{a+}$ such that $(-w)_0 \geqslant 0$ and hence $w\cdot w = (-w)\cdot(-w) < 0$.

Solution 3.12 "Part (a)": For every $\Lambda \in \mathcal{L}$ and $v \in H_{1a}$, we have that

$$-a^2 = v \cdot v = (\Lambda v) \cdot (\Lambda v)$$

and hence that

$$\Lambda H_{1a} = H_{1a} ,$$

for every $\Lambda \in \mathcal{L}$. For every $v \in H_{1a+}$, it follows that

$$v_0^2 = v \cdot v + v_1^2 + v_2^2 + v_3^2 = -a^2 + v_1^2 + v_2^2 + v_3^2$$

and as well as that

$$v_0 = (v_1^2 + v_2^2 + v_3^2 - a^2)^{1/2} .$$

In particular, for every $s \in \mathbb{R}$, we have that

$$\Lambda_s := \begin{pmatrix} \cosh(s) & 0 & 0 & \sinh(s) \\ 0 & 1 & 0 & 0 \\ 0 & 0 & 1 & 0 \\ \sinh(s) & 0 & 0 & \cosh(s) \end{pmatrix} \in \mathcal{L}_+^{\uparrow} ,$$

and that

$$v := a\,{}^t(0, 0, 0, -1) \in H_{1a+} .$$

Further, for every $s \in \mathbb{R}$, we have that

$$(\Lambda_s v)_{00} = (\Lambda_s)_{00} v_0 + (\Lambda_s)_{01} v_1 + (\Lambda_s)_{02} v_2 + (\Lambda_s)_{03} v_3 = -\sinh(s) < 0 ,$$

for every $s > 0$. Hence, $\Lambda_s v \notin H_{1a+}$, for every $s > 0$.
"Part (b)": We define $\gamma_1 : (0, \infty) \to H_{1a+}$ by

$$\gamma_1(s) = {}^t(s, 0, 0, (s^2 + a^2)^{1/2}) ,$$

for every $s > 0$. Then,

$$\gamma_1'(s) = {}^t(1, 0, 0, s\,(s^2 + a^2)^{-1/2}) ,$$

and

$$\gamma_1'(s) \cdot \gamma_1'(s) = 1 - \frac{s^2}{s^2 + a^2} = \frac{a^2}{s^2 + a^2} > 0 ,$$

for every $s > 0$. Further, we define $\gamma_2 : \mathbb{R} \to H_{1a+}$ by

$$\gamma_2(s) = {}^t(0, a\cos(s), a\sin(s), 0) ,$$

for every $s \in \mathbb{R}$. Then,

$$\gamma_2'(s) = {}^t(0, -a\sin(s), a\cos(s), 0) ,$$

and

$$\gamma_1'(s) \cdot \gamma_1'(s) = -[a^2 \sin^2(s) + a^2 \cos^2(s)] = -a^2 < 0 ,$$

for every $s \in \mathbb{R}$.

Solution 3.13 For the proof, let $a \geqslant 0$ and $\| \ \|_2$ the norm on $L^2_{\mathbb{C}}(\mathbb{R}^3, \varphi_a)$. Then $f \in C_0^\infty(\mathbb{R}^3, \mathbb{C})$ is everywhere defined on $\mathbb{R}^3$ and, as a continuous function, φ_a-measurable. Further, $|f|^2$ is φ_a-measurable and dominated by the φ_a-integrable function

$$\|f\|^2_\infty \cdot \chi_{U_R(0)} ,$$

where we use that $(v_a^0)^{-1}$ is locally Lebesgue integrable and $R > 0$ is such that $\text{supp}(f) \subset U_R(0)$. Hence, it follows from Lebesgue's dominated convergence theorem that $|f|^2$ is φ_a-integrable and therefore that $f \in L^2_{\mathbb{C}}(\mathbb{R}^3, \varphi_a)$. As a consequence,

$$C_0^\infty(\mathbb{R}^3, \mathbb{C}) \subset L^2_{\mathbb{C}}(\mathbb{R}^3, \varphi_a) .$$

Also, if $f \in L^2_{\mathbb{C}}(\mathbb{R}^3, \varphi_a)$, then f is φ_a-a.e. defined on $\mathbb{R}^3$ as well as φ_a-measurable and hence also v^3-a.e. defined on $\mathbb{R}^3$ and v^3-measurable. Since $(v_a^0)^{-1/2}$ is v^3-a.e. continuous, as a consequence, $(v_a^0)^{-1/2} f$ is v^3-a.e. defined on $\mathbb{R}^3$ and v^3-measurable. Further, $|(v_a^0)^{-1/2} f|^2$ is integrable. Hence $(v_a^0)^{-1/2} f \in L^2_{\mathbb{C}}(\mathbb{R}^3, \varphi_a)$. Since for every $p \in [1, \infty)$, $n \in \mathbb{N}^*$ and every non-empty subset Ω of $\mathbb{R}^n$, we have that $C_0^\infty(\mathbb{R}^n, \mathbb{C})$ is dense in $L^2_{\mathbb{C}}(\Omega)$ and $\{0\} \subset \mathbb{R}^3$ is a zero set, there is a canonical Hilbert space isomorphism between $L^2_{\mathbb{C}}(\mathbb{R}^3 \backslash \{0\})$ and $L^2_{\mathbb{C}}(\mathbb{R}^3)$ and hence

$$V := \{f \in C_0^\infty(\mathbb{R}^3, \mathbb{C}) : \text{supp}(f) \subset \mathbb{R}^3 \backslash \{0\}\}$$

is dense in $L^2_{\mathbb{C}}(\mathbb{R}^3)$. Since, for $f \in L^2_{\mathbb{C}}(\mathbb{R}^3, \varphi_a)$, we have that $(v_a^0)^{-1/2} f \in L^2_{\mathbb{C}}(\mathbb{R}^3)$, there is a sequence $g_1, g_2, \dots$ in V such that

$$\lim_{\nu \to \infty} \int_{\mathbb{R}^3} |g_\nu - (v_a^0)^{-1/2} f|^2 \, dv^3 = \lim_{\nu \to \infty} \int_{\mathbb{R}^3} |(v_a^0)^{1/2} g_\nu - f|^2 \, (v_a^0)^{-1} \, dv^3 = 0 .$$

Since $(v_a^0)^{1/2} g_\nu \in V$, for $\nu \in \mathbb{N}^*$, it follows that

$$\lim_{\nu \to \infty} \|(v_a^0)^{1/2} g_\nu - f\|_{2,a} = 0 .$$

Solution 3.14 For the proof, first, we prove by induction that

$$A^{2n-1} = (-1)^{n-1} A , \ A^{2n} = (-1)^n \begin{pmatrix} 0 & 0 & 0 & 0 \\ 0 & 1 & 0 & 0 \\ 0 & 0 & 1 & 0 \\ 0 & 0 & 0 & 0 \end{pmatrix} , \tag{A.9}$$

for every $n \in \mathbb{N}^*$. Since $A = A^{2 \cdot 1 - 1} = (-1)^{1-1} A$, and

$$A^2 = \begin{pmatrix} 0 & 0 & 0 & 0 \\ 0 & 0 & 1 & 0 \\ 0 & -1 & 0 & 0 \\ 0 & 0 & 0 & 0 \end{pmatrix} \cdot \begin{pmatrix} 0 & 0 & 0 & 0 \\ 0 & 0 & 1 & 0 \\ 0 & -1 & 0 & 0 \\ 0 & 0 & 0 & 0 \end{pmatrix} = \begin{pmatrix} 0 & 0 & 0 & 0 \\ 0 & -1 & 0 & 0 \\ 0 & 0 & -1 & 0 \\ 0 & 0 & 0 & 0 \end{pmatrix},$$

(A.9) is true, for $n = 1$. If (A.9) is true for $n \in \mathbb{N}^*$, then

$$A^{2(n+1)} = A^{2n} A^2 = (-1)^n \begin{pmatrix} 0 & 0 & 0 & 0 \\ 0 & 1 & 0 & 0 \\ 0 & 0 & 1 & 0 \\ 0 & 0 & 0 & 0 \end{pmatrix} \cdot \begin{pmatrix} 0 & 0 & 0 & 0 \\ 0 & -1 & 0 & 0 \\ 0 & 0 & -1 & 0 \\ 0 & 0 & 0 & 0 \end{pmatrix}$$

$$= (-1)^n \begin{pmatrix} 0 & 0 & 0 & 0 \\ 0 & -1 & 0 & 0 \\ 0 & 0 & -1 & 0 \\ 0 & 0 & 0 & 0 \end{pmatrix} = (-1)^{n+1} \begin{pmatrix} 0 & 0 & 0 & 0 \\ 0 & 1 & 0 & 0 \\ 0 & 0 & 1 & 0 \\ 0 & 0 & 0 & 0 \end{pmatrix},$$

and

$$A^{2(n+1)-1} = A^{2n+1} = A^{2n} A = (-1)^n \begin{pmatrix} 0 & 0 & 0 & 0 \\ 0 & 1 & 0 & 0 \\ 0 & 0 & 1 & 0 \\ 0 & 0 & 0 & 0 \end{pmatrix} \cdot \begin{pmatrix} 0 & 0 & 0 & 0 \\ 0 & 0 & 1 & 0 \\ 0 & -1 & 0 & 0 \\ 0 & 0 & 0 & 0 \end{pmatrix}$$

$$= (-1)^n \begin{pmatrix} 0 & 0 & 0 & 0 \\ 0 & 0 & 1 & 0 \\ 0 & -1 & 0 & 0 \\ 0 & 0 & 0 & 0 \end{pmatrix} = (-1)^n A .$$

Hence, (A.9) is true for the subsequent case and therefore true for all $n \in \mathbb{N}^*$. From (A.9), we infer that

$$\exp(sA) = \sum_{n=0}^{\infty} \frac{s^n}{n!} A^n = E + \sum_{n=1}^{\infty} \frac{s^{2n}}{(2n)!} A^{2n} + \sum_{n=1}^{\infty} \frac{s^{2n-1}}{(2n-1)!} A^{2n-1}$$

$$= E + \left[\sum_{n=1}^{\infty} (-1)^n \frac{s^{2n}}{(2n)!} \right] \begin{pmatrix} 0 & 0 & 0 & 0 \\ 0 & 1 & 0 & 0 \\ 0 & 0 & 1 & 0 \\ 0 & 0 & 0 & 0 \end{pmatrix} + \left[\sum_{n=1}^{\infty} (-1)^{n-1} \frac{s^{2n-1}}{(2n-1)!} \right] A$$

$$= \begin{pmatrix} 1 & 0 & 0 & 0 \\ 0 & 0 & 0 & 0 \\ 0 & 0 & 0 & 0 \\ 0 & 0 & 0 & 1 \end{pmatrix} + \left[\sum_{n=0}^{\infty} (-1)^n \frac{s^{2n}}{(2n)!} \right] \begin{pmatrix} 0 & 0 & 0 & 0 \\ 0 & 1 & 0 & 0 \\ 0 & 0 & 1 & 0 \\ 0 & 0 & 0 & 0 \end{pmatrix}$$

$$+\left[\sum_{n=0}^{\infty}(-1)^n \frac{s^{2n+1}}{(2n+1)!}\right] A$$

$$= \begin{pmatrix} 1&0&0&0\\ 0&0&0&0\\ 0&0&0&0\\ 0&0&0&1 \end{pmatrix} + \cos(s) \begin{pmatrix} 0&0&0&0\\ 0&1&0&0\\ 0&0&1&0\\ 0&0&0&0 \end{pmatrix} + \sin(s) \begin{pmatrix} 0&0&0&0\\ 0&0&1&0\\ 0&-1&0&0\\ 0&0&0&0 \end{pmatrix}$$

$$= \begin{pmatrix} 1&0&0&0\\ 0&\cos(s)&\sin(s)&0\\ 0&-\sin(s)&\cos(s)&0\\ 0&0&0&1 \end{pmatrix} = M_3(s) \ ,$$

for every $s \in \mathbb{R}$.

Solution 3.15 We note that

$$(U_1x)\cdot(U_1x) = {}^t(x_0, x_3, -x_2, x_1)\cdot{}^t(x_0, x_3, -x_2, x_1) = x\cdot x \ ,$$
$$(U_2x)\cdot(U_2x) = {}^t(x_0, -x_1, x_3, x_2)\cdot{}^t(x_0, -x_1, x_3, x_2) = x\cdot x \ ,$$

for all $x \in \mathbb{R}^4$. Since $\det(U_1) = \det(U_2) = 1$ and $(U_1)_{00} = (U_2)_{00} = 1 \geqslant 0$, it follows that $U_1, U_2 \in \mathcal{L}_+^{\uparrow}$. In addition, we note that

$$U_1^2 = \begin{pmatrix} 1&0&0&0\\ 0&0&0&1\\ 0&0&-1&0\\ 0&1&0&0 \end{pmatrix} \cdot \begin{pmatrix} 1&0&0&0\\ 0&0&0&1\\ 0&0&-1&0\\ 0&1&0&0 \end{pmatrix} = E \ ,$$

$$U_2^2 = \begin{pmatrix} 1&0&0&0\\ 0&-1&0&0\\ 0&0&0&1\\ 0&0&1&0 \end{pmatrix} \cdot \begin{pmatrix} 1&0&0&0\\ 0&-1&0&0\\ 0&0&0&1\\ 0&0&1&0 \end{pmatrix} = E \ ,$$

where E denotes the 4×4 unit matrix. As a consequence,

$$U_1^{-1} = U_1 \ , \ U_2^{-1} = U_2 \ .$$

Further, for $s \in \mathbb{R}$, we have that

$$
\begin{aligned}
&U_1 M_3(s) U_1 \\
&= \begin{pmatrix} 1 & 0 & 0 & 0 \\ 0 & 0 & 0 & 1 \\ 0 & 0 & -1 & 0 \\ 0 & 1 & 0 & 0 \end{pmatrix} \cdot \begin{pmatrix} 1 & 0 & 0 & 0 \\ 0 & \cos(s) & \sin(s) & 0 \\ 0 & -\sin(s) & \cos(s) & 0 \\ 0 & 0 & 0 & 1 \end{pmatrix} \cdot \begin{pmatrix} 1 & 0 & 0 & 0 \\ 0 & 0 & 0 & 1 \\ 0 & 0 & -1 & 0 \\ 0 & 1 & 0 & 0 \end{pmatrix} \\
&= \begin{pmatrix} 1 & 0 & 0 & 0 \\ 0 & 0 & 0 & 1 \\ 0 & 0 & -1 & 0 \\ 0 & 1 & 0 & 0 \end{pmatrix} \cdot \begin{pmatrix} 1 & 0 & 0 & 0 \\ 0 & 0 & -\sin(s) & \cos(s) \\ 0 & 0 & -\cos(s) & -\sin(s) \\ 0 & 1 & 0 & 0 \end{pmatrix} \\
&= \begin{pmatrix} 1 & 0 & 0 & 0 \\ 0 & 1 & 0 & 0 \\ 0 & 0 & \cos(s) & \sin(s) \\ 0 & 0 & -\sin(s) & \cos(s) \end{pmatrix} = M_1(s)
\end{aligned}
$$

and that

$$
\begin{aligned}
&U_2 M_3(s) U_2 \\
&= \begin{pmatrix} 1 & 0 & 0 & 0 \\ 0 & -1 & 0 & 0 \\ 0 & 0 & 0 & 1 \\ 0 & 0 & 1 & 0 \end{pmatrix} \cdot \begin{pmatrix} 1 & 0 & 0 & 0 \\ 0 & \cos(s) & \sin(s) & 0 \\ 0 & -\sin(s) & \cos(s) & 0 \\ 0 & 0 & 0 & 1 \end{pmatrix} \cdot \begin{pmatrix} 1 & 0 & 0 & 0 \\ 0 & -1 & 0 & 0 \\ 0 & 0 & 0 & 1 \\ 0 & 0 & 1 & 0 \end{pmatrix} \\
&= \begin{pmatrix} 1 & 0 & 0 & 0 \\ 0 & -1 & 0 & 0 \\ 0 & 0 & 0 & 1 \\ 0 & 0 & 1 & 0 \end{pmatrix} \cdot \begin{pmatrix} 1 & 0 & 0 & 0 \\ 0 & -\cos(s) & 0 & \sin(s) \\ 0 & \sin(s) & 0 & \cos(s) \\ 0 & 0 & 1 & 0 \end{pmatrix} \\
&= \begin{pmatrix} 1 & 0 & 0 & 0 \\ 0 & \cos(s) & 0 & -\sin(s) \\ 0 & 0 & 1 & 0 \\ 0 & \sin(s) & 0 & \cos(s) \end{pmatrix} = M_2(s) \ .
\end{aligned}
$$

Hence, we have that

$$
\begin{aligned}
M_1(s) &= U_1 M_3(s) U_1 = U_1 \exp(sA) U_1 = \exp(s U_1 A U_1) = \exp(sB) \ , \\
M_2(s) &= U_2 M_3(s) U_2 = U_2 \exp(sA) U_2 = \exp(s U_2 A U_2) = \exp(sC) \ ,
\end{aligned}
$$

where we note that

$$U_1 A U_1 = \begin{pmatrix} 1 & 0 & 0 & 0 \\ 0 & 0 & 0 & 1 \\ 0 & 0 & -1 & 0 \\ 0 & 1 & 0 & 0 \end{pmatrix} \cdot \begin{pmatrix} 0 & 0 & 0 & 0 \\ 0 & 0 & 1 & 0 \\ 0 & -1 & 0 & 0 \\ 0 & 0 & 0 & 0 \end{pmatrix} \cdot \begin{pmatrix} 1 & 0 & 0 & 0 \\ 0 & 0 & 0 & 1 \\ 0 & 0 & -1 & 0 \\ 0 & 1 & 0 & 0 \end{pmatrix}$$

$$= \begin{pmatrix} 1 & 0 & 0 & 0 \\ 0 & 0 & 0 & 1 \\ 0 & 0 & -1 & 0 \\ 0 & 1 & 0 & 0 \end{pmatrix} \cdot \begin{pmatrix} 0 & 0 & 0 & 0 \\ 0 & 0 & -1 & 0 \\ 0 & 0 & 0 & -1 \\ 0 & 0 & 0 & 0 \end{pmatrix} = \begin{pmatrix} 0 & 0 & 0 & 0 \\ 0 & 0 & 0 & 0 \\ 0 & 0 & 0 & 1 \\ 0 & 0 & -1 & 0 \end{pmatrix} = B$$

and that

$$U_2 A U_2 = \begin{pmatrix} 1 & 0 & 0 & 0 \\ 0 & -1 & 0 & 0 \\ 0 & 0 & 0 & 1 \\ 0 & 0 & 1 & 0 \end{pmatrix} \cdot \begin{pmatrix} 0 & 0 & 0 & 0 \\ 0 & 0 & 1 & 0 \\ 0 & -1 & 0 & 0 \\ 0 & 0 & 0 & 0 \end{pmatrix} \cdot \begin{pmatrix} 1 & 0 & 0 & 0 \\ 0 & -1 & 0 & 0 \\ 0 & 0 & 0 & 1 \\ 0 & 0 & 1 & 0 \end{pmatrix}$$

$$= \begin{pmatrix} 1 & 0 & 0 & 0 \\ 0 & -1 & 0 & 0 \\ 0 & 0 & 0 & 1 \\ 0 & 0 & 1 & 0 \end{pmatrix} \cdot \begin{pmatrix} 0 & 0 & 0 & 0 \\ 0 & 0 & 0 & 1 \\ 0 & 1 & 0 & 0 \\ 0 & 0 & 0 & 0 \end{pmatrix} = \begin{pmatrix} 0 & 0 & 0 & 0 \\ 0 & 0 & 0 & -1 \\ 0 & 0 & 0 & 0 \\ 0 & 1 & 0 & 0 \end{pmatrix} = C\ .$$

Solution 3.16 With the help of the result of Exercise 3.15, it follows for $j \in \{1, 2\}$ and $s \in \mathbb{R}$ that

$$U_a(M_j(s)) = U_a(U_j^{-1} \cdot M_3(s) \cdot U_j) = U_a(U_j \cdot M_3(s) \cdot U_j)$$
$$= U_a(U_j) \circ U_a(M_3(s)) \circ U_a(U_j) = U_a(U_j) \circ \exp\Big(i \, \frac{s}{\hbar} \, \hat{L}_3\Big) \circ U_a(U_j)$$
$$= \exp\Big(i \, \frac{s}{\hbar} \, U_a(U_j) \circ \hat{L}_3 \circ U_a(U_j)\Big)\ .$$

Further, for every $v = {}^t(v_1, v_2, v_3) \in \mathbb{R}^3$, we have that

$$U_1^{-1} \cdot p_a(v) = U_1 \cdot p_a(v) = {}^t((a^2 + |v|^2)^{1/2}, v_3, -v_2, v_1)\ ,$$
$$U_2^{-1} \cdot p_a(v) = U_2 \cdot p_a(v) = {}^t((a^2 + |v|^2)^{1/2}, -v_1, v_3, v_2)\ ,$$

and hence that

$$h_{aU_1}(v) = p_a^{-1}(U_1 \cdot p_a(v)) = {}^t(v_3, -v_2, v_1)\ ,$$
$$h_{aU_2}(v) = p_a^{-1}(U_2 \cdot p_a(v)) = {}^t(-v_1, v_3, v_2)\ .$$

As a consequence, for every $f \in X$, it follows that

$$[U_a(U_1)f](v) = (f \circ h_{aU_1})(v) = f(v_3, -v_2, v_1) \ ,$$
$$[U_a(U_2)f](v) = (f \circ h_{aU_2})(v) = f(-v_1, v_3, v_2)$$

and almost all $v = {}^t(v_1, v_2, v_3) \in \mathbb{R}^3$. Further, we note for $j \in \{1, 2\}$ that $U_a(U_j)C_0^1(\mathbb{R}^3, \mathbb{C}) = C_0^1(\mathbb{R}^3, \mathbb{C})$ and for $f \in C_0^1(\mathbb{R}^3, \mathbb{C})$ that

$$\begin{aligned}
&[\hat{L}_3 U_a(U_1)f](v) = [\hat{L}_3(f \circ h_1)](v) \\
&= \frac{\hbar}{i}\left[v_1\,(f \circ h_1)'(v) \cdot e_2 - v_2\,(f \circ h_1)'(v) \cdot e_1\right] \\
&= \frac{\hbar}{i}\left[v_1\,f'(h_1(v)) \cdot h_1'(v) \cdot e_2 - v_2\,f'(h_1(v)) \cdot h_1'(v) \cdot e_1\right] \\
&= \frac{\hbar}{i}\left[-v_1\,f'(h_1(v)) \cdot e_2 - v_2\,f'(h_1(v)) \cdot e_3\right] \\
&= \frac{\hbar}{i}\left[-(h_1(v))_3\,f'(h_1(v)) \cdot e_2 + (h_1(v))_2\,f'(h_1(v)) \cdot e_3\right] \\
&= \left[\frac{\hbar}{i}\left(v_2\,\frac{\partial f}{\partial v_3} - v_3\,\frac{\partial f}{\partial v_2}\right)\right](h_1(v)) = \left[U_a(U_1)\,\frac{\hbar}{i}\left(v_2\,\frac{\partial f}{\partial v_3} - v_3\,\frac{\partial f}{\partial v_2}\right)\right](v)
\end{aligned}$$

and that

$$\begin{aligned}
&[\hat{L}_3 U_a(U_2)f](v) = [\hat{L}_3(f \circ h_2)](v) \\
&= \frac{\hbar}{i}\left[v_1\,(f \circ h_2)'(v) \cdot e_2 - v_2\,(f \circ h_2)'(v) \cdot e_1\right] \\
&= \frac{\hbar}{i}\left[v_1\,f'(h_2(v)) \cdot h_2'(v) \cdot e_2 - v_2\,f'(h_2(v)) \cdot h_2'(v) \cdot e_1\right] \\
&= \frac{\hbar}{i}\left[v_1\,f'(h_2(v)) \cdot e_3 + v_2\,f'(h_2(v)) \cdot e_1\right] \\
&= \frac{\hbar}{i}\left[-(h_2(v))_1\,f'(h_2(v)) \cdot e_3 + (h_2(v))_3\,f'(h_2(v)) \cdot e_1\right] \\
&= \left[\frac{\hbar}{i}\left(v_3\,\frac{\partial f}{\partial v_1} - v_1\,\frac{\partial f}{\partial v_3}\right)\right](h_2(v)) = \left[U_a(U_2)\,\frac{\hbar}{i}\left(v_3\,\frac{\partial f}{\partial v_1} - v_1\,\frac{\partial f}{\partial v_3}\right)\right](v) \ ,
\end{aligned}$$

for every $v = {}^t(v_1, v_2, v_3) \in \mathbb{R}^3$, where $h_1 : \mathbb{R}^3 \to \mathbb{R}^3$ and $h_2 : \mathbb{R}^3 \to \mathbb{R}^3$ are defined by

$$h_1(v) := (v_3, -v_2, v_1) \ , \ \ h_2(v) := (-v_1, v_3, v_2) \ ,$$

for every $v = {}^t(v_1, v_2, v_3) \in \mathbb{R}^3$, and e_1, e_2, e_3 is the canonical basis of $\mathbb{R}^3$. The latter implies that

$$h_1'(v) = \begin{pmatrix} 0 & 0 & 1 \\ 0 & -1 & 0 \\ 1 & 0 & 0 \end{pmatrix} \ , \ \ h_2'(v) = \begin{pmatrix} -1 & 0 & 0 \\ 0 & 0 & 1 \\ 0 & 1 & 0 \end{pmatrix} \ ,$$

for every $v \in \mathbb{R}^3$. Hence, it follows that

$$U_a(U_1) \circ \hat{L}_3 \circ U_a(U_1) f = \frac{\hbar}{i} \left(v_2 \frac{\partial f}{\partial v_3} - v_3 \frac{\partial f}{\partial v_2} \right) ,$$

$$U_a(U_2) \circ \hat{L}_3 \circ U_a(U_2) f = \frac{\hbar}{i} \left(v_3 \frac{\partial f}{\partial v_1} - v_1 \frac{\partial f}{\partial v_3} \right) .$$

Solution 3.17 For the proof, let $a \geqslant 0$ and $\| \ \|_2$ the norm on $L^2_{\mathbb{C}}(\mathbb{R}^3, \varphi_a)$. If $f \in L^2_{\mathbb{C}}(\mathbb{R}^3, \varphi_a)$, then f is φ_a-a.e. defined on $\mathbb{R}^3$ as well as φ_a-measurable and hence also v^3-a.e. defined on $\mathbb{R}^3$ and v^3-measurable. Since $(v^0_a)^{-1/2}$ is φ_a-a.e. continuous, as a consequence, $(v^0_a)^{-1/2} f$ is a.e. defined on $\mathbb{R}^3$ and v^3-measurable. Further, $|f|^2$ is φ_a-integrable and hence $(v^0_a)^{-1}|f|^2 = |(v^0_a)^{-1/2} f|^2$ is v^3-integrable such that

$$\|f\|_2^2 = \int_{\mathbb{R}^3} |f|^2 \, d\varphi_a = \int_{\mathbb{R}^3} |f|^2 \cdot (v^0_a)^{-1} \, dv^3 = \int_{\mathbb{R}^3} |(v^0_a)^{-1/2} f|^2 \, dv^3 .$$

Hence, it follows that $(v^0_a)^{-1/2} f \in L^2_{\mathbb{C}}(\mathbb{R}^3)$. Hence, the map V_a is well-defined and isometric. Further, V_a is obviously linear, and it follows from the polarization identity for complex sesquilinear forms that V_a preserves scalar products. If $f \in L^2_{\mathbb{C}}(\mathbb{R}^3)$, then f is v^3-a.e. defined on $\mathbb{R}^3$ as well as v^3-measurable and hence also φ_a-a.e. defined on $\mathbb{R}^3$ as well as φ_a-measurable. Since $(v^0_a)^{1/2}$ is φ_a-a.e. continuous, $(v^0_a)^{1/2} f$ is φ_a-a.e. defined on $\mathbb{R}^3$ and φ_a-measurable. Further, $(v^0_a)^{-1}|(v^0_a)^{1/2} f|^2 = |f|^2$ is v^3-integrable and therefore $|(v^0_a)^{1/2} f|^2$ is φ_a-integrable. As a consequence, $(v^0_a)^{1/2} f \in L^2_{\mathbb{C}}(\mathbb{R}^3, \varphi_a)$ such that $V_a \, (v^0_a)^{1/2} f = f$. Hence, V_a is a Hilbert space isomorphism and $V_a^{-1} f = (v^0_a)^{1/2} f$.

Solution 3.18 For the proof, first, we prove by induction that

$$A^{2n} = \begin{pmatrix} 1&0&0&0 \\ 0&0&0&0 \\ 0&0&0&0 \\ 0&0&0&1 \end{pmatrix} , \ A^{2n-1} = A , \tag{A.10}$$

for every $n \in \mathbb{N}^*$. Since,

$$A^2 = \begin{pmatrix} 0&0&0&1 \\ 0&0&0&0 \\ 0&0&0&0 \\ 1&0&0&0 \end{pmatrix} \cdot \begin{pmatrix} 0&0&0&1 \\ 0&0&0&0 \\ 0&0&0&0 \\ 1&0&0&0 \end{pmatrix} = \begin{pmatrix} 1&0&0&0 \\ 0&0&0&0 \\ 0&0&0&0 \\ 0&0&0&1 \end{pmatrix} ,$$

(A.10) is true, for $n = 1$. If (A.10) is true for $n \in \mathbb{N}^*$, then

$$A^{2(n+1)} = A^{2n} A^2 = \begin{pmatrix} 1&0&0&0 \\ 0&0&0&0 \\ 0&0&0&0 \\ 0&0&0&1 \end{pmatrix} \cdot \begin{pmatrix} 1&0&0&0 \\ 0&0&0&0 \\ 0&0&0&0 \\ 0&0&0&1 \end{pmatrix} = \begin{pmatrix} 1&0&0&0 \\ 0&0&0&0 \\ 0&0&0&0 \\ 0&0&0&1 \end{pmatrix} ,$$

and

$$A^{2(n+1)-1} = A^{2n+1} = A^{2n}A = \begin{pmatrix} 1&0&0&0 \\ 0&0&0&0 \\ 0&0&0&0 \\ 0&0&0&1 \end{pmatrix} \cdot \begin{pmatrix} 0&0&0&1 \\ 0&0&0&0 \\ 0&0&0&0 \\ 1&0&0&0 \end{pmatrix}$$

$$= \begin{pmatrix} 0&0&0&1 \\ 0&0&0&0 \\ 0&0&0&0 \\ 1&0&0&0 \end{pmatrix} = A \ .$$

Hence, (A.10) is true for the subsequent case and therefore true for all $n \in \mathbb{N}^*$. From (A.10), we infer that

$$\begin{aligned} \exp(sA) &= \sum_{n=0}^{\infty} \frac{s^n}{n!} A^n = E + \sum_{n=1}^{\infty} \frac{s^{2n}}{(2n)!} A^{2n} + \sum_{n=1}^{\infty} \frac{s^{2n-1}}{(2n-1)!} A^{2n-1} \\ &= E + \left[\sum_{n=1}^{\infty} \frac{s^{2n}}{(2n)!} \right] A^2 + \left[\sum_{n=1}^{\infty} \frac{s^{2n-1}}{(2n-1)!} \right] A \\ &= E + [\cosh(s) - 1] A^2 + \sinh(s) A \\ &= \begin{pmatrix} 1&0&0&0 \\ 0&1&0&0 \\ 0&0&1&0 \\ 0&0&0&1 \end{pmatrix} + \begin{pmatrix} \cosh(s)-1 & 0&0 & 0 \\ 0 & 0&0 & 0 \\ 0 & 0&0 & 0 \\ 0 & 0&0 & \cosh(s)-1 \end{pmatrix} \\ &\quad + \begin{pmatrix} 0 & 0&0 & \sinh(s) \\ 0 & 0&0 & 0 \\ 0 & 0&0 & 0 \\ \sinh(s) & 0&0 & 0 \end{pmatrix} \\ &= \begin{pmatrix} \cosh(s) & 0&0 & \sinh(s) \\ 0 & 1&0 & 0 \\ 0 & 0&1 & 0 \\ \sinh(s) & 0&0 & \cosh(s) \end{pmatrix} = M_{03}(s) \ . \end{aligned}$$

Solution 3.19 We note that

$$\begin{aligned} (U_{13}x) \cdot (U_{13}x) &= {}^t(x_0, x_3, -x_2, x_1) \cdot {}^t(x_0, x_3, -x_2, x_1) = x \cdot x \ , \\ (U_{23}x) \cdot (U_{23}x) &= {}^t(x_0, -x_1, x_3, x_2) \cdot {}^t(x_0, -x_1, x_3, x_2) = x \cdot x \ , \end{aligned}$$

for all $x \in \mathbb{R}^4$. Since $\det(U_{13}) = \det(U_{23}) = 1$ and $(U_{13})_{00} = (U_{23})_{00} = 1 \geqslant 0$, it follows that $U_{13}, U_{23} \in \mathcal{L}_+^\uparrow$. In addition, we note that

$$U_{13}^2 = \begin{pmatrix} 1 & 0 & 0 & 0 \\ 0 & 0 & 0 & 1 \\ 0 & 0 & -1 & 0 \\ 0 & 1 & 0 & 0 \end{pmatrix} \cdot \begin{pmatrix} 1 & 0 & 0 & 0 \\ 0 & 0 & 0 & 1 \\ 0 & 0 & -1 & 0 \\ 0 & 1 & 0 & 0 \end{pmatrix} = E \ ,$$

$$U_{23}^2 = \begin{pmatrix} 1 & 0 & 0 & 0 \\ 0 & -1 & 0 & 0 \\ 0 & 0 & 0 & 1 \\ 0 & 0 & 1 & 0 \end{pmatrix} \cdot \begin{pmatrix} 1 & 0 & 0 & 0 \\ 0 & -1 & 0 & 0 \\ 0 & 0 & 0 & 1 \\ 0 & 0 & 1 & 0 \end{pmatrix} = E \ ,$$

where E denotes the 4×4 unit matrix. As a consequence,

$$U_{13}^{-1} = U_{13} \ , \ U_{23}^{-1} = U_{23} \ .$$

Further, for $s \in \mathbb{R}$, we have that

$$\begin{aligned} & U_{13} M_{03}(s) U_{13} \\ &= \begin{pmatrix} 1 & 0 & 0 & 0 \\ 0 & 0 & 0 & 1 \\ 0 & 0 & -1 & 0 \\ 0 & 1 & 0 & 0 \end{pmatrix} \cdot \begin{pmatrix} \cosh(s) & 0 & 0 & \sinh(s) \\ 0 & 1 & 0 & 0 \\ 0 & 0 & 1 & 0 \\ \sinh(s) & 0 & 0 & \cosh(s) \end{pmatrix} \cdot \begin{pmatrix} 1 & 0 & 0 & 0 \\ 0 & 0 & 0 & 1 \\ 0 & 0 & -1 & 0 \\ 0 & 1 & 0 & 0 \end{pmatrix} \\ &= \begin{pmatrix} 1 & 0 & 0 & 0 \\ 0 & 0 & 0 & 1 \\ 0 & 0 & -1 & 0 \\ 0 & 1 & 0 & 0 \end{pmatrix} \cdot \begin{pmatrix} \cosh(s) & \sinh(s) & 0 & 0 \\ 0 & 0 & 0 & 1 \\ 0 & 0 & -1 & 0 \\ \sinh(s) & \cosh(s) & 0 & 0 \end{pmatrix} \\ &= \begin{pmatrix} \cosh(s) & \sinh(s) & 0 & 0 \\ \sinh(s) & \cosh(s) & 0 & 0 \\ 0 & 0 & 1 & 0 \\ 0 & 0 & 0 & 1 \end{pmatrix} = M_{01}(s) \end{aligned}$$

and that

$$\begin{aligned} & U_{23} M_{03}(s) U_{23} \\ &= \begin{pmatrix} 1 & 0 & 0 & 0 \\ 0 & -1 & 0 & 0 \\ 0 & 0 & 0 & 1 \\ 0 & 0 & 1 & 0 \end{pmatrix} \cdot \begin{pmatrix} \cosh(s) & 0 & 0 & \sinh(s) \\ 0 & 1 & 0 & 0 \\ 0 & 0 & 1 & 0 \\ \sinh(s) & 0 & 0 & \cosh(s) \end{pmatrix} \cdot \begin{pmatrix} 1 & 0 & 0 & 0 \\ 0 & -1 & 0 & 0 \\ 0 & 0 & 0 & 1 \\ 0 & 0 & 1 & 0 \end{pmatrix} \\ &= \begin{pmatrix} 1 & 0 & 0 & 0 \\ 0 & -1 & 0 & 0 \\ 0 & 0 & 0 & 1 \\ 0 & 0 & 1 & 0 \end{pmatrix} \cdot \begin{pmatrix} \cosh(s) & 0 & \sinh(s) & 0 \\ 0 & -1 & 0 & 0 \\ 0 & 0 & 0 & 1 \\ \sinh(s) & 0 & \cosh(s) & 0 \end{pmatrix} \end{aligned}$$

$$= \begin{pmatrix} \cosh(s) & 0 & \sinh(s) & 0 \\ 0 & 1 & 0 & 0 \\ \sinh(s) & 0 & \cosh(s) & 0 \\ 0 & 0 & 0 & 1 \end{pmatrix} = M_{02}(s) \ .$$

Hence, we have that

$$M_{01}(s) = U_{13} M_{03}(s) U_{13} = U_{13} \exp(sA) U_{13} = \exp(s U_{13} A U_{13}) = \exp(sB) \ ,$$
$$M_{02}(s) = U_{23} M_{03}(s) U_{23} = U_{23} \exp(sA) U_{23} = \exp(s U_{23} A U_{23}) = \exp(sC) \ ,$$

where we note that

$$U_{13} A U_{13} = \begin{pmatrix} 1 & 0 & 0 & 0 \\ 0 & 0 & 0 & 1 \\ 0 & 0 & -1 & 0 \\ 0 & 1 & 0 & 0 \end{pmatrix} \cdot \begin{pmatrix} 0 & 0 & 0 & 1 \\ 0 & 0 & 0 & 0 \\ 0 & 0 & 0 & 0 \\ 1 & 0 & 0 & 0 \end{pmatrix} \cdot \begin{pmatrix} 1 & 0 & 0 & 0 \\ 0 & 0 & 0 & 1 \\ 0 & 0 & -1 & 0 \\ 0 & 1 & 0 & 0 \end{pmatrix}$$
$$= \begin{pmatrix} 1 & 0 & 0 & 0 \\ 0 & 0 & 0 & 1 \\ 0 & 0 & -1 & 0 \\ 0 & 1 & 0 & 0 \end{pmatrix} \cdot \begin{pmatrix} 0 & 1 & 0 & 0 \\ 0 & 0 & 0 & 0 \\ 0 & 0 & 0 & 0 \\ 1 & 0 & 0 & 0 \end{pmatrix} = \begin{pmatrix} 0 & 1 & 0 & 0 \\ 1 & 0 & 0 & 0 \\ 0 & 0 & 0 & 0 \\ 0 & 0 & 0 & 0 \end{pmatrix} = B$$

and that

$$U_{23} A U_{23} = \begin{pmatrix} 1 & 0 & 0 & 0 \\ 0 & -1 & 0 & 0 \\ 0 & 0 & 0 & 1 \\ 0 & 0 & 1 & 0 \end{pmatrix} \cdot \begin{pmatrix} 0 & 0 & 0 & 1 \\ 0 & 0 & 0 & 0 \\ 0 & 0 & 0 & 0 \\ 1 & 0 & 0 & 0 \end{pmatrix} \cdot \begin{pmatrix} 1 & 0 & 0 & 0 \\ 0 & -1 & 0 & 0 \\ 0 & 0 & 0 & 1 \\ 0 & 0 & 1 & 0 \end{pmatrix}$$
$$= \begin{pmatrix} 1 & 0 & 0 & 0 \\ 0 & -1 & 0 & 0 \\ 0 & 0 & 0 & 1 \\ 0 & 0 & 1 & 0 \end{pmatrix} \cdot \begin{pmatrix} 0 & 0 & 1 & 0 \\ 0 & 0 & 0 & 0 \\ 0 & 0 & 0 & 0 \\ 1 & 0 & 0 & 0 \end{pmatrix} = \begin{pmatrix} 0 & 0 & 1 & 0 \\ 0 & 0 & 0 & 0 \\ 1 & 0 & 0 & 0 \\ 0 & 0 & 0 & 0 \end{pmatrix} = C \ .$$

Solution 3.20 With the help of Exercise 3.19, it follows for $j \in \{1, 2\}$ and $s \in \mathbb{R}$ that

$$U_a(M_{0j}(s)) = U_a(U_{j3}^{-1} \cdot M_{03}(s) \cdot U_{j3}) = U_a(U_{j3} \cdot M_{03}(s) \cdot U_{j3})$$
$$= U_a(U_{j3}) \circ U_a(M_{03}(s)) \circ U_a(U_{j3}) = U_a(U_{j3}) \circ \exp\Big(i \, \frac{s}{\hbar} \, \hat{L}_{03}\Big) \circ U_a(U_{j3})$$
$$= \exp\Big(i \, \frac{s}{\hbar} \, U_a(U_{j3}) \circ \hat{L}_{03} \circ U_a(U_{j3})\Big) \ .$$

Further, for every $v = {}^t(v_1, v_2, v_3) \in \mathbb{R}^3$, we have that

$$U_{13}^{-1} \cdot p_a(v) = U_{13} \cdot p_a(v) = {}^t((a^2 + |v|^2)^{1/2}, v_3, -v_2, v_1) \, ,$$
$$U_{23}^{-1} \cdot p_a(v) = U_{23} \cdot p_a(v) = {}^t((a^2 + |v|^2)^{1/2}, -v_1, v_3, v_2) \, ,$$

and hence that

$$h_{aU_{13}}(v) = p_a^{-1}(U_{13} \cdot p_a(v)) = {}^t(v_3, -v_2, v_1) \, ,$$
$$h_{aU_{23}}(v) = p_a^{-1}(U_{23} \cdot p_a(v)) = {}^t(-v_1, v_3, v_2) \, .$$

As a consequence, for every $f \in X$, it follows that

$$[U_a(U_{13})f](v) = (f \circ h_{aU_{13}})(v) = f(v_3, -v_2, v_1) \, ,$$
$$[U_a(U_{23})f](v) = (f \circ h_{aU_{23}})(v) = f(-v_1, v_3, v_2)$$

and almost all $v = {}^t(v_1, v_2, v_3) \in \mathbb{R}^3$. Further, we note for $j \in \{1, 2\}$ that $U_a(U_{j3})C_0^1(\mathbb{R}^3, \mathbb{C}) = C_0^1(\mathbb{R}^3, \mathbb{C})$ and for $f \in C_0^1(\mathbb{R}^3, \mathbb{C})$ that

$$\begin{aligned}
[\hat{L}_{03}U_a(U_{13})f](v) &= [\hat{L}_{03}(f \circ h_1)](v) = i\hbar\, v_a^0(v) \cdot (f \circ h_1)'(v) \cdot e_3 \\
&= i\hbar\, v_a^0(v) \cdot f'(h_1(v)) \cdot h_1'(v) \cdot e_3 \\
&= i\hbar\, v_a^0(h_1(v)) \cdot f'(h_1(v)) \cdot e_1 \\
&= \left(i\hbar\, v_a^0 \frac{\partial f}{\partial v_1}\right)(h_1(v)) = \left[U_a(U_{13}) \left(i\hbar\, v_a^0 \frac{\partial f}{\partial v_1}\right)\right](v) \, , \\
[\hat{L}_{03}U_a(U_{23})f](v) &= [\hat{L}_{03}(f \circ h_2)](v) = i\hbar\, v_a^0(v) \cdot (f \circ h_2)'(v) \cdot e_3 \\
&= i\hbar\, v_a^0(v) \cdot f'(h_2(v)) \cdot h_2'(v) \cdot e_3 \\
&= i\hbar\, v_a^0(h_2(v)) \cdot f'(h_2(v)) \cdot e_2 \\
&= \left(i\hbar\, v_a^0 \frac{\partial f}{\partial v_2}\right)(h_2(v)) = \left[U_a(U_{23}) \left(i\hbar\, v_a^0 \frac{\partial f}{\partial v_2}\right)\right](v) \, ,
\end{aligned}$$

for every $v = {}^t(v_1, v_2, v_3) \in \mathbb{R}^3$, where $h_1 : \mathbb{R}^3 \to \mathbb{R}^3$ and $h_2 : \mathbb{R}^3 \to \mathbb{R}^3$ are defined by

$$h_1(v) := (v_3, -v_2, v_1) \, , \quad h_2(v) := (-v_1, v_3, v_2) \, ,$$

for every $v = {}^t(v_1, v_2, v_3) \in \mathbb{R}^3$, and e_1, e_2, e_3 is the canonical basis of $\mathbb{R}^3$. The latter implies that

$$h_1'(v) = \begin{pmatrix} 0 & 0 & 1 \\ 0 & -1 & 0 \\ 1 & 0 & 0 \end{pmatrix} \, , \quad h_2'(v) = \begin{pmatrix} -1 & 0 & 0 \\ 0 & 0 & 1 \\ 0 & 1 & 0 \end{pmatrix} \, ,$$

for every $v \in \mathbb{R}^3$. Hence, it follows that

$$U_a(U_{13}) \circ \hat{L}_{03} \circ U_a(U_{13})f = i\hbar\, v_a^0 \frac{\partial f}{\partial v_1} \, , \quad U_a(U_{23}) \circ \hat{L}_{03} \circ U_a(U_{23})f = i\hbar\, v_a^0 \frac{\partial f}{\partial v_2} \, .$$

Solution 3.21 Part (i): If $a > 0$, then $\hbar_a$ and $\hbar_a^{-1}$ are C^∞-functions such that

$$\hbar_a^{-1}(\hbar_a(w)) = {}^t\!\left([\hbar_a(w)]_1, [\hbar_a(w)]_2, \operatorname{arsinh}\left(\frac{[\hbar_a(w)]_3}{([\hbar_a(w)]_1^2 + [\hbar_a(w)]_2^2 + a^2)^{1/2}}\right)\right)$$
$$= {}^t\!\left(w_1, w_2, \operatorname{arsinh}\left(\frac{(w_1^2 + w_2^2 + a^2)^{1/2}\sinh(w_3)}{(w_1^2 + w_2^2 + a^2)^{1/2}}\right)\right) = w\ ,$$

for every $w \in \mathbb{R}^3$, and

$$\hbar_a(\hbar_a^{-1}(v))$$
$$= {}^t([\hbar_a^{-1}(v)]_1, [\hbar_a^{-1}(v)]_2, ([\hbar_a^{-1}(v)]_1^2 + [\hbar_a^{-1}(v)]_2^2 + a^2)^{1/2}\sinh([\hbar_a^{-1}(v)]_3))$$
$$= {}^t(v_1, v_2, (v_1^2 + v_2^2 + a^2)^{1/2}\,\frac{v_3}{(v_1^2 + v_2^2 + a^2)^{1/2}}) = v\ ,$$

for every $v \in \mathbb{R}^3$. If $a = 0$, then $\hbar_a$ and $\hbar_a^{-1}$ are well-defined as well as C^∞ such that

$$\hbar_a^{-1}(\hbar_a(w)) = {}^t\!\left([\hbar_a(w)]_1, [\hbar_a(w)]_2, \operatorname{arsinh}\left(\frac{[\hbar_a(w)]_3}{([\hbar_a(w)]_1^2 + [\hbar_a(w)]_2^2 + a^2)^{1/2}}\right)\right)$$
$$= {}^t\!\left(w_1, w_2, \operatorname{arsinh}\left(\frac{(w_1^2 + w_2^2 + a^2)^{1/2}\sinh(w_3)}{(w_1^2 + w_2^2 + a^2)^{1/2}}\right)\right) = w\ ,$$

for every $w \in \mathbb{R}^3 \setminus (\{0\} \times \{0\} \times \mathbb{R})$, and

$$\hbar_a(\hbar_a^{-1}(v))$$
$$= {}^t([\hbar_a^{-1}(v)]_1, [\hbar_a^{-1}(v)]_2, ([\hbar_a^{-1}(v)]_1^2 + [\hbar_a^{-1}(v)]_2^2 + a^2)^{1/2}\sinh([\hbar_a^{-1}(v)]_3))$$
$$= {}^t(v_1, v_2, (v_1^2 + v_2^2 + a^2)^{1/2}\,\frac{v_3}{(v_1^2 + v_2^2 + a^2)^{1/2}}) = v\ ,$$

for every $v \in \mathbb{R}^3 \setminus (\{0\} \times \{0\} \times \mathbb{R})$. In both cases, $a > 0$ and $a = 0$, we have that

$$\hbar_a'(w) = \begin{pmatrix} 1 & 0 & 0 \\ 0 & 1 & 0 \\ w_1\,(w_1^2 + w_2^2 + a^2)^{-1/2}\sinh(w_3) & w_2\,(w_1^2 + w_2^2 + a^2)^{-1/2}\sinh(w_3) & (w_1^2 + w_2^2 + a^2)^{1/2}\cosh(w_3) \end{pmatrix}$$

and hence that

$$\det(\hbar_a'(w)) = (w_1^2 + w_2^2 + a^2)^{1/2}\cosh(w_3)\ ,$$

for every $w \in \mathcal{U}_a$. Finally, since

$$
\begin{aligned}
v_a^0(\hbar_a(w)) &= (|\hbar_a(w)|^2 + a^2)^{1/2} \\
&= [w_1^2 + w_2^2 + (w_1^2 + w_2^2 + a^2)\sinh^2(w_3) + a^2]^{1/2} \\
&= (w_1^2 + w_2^2 + a^2)^{1/2}\cosh(w_3)\ ,
\end{aligned}
$$

for every $w \in \mathcal{U}_a$, we have that

$$
(v_a^0 \circ \hbar_a)^{-1}\ \det(\hbar_a') = 1\ .
$$

Part (ii): We reason as follows. If $f \in L^2_{\mathbb{C}}(\mathbb{R}^3, \varphi_a)$, then f is φ_a-a.e. defined on $\mathbb{R}^3$ as well as φ_a-measurable and hence also v^3-a.e. defined on $\mathbb{R}^3$ and v^3-measurable. Further, since $|f|^2$ is φ_a-integrable, it follows that $(v_a^0)^{-1}|f|^2$ is v^3-integrable and that

$$
\|f\|_2^2 = \int_{\mathbb{R}^3} |f|^2\, d\varphi_a = \int_{\mathbb{R}^3} |f|^2 \cdot (v_a^0)^{-1}\, dv^3 = \int_{\mathcal{U}_a} |f|^2 \cdot (v_a^0)^{-1}\, dv^3\ ,
$$

where $\|\ \|_2$ denotes the norm on $L^2_{\mathbb{C}}(\mathbb{R}^3, \varphi_a)$. Since $\hbar_a : \mathcal{U}_a \to \mathcal{U}_a$ is a C^1-diffeomorphism, it follows from the change of variable theorem for the Lebesgue integral that $f \circ \hbar_a$ is v^3-a.e. defined on $\mathbb{R}^3$ and v^3-measurable. Further, from the same theorem, it follows that

$$
\begin{aligned}
&\int_{\mathcal{U}_a} |f|^2 \cdot (v_a^0)^{-1}\, dv^3 = \int_{\mathcal{U}_a} |f \circ \hbar_a|^2 \cdot (v_a^0 \circ \hbar_a)^{-1}\, |\det(\hbar_a')|\, dv^3 \\
&= \int_{\mathcal{U}_a} |f \circ \hbar_a|^2\, dv^3 = \int_{\mathbb{R}^3} |f \circ \hbar_a|^2\, dv^3\ .
\end{aligned}
$$

Hence, we have that $f \circ \hbar_a \in L^2_{\mathbb{C}}(\mathbb{R}^3)$ and that

$$
\|f\|_2^2 = \int_{\mathbb{R}^3} |f \circ \hbar_a|^2\, dv^3\ .
$$

Therefore, the map $\mathcal{W}_a$ is well-defined and isometric. Further, $\mathcal{W}_a$ is obviously linear, and it follows from the polarization identity for complex sesquilinear forms that $\mathcal{W}_a$ preserves scalar products. Furthermore, if $f \in L^2_{\mathbb{C}}(\mathbb{R}^3)$, then f is v^3-a.e. defined on $\mathbb{R}^3$ as well as v^3-measurable. Since, $\hbar_a^{-1} : \mathcal{U}_a \to \mathcal{U}_a$ is a C^1-diffeomorphism, it follows from the change of variable theorem for the Lebesgue integral that $f \circ \hbar_a^{-1}$ is v^3-a.e. defined on $\mathbb{R}^3$ and v^3-measurable, where we note that the latter implies that $f \circ \hbar_a^{-1}$ is φ_a-a.e. defined on $\mathbb{R}^3$ and φ_a-measurable, as well as that

$$
\int_{\mathbb{R}^3} |f|^2\, dv^3 = \int_{\mathcal{U}_a} |f|^2\, dv^3 = \int_{\mathcal{U}_a} |f \circ \hbar_a^{-1}|^2 \cdot |\det\big((\hbar_a^{-1})'\big)|\, dv^3\ .
$$

Since

$$
E = (\hbar_a^{-1} \circ \hbar_a)'(w) = [(\hbar_a^{-1})' \circ \hbar_a](w) \cdot \hbar_a'(w) = (\hbar_a^{-1})'(\hbar_a(w)) \cdot \hbar_a'(w)\ ,
$$

for every $w \in \mathcal{U}_a$, we infer that

$$E = (\hbar_a^{-1})'(v) \cdot \hbar_a'(\hbar_a^{-1}(v)) \ ,$$

and that

$$1 = \det\big((\hbar_a^{-1})'(v)\big) \cdot \det(\hbar_a'(\hbar_a^{-1}(v))) = \det\big((\hbar_a^{-1})'(v)\big) \cdot v_a^0(v) \ ,$$

for every $v \in \mathcal{U}_a$. The latter implies that

$$\det\big((\hbar_a^{-1})'\big) = (v_a^0)^{-1} \ .$$

Hence,

$$\begin{aligned}\int_{\mathcal{U}_a} |f \circ \hbar_a^{-1}|^2 \cdot |\det\big((\hbar_a^{-1})'\big)| \, dv^3 &= \int_{\mathcal{U}_a} |f \circ \hbar_a^{-1}|^2 \cdot (v_a^0)^{-1} \, dv^3 \\ &= \int_{\mathbb{R}^3} |f \circ \hbar_a^{-1}|^2 \cdot (v_a^0)^{-1} \, dv^3 \ .\end{aligned}$$

In particular, it follows that $f \circ \hbar_a^{-1} \in L^2_{\mathbb{C}}(\mathbb{R}^3, \varphi_a)$ and that

$$\int_{\mathbb{R}^3} |f|^2 \, dv^3 = \|f \circ \hbar_a^{-1}\|_2^2 \ .$$

As a consequence, the map $\mathcal{W}_a^{-1}$ is well-defined and isometric. Further, $\mathcal{W}_a^{-1}$ is obviously linear, and it follows from the polarization identity for complex sesquilinear forms that $\mathcal{W}_a^{-1}$ preserves scalar products. Since $\mathcal{W}_a \mathcal{W}_a^{-1} f = f$, for every $f \in L^2_{\mathbb{C}}(\mathbb{R}^3)$, we infer that $\mathcal{W}_a$ is surjective with inverse $\mathcal{W}_a^{-1}$.

Solution 3.22 If $(m_1, \Lambda_1), (m_2, \Lambda_2), \dots$ is a sequence in $\mathcal{P}$ that is component-wise convergent to $(m, \Lambda) \in \mathbb{R}^4 \times \mathrm{M}(4, \mathbb{R})$, it follows that the sequence $\Lambda_1, \Lambda_2, \dots$ is component-wise convergent to Λ and hence, since $\mathcal{L}$ is a closed subset of $\mathrm{M}(4, \mathbb{R})$, that $\Lambda \in \mathcal{L}$. Hence, $(m, \Lambda) \in \mathcal{P}$. Further, for every $\nu \in \mathbb{N}$, we have that $({}^t(\nu, 0, 0, 0), E) \in \mathcal{P}$ and that $\|({}^t(\nu, 0, 0, 0), E)\| = (1 + \nu^2)^{1/2} \geqslant \nu$. Hence, $\mathcal{P}$ is unbounded.

Solution 3.23 If $(m_1, \Lambda_1), (m_2, \Lambda_2), \dots$ is a sequence in $\mathcal{P}_+^{\uparrow}$ that is component-wise convergent to $(m, \Lambda) \in \mathbb{R}^4 \times \mathrm{M}(4, \mathbb{R})$, it follows that the sequence $\Lambda_1, \Lambda_2, \dots$ is component-wise convergent to Λ and hence, since $\mathcal{L}_+^{\uparrow}$ is a closed subset of $\mathrm{M}(4, \mathbb{R})$, that $\Lambda \in \mathcal{L}_+^{\uparrow}$. Hence, $(m, \Lambda) \in \mathcal{P}$. Further, for every $\nu \in \mathbb{N}$, we have that $({}^t(\nu, 0, 0, 0), E) \in \mathcal{P}_+^{\uparrow}$ and that $\|({}^t(\nu, 0, 0, 0), E)\| = (1 + \nu^2)^{1/2} \geqslant \nu$. Hence, $\mathcal{P}_+^{\uparrow}$ is unbounded. Since $\mathbb{R}^4$ and $\mathcal{L}_+^{\uparrow}$ are path-connected, for $(m_1, \Lambda_1), (m_2, \Lambda_2) \in \mathcal{P}_+^{\uparrow}$, it follows the existence of continuous paths $c_1 : [0, 1] \to \mathbb{R}^4$ and $c_2 : [0, 1] \to \mathcal{L}_+^{\uparrow}$ such that $c_1(0) = m_1$, $c_1(1) = m_2$ and $c_2(0) = \Lambda_1$, $c_2(1) = \Lambda_2$. Hence, $c : [0, 1] \to \mathcal{P}$, defined by $c(s) := (c_1(s), c_2(s))$, for every $s \in [0, 1]$, is component-wise continuous and hence continuous, and such that $c(0) = (m_1, \Lambda_1)$, $c(0) = (m_2, \Lambda_2)$. Hence, $\mathcal{P}_+^{\uparrow}$ is path-connected.

Solution 3.24 Part (a): We note that, since for every $(a, \Lambda) \in \mathscr{P}$, $f, g \in (\mathbb{R}^4)^{\mathbb{R}^4}$, $\lambda \in \mathbb{R}$ and $x \in \mathbb{R}^4$, we have that

$$\begin{aligned}
&[R((a,\Lambda))(f+g)](x) = (f+g)(\Lambda^{-1}\cdot(x-a)) \\
&= f(\Lambda^{-1}\cdot(x-a)) + g(\Lambda^{-1}\cdot(x-a)) \\
&= [R((a,\Lambda))f](x) + [R((a,\Lambda))g](x) \\
&= [R((a,\Lambda))f + R((a,\Lambda))g](x)\ , \\
&[R((a,\Lambda))(\lambda f)](x) = (\lambda f)(\Lambda^{-1}\cdot(x-a)) = \lambda f(\Lambda^{-1}\cdot(x-a)) \\
&= \lambda\,[R((a,\Lambda))f](x) = [\lambda\, R((a,\Lambda))f](x)\ ,
\end{aligned}$$

R is well-defined. Further, for $(a_1, \Lambda_1), (a_2, \Lambda_2) \in \mathscr{P}$, $f \in (\mathbb{R}^4)^{\mathbb{R}^4}$ and $x \in \mathbb{R}^4$, it follows that

$$\begin{aligned}
&[R((a_1,\Lambda_1)\cdot(a_2,\Lambda_2))f](x) = [R((a_1+\Lambda_1 a_2, \Lambda_1\Lambda_2))f](x) \\
&= f((\Lambda_1\Lambda_2)^{-1}\cdot(x-(a_1+\Lambda_1 a_2))) = f(\Lambda_2^{-1}\Lambda_1^{-1}\cdot(x-(a_1+\Lambda_1 a_2))) \\
&= f(\Lambda_2^{-1}\Lambda_1^{-1}\cdot(x-a_1) - \Lambda_2^{-1}a_2) = f(\Lambda_2^{-1}\cdot(\Lambda_1^{-1}\cdot(x-a_1)-a_2)) \\
&= [R(a_2,\Lambda_2)f](\Lambda_1^{-1}\cdot(x-a_1)) = [R(a_1,\Lambda_1)[R(a_2,\Lambda_2)f]](x)\ , \\
&[R((0,E))f](x) = f(E^{-1}\cdot(x-0)) = f(x)
\end{aligned}$$

and hence that

$$R((a_1,\Lambda_1)\cdot(a_2,\Lambda_2)) = R(a_1,\Lambda_1)\circ R(a_2,\Lambda_2)\ ,\quad R((0,E)) = \mathrm{id}_{(\mathbb{R}^4)^{\mathbb{R}^4}}\ .$$

As a consequence, R is a representation of $\mathscr{P}$.
Part (b): If $(a, \Lambda) \in \mathscr{P}$, then it follows for $(m, M) \in \mathbb{R}^4 \times \mathrm{M}(4, \mathbb{R})$ that

$$\begin{aligned}
&[R((a,\Lambda))(m + M\cdot \mathrm{id}_{\mathbb{R}^4})](x) = (m + M\cdot\mathrm{id}_{\mathbb{R}^4})(\Lambda^{-1}\cdot(x-a)) \\
&= m + M\cdot\Lambda^{-1}\cdot(x-a) = m - M\cdot\Lambda^{-1}\cdot a + (M\cdot\Lambda^{-1})\cdot x\ ,
\end{aligned}$$

for every $x \in \mathbb{R}^4$ and hence that

$$R((a,\Lambda))(m + M\cdot\mathrm{id}_{\mathbb{R}^4}) = m - M\cdot\Lambda^{-1}\cdot a + (M\cdot\Lambda^{-1})\cdot\mathrm{id}_{\mathbb{R}^4} \in \mathrm{A}(\mathbb{R}^4)\ .$$

As a consequence, $R((a, \Lambda))$ leaves $\mathrm{A}(\mathbb{R}^4)$ invariant and through restriction of $R((a, \Lambda))$ in domain and in range to $\mathrm{A}(\mathbb{R}^4)$, for every $(a, \Lambda) \in \mathscr{P}$, we arrive at representation $R_0 : \mathscr{P} \to \mathrm{L}(\mathrm{A}(\mathbb{R}^4), \mathrm{A}(\mathbb{R}^4))$ of $\mathscr{P}$.
Part (c): T is well-defined, since

$$
\begin{aligned}
&[T((a,\Lambda))]((m_1,M_1)+(m_2,M_2)) = [T((a,\Lambda))]((m_1+m_2, M_1+M_2))\\
&= (m_1+m_2-(M_1+M_2)\,\Lambda^{-1}a, (M_1+M_2)\,\Lambda^{-1})\\
&= (m_1-M_1\Lambda^{-1}a, M_1\Lambda^{-1}) + (m_2-M_2\Lambda^{-1}a, M_2\Lambda^{-1})\\
&= [T((a,\Lambda))]((m_1,M_1)) + [T((a,\Lambda))]((m_2,M_2))\ ,\\
&[T((a,\Lambda))](\lambda.(m,M)) = [T((a,\Lambda))]((\lambda.m, \lambda.M))\\
&= (\lambda.m-(\lambda.M)\,\Lambda^{-1}a, (\lambda.M)\,\Lambda^{-1}) = \lambda.(m-M\Lambda^{-1}a, M\Lambda^{-1})\\
&= \lambda.[T((a,\Lambda))]((m,M))\ ,
\end{aligned}
$$

for every $(a, \Lambda) \in \mathcal{P}$ and all $(m_1, M_1), (m_2, M_2) \in \mathbb{R}^4 \times \mathrm{M}(4, \mathbb{R})$. In addition, we have that

$$
\begin{aligned}
&[T((a_1,\Lambda_1)\cdot(a_2,\Lambda_2))]((m,M)) = [T((a_1+\Lambda_1 a_2, \Lambda_1\Lambda_2))]((m,M))\\
&= (m-M\,(\Lambda_1\Lambda_2)^{-1}(a_1+\Lambda_1 a_2), M\,(\Lambda_1\Lambda_2)^{-1})\\
&= (m-M\,\Lambda_2^{-1}\Lambda_1^{-1}\,(a_1+\Lambda_1 a_2), M\,\Lambda_2^{-1}\Lambda_1^{-1})\\
&= (m-M\,(\Lambda_2^{-1}\Lambda_1^{-1}\,a_1+\Lambda_2^{-1}a_2), M\,\Lambda_2^{-1}\Lambda_1^{-1})\ ,\\
&[T((a_1,\Lambda_1))]([T((a_2,\Lambda_2))](m,M))\\
&= [T((a_1,\Lambda_1))]((m-M\Lambda_2^{-1}a_2, M\Lambda_2^{-1}))\\
&= ((m-M\Lambda_2^{-1}a_2)-M\Lambda_2^{-1}\Lambda_1^{-1}a_1, M\Lambda_2^{-1}\Lambda_1^{-1})\\
&= (m-M\,(\Lambda_2^{-1}\Lambda_1^{-1}\,a_1+\Lambda_2^{-1}a_2), M\,\Lambda_2^{-1}\Lambda_1^{-1})\ ,\\
&[T((0,E))]((m,M)) = (m-M\cdot E\cdot 0, M\cdot E) = (m,M)\ ,
\end{aligned}
$$

for $(a_1, \Lambda_1), (a_2, \Lambda_2) \in \mathcal{P}$ and all $(m, M) \in \mathbb{R}^4 \times \mathrm{M}(4, \mathbb{R})$. Further, if $(a_1, \Lambda_1), (a_2, \Lambda_2) \in \mathcal{P}$ are such that

$$[T((a_1,\Lambda_1))]((m,M)) = [T((a_2,\Lambda_2))]((m,M))$$

for every $(m, M) \in \mathbb{R}^4 \times \mathrm{M}(4, \mathbb{R})$, then

$$
\begin{aligned}
&[T((a_1,\Lambda_1))]((m,E)) = (m-\Lambda_1^{-1}a_1, \Lambda_1^{-1}) = (m-\Lambda_2^{-1}a_2, \Lambda_2^{-1})\\
&= [T((a_2,\Lambda_2))]((m,E))\ .
\end{aligned}
$$

Hence

$$m-\Lambda_1^{-1}a_1 = m-\Lambda_2^{-1}a_2\ ,\quad \Lambda_1^{-1} = \Lambda_2^{-1}\ ,$$

and therefore $(a_1, \Lambda_1) = (a_2, \Lambda_2)$.

Solution 3.25 If $(a, \Lambda) \in \mathcal{P}_+^{\uparrow}$, then $(\Lambda_0 a, \Lambda_0\Lambda) \in \mathcal{P}_+^{\uparrow}$ is such that $(0, \Lambda_0) \cdot (\Lambda_0 a, \Lambda_0\Lambda) = (a, \Lambda)$; if $(a, \Lambda) \in \mathcal{P}_+^{\downarrow}$, then $(\Lambda_{PT} a, \Lambda_{PT}\Lambda) \in \mathcal{P}_+^{\uparrow}$ is such that $(0, \Lambda_{PT}) \cdot (\Lambda_{PT} a, \Lambda_{PT}\Lambda) = (a, \Lambda)$; if $(a, \Lambda) \in \mathcal{P}_-^{\uparrow}$, then $(\Lambda_P a, \Lambda_P\Lambda) \in \mathcal{P}_+^{\uparrow}$ is such that $(0, \Lambda_P) \cdot (\Lambda_P a, \Lambda_P\Lambda) = (a, \Lambda)$; if $(a, \Lambda) \in \mathcal{P}_-^{\downarrow}$, then $(\Lambda_T a, \Lambda_T\Lambda) \in \mathcal{P}_+^{\uparrow}$ is such that $(0, \Lambda_T) \cdot (\Lambda_T a, \Lambda_T\Lambda) =$

(a, Λ). Moreover, such $(\bar{a}, \bar{\Lambda})$, with the prescribed properties, are obviously uniquely determined.

Solution 3.26 As a a square of densely-defined, linear and self-adjoint operator in a complex Hilbert space, the operator $\hat{H}^2$ is densely-defined, linear and self-adjoint. Further, we have that $(\hbar\kappa c)^2\, T_{|\ |^2+a^2}$ is an extension of $\hat{H}^2$. Since densely-defined, linear and self-adjoint operators in complex Hilbert spaces are maximal, in the sense that they have no proper symmetric extensions, and $(\hbar\kappa c)^2\, T_{|\ |^2+a^2}$ is densely-defined, linear and self-adjoint, this implies that $\hat{H}^2 = (\hbar\kappa c)^2\, T_{|\ |^2+a^2}$.

References

1. Baez JC, Segal IE, Zhou Z (1992) Introduction to algebraic and constructive quantum field theory. Princeton, Princeton University Press
2. Bargmann V, Wigner EP (1948) Group theoretical discussion of relativistic wave equations. Proc Natl Acad Sci USA, series 34:211–223
3. Barut AO, Raczka R (1980) Theory of group representations and applications, 2nd edn. Polish Scientific Publishers, Warszawa
4. Beyer HR (2007) Beyond partial differential equations: On linear and quasi-linear abstract hyperbolic evolution equations, vol 1898. Lecture Notes in Mathematics. Springer, Berlin
5. Beyer HR (1991) Remarks on Fulling's quantization. Class. Quantum Grav. series 8:1091–1112
6. Beyer H (2022) The reasoning of quantum mechanics: Operator theory and the harmonic oscillator. Cham, Springer Nature
7. Beyer H (2023) Introduction to quantum mechanics: Physics and Operator Theory. Cham, Springer Nature
8. Beyer HR (2010) Calculus and analysis: A combined approach. Wiley, New York
9. Bishop RL, Crittenden RJ (1964) Geometry of manifolds. Academic Press, New York
10. Bjorken JD, Drell SD (1964) Relativistic quantum mechanics. McGraw-Hill, New York
11. Bjorken JD, Drell SD (1965) Relativistic quantum fields. McGraw-Hill, New York
12. Brezis H (1983) Analyse fonctionnelle: Théorie et applications. Collection Mathématiques Appliquées pour la Maîtrise, Masson, Paris
13. Bonanos S 2003, *Capabilities of the Mathematica package 'Riemannian Geometry and Tensor Calculus'*, Recent Developments in Gravity, 174–182
14. Buchholz D 2000, *Algebraic quantum field theory: A status report*, Plenary talk given at XIIIth International Congress on Mathematical Physics, London, http://xxx.lanl.gov/abs/math-ph/0011044
15. Carmeli M, Malin M (2000) Theory of spinors: An introduction. World Scientific, Singapore
16. Davydov AS (1965) Quantum mechanics. Pergamon press, Oxford
17. Dixmier J (1977) $\mathbb{C}^*$ -Algebras. North-Holland, Amsterdam
18. Dunford N, Schwartz JT (1957) Linear operators, Part I: General theory. Wiley, New York
19. Dunford N, Schwartz JT (1963) Linear operators, Part II: Spectral theory: Self adjoint operators in Hilbert space theory. Wiley, New York
20. Engel K-J, Nagel R (2000) One-parameter semigroups for linear evolution equations. Springer, New York

21. Fulling SA (1989) Aspects of quantum field theory in curved spacetime. Cambridge University Press, Cambridge
22. Gelfand IM, Naimark MA (1957) Unitäre Darstellungen der klassischen Gruppen. Akademie-Verlag, Berlin
23. Gallier J, Quaintance J (2020) Differential geometry and Lie groups. Cham, Springer Nature
24. Goldberg S 1985, *Unbounded linear operators* Dover: New York
25. Goldstein JA (1985) Semigroups of linear operators and applications. Oxford University Press, New York
26. Gromoll D, Klingenberg W, Meyer W (1975) Riemannsche Geometrie im Großen. Springer, Berlin
27. Haag R (1996) Local quantum physics: Fields, particles, algebras. Springer, New York
28. Hamermesh M (1989) Group theory and its applications to physical problems. Dover, New York
29. Hall BC (2015) Lie groups, Lie algebras and representations, 2nd edn. Cham, Springer Nature
30. Helgason S (1978) Differential geometry, Lie groups, and symmetric spaces. Academic Press, New York
31. Hille E, Phillips RS (1957) Functional analysis and semi-groups, Revised. Providence, AMS
32. Hirzebruch F, Scharlau W (1971) Einführung in die Funktionalanalysis. Mannheim, BI
33. Hladik J (1999) Spinors in physics. Springer, New York
34. Kaku M (1993) Quantum field theory: A modern introduction. Oxford University Press, New York
35. Kato T (1966) Perturbation theory for linear operators. Springer, New York
36. Kosmann-Schwarzbach Y (2022) Groups and symmetries, 2nd edn. Cham, Springer Nature
37. Lang S (1972) Differentiable manifolds. Addison-Wesley, Reading
38. Lang S 1996, *Real and functional analysis*, 3rd ed., Springer: New York
39. Lang S (1997) Undergraduate analysis, 2nd edn. Springer, New York
40. Mackey GW (1951) On Induced representations of groups. American Journal of Mathematics, series 73:576–592
41. Mackey GW (2004) Mathematical foundations of quantum mechanics. Dover, Dover Publications
42. Maggiore M (2005) A Modern introduction to quantum field theory. Oxford University Press, Oxford
43. Messiah A (2014) Quantum mechanics. Dover Publication, New York
44. Von Neumann J 1930, *Allgemeine Eigenwerttheorie hermitescher Funktionaloperatoren*, Mathematische Annalen, Vol. **102**, 49–131
45. Von Neumann J (1932) Mathematische Grundlagen der Quantenmechanik. Springer, Berlin
46. Ohnuki Y (1988) Unitary representations of the Poincaré group and relativistic wave equations. World Scientific, Singapore
47. Pazy A (1983) Semigroups of linear operators and applications to partial differential equations. Springer, New York
48. Penrose R, Rindler W (1987) Spinors and space-time, vol 1. Cambridge University Press, Cambridge
49. Peskin ME, Schroeder DV (2018) An Introduction to quantum field theory. CRC Press, Boca Raton
50. Prugovecki E (1981) Quantum mechanics in Hilbert space. Academic Press, New York
51. Ratcliffe J G 2019, *Foundations of hyperbolic manifolds*, 3rd ed., Springer: New York
52. Rossmann W (2002) Lie groups: An introduction through linear groups. Oxford University Press, New York
53. Reed M and Simon B, 1980, 1975, 1979, 1978, *Methods of modern mathematical physics*, Volume I, II, III, IV, Academic: New York

54. Riesz F, Sz-Nagy B (1955) Functional analysis. Unger, New York
55. Rudin W (1991) Functional analysis, 2nd edn. MacGraw-Hill, New York
56. Ryder LH (1996) Quantum field theory, 2nd edn. Cambridge University Press, Cambridge
57. Sakurai JJ (1967) Advanced quantum mechanics. Addison-Wesley, Reading
58. Schechter M (2003) Operator methods in quantum mechanics. Dover Publication, New York
59. Schiff LI (1968) Quantum mechanics, 3Rev edn. McGraw-Hill Education, New York
60. Schroer B 2001, *Lectures on algebraic quantum field theory and operator algebras*, http://xxx.lanl.gov/abs/math-ph/0102018
61. Sexl RU, Urbantke HK (1976) Relativität Gruppen Teilchen. Springer, Wien
62. Simon B (2015) A comprehensive course in analysis Part 4: Operator theory. Providence, AMS
63. Sternberg S (1995) Group theory and physics. Cambridge University Press, Cambridge
64. Streater RF, Wightman AS (2000) PCT, spin and statistics, and all that. Princeton, Princeton University Press
65. Thirring W (1981) A course in mathematical physics 3: Quantum Mechanics of Atoms and Molecules. Springer, New York
66. Thorpe JA (1979) Elementary topics in differential geometry. Springer, New York
67. Tomonaga S (1997) The story of spin. The university of Chicago press, Chicago
68. Tung W-K (1985) Group theory in physics. World Scientific, Singapore
69. Wald RM (1994) Quantum field theory in curved spacetime and black hole thermodynamics. University of Chicago Press, Chicago
70. Warner FW (2010) Foundations of differentiable manifolds and Lie groups. Springer, New York
71. Weidmann J (1980) Linear operators in Hilbert spaces. Springer, New York
72. Weidmann J (2000) Lineare Operatoren in Hilberträumen: Teil I: Grundlagen. Teubner, Stuttgart
73. Weidmann J (2003) Lineare Operatoren in Hilberträumen: Teil II: Anwendungen. Teubner, Stuttgart
74. Weinberg S (1995) The quantum theory of fields, vol I. University of Cambridge, Cambridge
75. Weyl H (1950) Group theory and quantum mechanics. Dover, New York
76. Weyl H (1953) The classical groups: Their invariants and representations, 2nd edn. Princeton, Princeton University Press
77. Wigner E (1939) On unitary representations of the inhomogeneous Lorentz group. Annals of Mathematics 40:149–204
78. Yosida K (1968) Functional analysis, 2nd edn. Springer, Berlin

Index

D
Conventions, ix

D
Double covering
  of SO(3), 76
    particular images, 79

E
Equations
  Klein-Gordon, 161
Exponential map, 17
  $T_E GL(n, \mathbb{C}) \to GL(n, \mathbb{C})$, 47
    surjective, 47
  $T_E GL_+(2, \mathbb{R}) \to GL_+(2, \mathbb{R})$, 37
    not surjective, 37
  $T_E SL(2, \mathbb{K}) \to SL(2, \mathbb{K})$, 37
    not surjective, 37
  $T_E \mathrm{SO}(2) \to \mathrm{SO}(2)$, 52
    surjective, 52
  $T_E \mathrm{SO}(3) \to \mathrm{SO}(3)$, 60
    surjective, 60
  $T_E SO(n) \to SO(n)$, 33
    surjective, 33
  $T_E SU(n) \to SU(n)$, 36
    surjective, 36
  $T_E \mathrm{U}(n) \to \mathrm{U}(n)$, 36
    surjective, 36
  $T_E \mathcal{L}_+^\uparrow \to \mathcal{L}_+^\uparrow$, 110
  $\to \mathrm{SO}(3)$
    injectivity, 61

G
Groups
  $GL(n, \mathbb{K})$, 3
    tangent space, 28
  $GL(n, \mathbb{R})$, 6
    connected components, 38
  $GL_+(n, \mathbb{R})$, 6
  O(3), 91
    unitary representation, 91
    tangent space, 28
  O(n), 6
    connected components, 12
    tangent space, 31
  $SL(n, \mathbb{K})$, 10
    tangent space, 29
  SO(2), 49
    isomorphic to S^1, 53
    parametrization, 50
  SO(3), 53
    double covering, 76
    parametrization, 53, 57
    proper Euler angles, 63
  $SO(n)$, 8

H. R. Beyer, *Quantum Spin and Representations of the Poincaré Group, Part I*, Synthesis Lectures on Engineering, Science, and Technology, https://doi.org/10.1007/978-3-031-84140-8

Symbol Index

Symbols

Ran f, range of a map f, ix
ker f, kernel of a linear map f, ix
id_S, identity map on a set S, ix
$\mathbb{N}$, set of natural numbers, ix
$\mathbb{R}$, set of real numbers, ix
$\mathbb{C}$, set of complex numbers, ix
$\mathbb{N}^*$, $\mathbb{N}\setminus\{0\}$, ix
$\mathbb{R}^*$, $\mathbb{R}\setminus\{0\}$, ix
$\mathbb{C}^*$, $\mathbb{C}\setminus\{0\}$, ix
positive, ix
negative, ix
strictly positive, ix
strictly negative, ix
S^n, unit sphere in Euclidean n-space, ix
$\mathbb{K}$, $\in \{\mathbb{R}, \mathbb{C}\}$, ix
$e_1, \dots, e_n$, canonical basis of $\mathbb{K}^n$, ix
$\mathrm{M}(n, \mathbb{K})$, $n \times n$ matrices, ix
$\mathrm{M}(n \times m, \mathbb{K})$, $n \times m$ matrices, ix
$E_{n \times n}$, $n \times n$ unit matrix, ix
E, unit matrix, ix
$\det(A)$, determinant of a matrix A, ix
$\mathrm{Tr}(A)$, trace of a matrix A, ix
$C^k(\Omega, \mathbb{C})$, ix
$C_0^k(\Omega, \mathbb{C})$, ix
$C^k(\overline{\Omega}, \mathbb{C})$, ix
$f'(x)$, derivative of f in x (matrix), ix
∇f, gradient of f, ix
$'$, ordinary derivative, ix
$BC(\mathbb{R}^n, \mathbb{C})$, ix
$C_\infty(\mathbb{R}^n, \mathbb{C})$, x
v^n, Lebesgue measure on $\mathbb{R}^n$, x
$L^p_{\mathbb{C}}(\Omega, \rho)$, weighted L^p-space, x
$\| \, \|_p$, L^p-norm, x
$\langle \, | \, \rangle_2$, L^2-scalar product, x
$L^\infty_{\mathbb{C}}(\Omega)$, x
$\| \, \|_\infty$, infinity-norm, x
$\mathrm{L}(X, Y)$, bounded linear operators, x
$\| \, \|_{\mathrm{op},X,Y}$, operator norm, x
$C(U, Y)$, xi
$| \, |$, canonical norm on $\mathbb{K}^n$, 2
$\mathrm{GL}(n, \mathbb{K})$, general linear group, 3
$\mathrm{GL}_+(n, \mathbb{R})$,$\{M \in \mathrm{GL}(n, \mathbb{R}) : \det(M) > 0\}$, 6
$\mathrm{O}(n)$, orthogonal group, 6
$\mathrm{SO}(n)$, special orthogonal group, 8
$\mathrm{U}(n)$, unitary group, 8
$\mathrm{SU}(n)$, special unitary group, 9
ε_{jkl}, Levi-Civita symbol, 75
$\hat{J}$, total angular momentum operator, 97
$\hbar$, reduced Planck's constant, 97
$\hat{S}$, intrinsic angular momentum operator, 97
$\hat{H}$, Hamilton operator, 99
q, charge, 99
B, magnetic field, 99
c, speed of light, 99
m, mass, 99
κ, scale factor, 114
$\lambda\!\!\!^-_{\mathrm{C}}$, reduced Compton wavelength, 158
$\hat{p}$, 4-momentum, 158

H. R. Beyer, *Quantum Spin and Representations of the Poincaré Group, Part I*, Synthesis Lectures on Engineering, Science, and Technology,
https://doi.org/10.1007/978-3-031-84140-8

The manufacturer's authorised representative in the EU is Springer Nature Customer Service Centre GmbH, Europaplatz 3, 69115 Heidelberg, Germany. If you have any concerns regarding our products, please contact ProductSafety@springernature.com

Printed and bound by CPI Group (UK) Ltd, Croydon, CR0 4YY
26/03/2026
02078939-0015